STATICS AND STRENGTH OF MATERIALS

STATICS AND STRENGTH OF MATERIALS

FOURTH EDITION

MILTON G. BASSIN, B.M.E., M.M.E., P.E.
York College

STANLEY M. BRODSKY, B.M.E., M.M.E., Ph.D., P.E.
N.Y.C. Technical College

HAROLD WOLKOFF, B.M.E., M.B.A., P.E.
N.Y.C. Technical College

Gregg Division
McGRAW-HILL BOOK COMPANY

New York   Atlanta   Dallas   St. Louis   San Francisco
Auckland   Bogotá   Guatemala   Hamburg   Lisbon
London   Madrid   Mexico   Milan   Montreal   New Delhi
Panama   Paris   San Juan   São Paulo   Singapore
Sydney   Tokyo   Toronto

Sponsoring Editor: D. Eugene Gilmore
Editing Supervisors: Alfred Bernardi and Larry Goldberg
Design and Art Supervisor/Cover Design: Annette Mastrolia-Tynan
Production Supervisor: Mirabel Flores

Text Designer: Edward A. Butler
Cover Photo Researcher: Peri E. Zules
Cover Photographer: © Robert Phillips/The Image Bank

Library of Congress Cataloging-in-Publication Data

Bassin, Milton G.
    Statics and strength of materials.

    Includes index.
    1. Mechanics, Applied.   2. Strength of materials.
I. Brodsky, Stanley M.   II. Wolkoff, Harold.
III. Title.
TA350.B32   1988   620.1'12   87-3168
ISBN 0-07-004023-0

Statics and Strength of Materials, Fourth Edition

2 3 4 5 6 7 8 9 0   DOCDOC   8 9 4 3 2 1 0 9

ISBN 0-07-004023-0

C O N T E N T S

L I S T O F T A B L E S

P R E F A C E

As the previous editions of *Statics and Strength of Materials* have done, this fourth edition focuses on courses in statics and strength of materials which do not require the use of calculus. The practical orientation of the text has been retained and emphasized to ensure comprehension of the underlying principles and their applications. The text can therefore also serve as a reference for courses in materials testing, materials selection, structural design, and machine design.

As before, the text consists of two main sections: Statics (Chapters 1 through 6) and Strength of Materials (Chapters 7 through 16). For those who have previously completed a course in statics, it is recommended that they be referred to the review of statics in Appendix A. The most frequently used tables are given in Appendix B, which also contains a list of important formulas for convenient reference. The symbols used throughout the book conform to the recommendations of the American National Standards Institute (ANSI).

A number of significant changes have been introduced in this edition. Among them are the following.

A new Chapter 17, Computer Programs, has been added. It incorporates 26 sample computer programs written in BASIC, as well as a run of each sample program. These programs demonstrate applications of computer problem solving for a selection of topics in this book, and they are cross-referenced to the corresponding chapter, topic, and problem as appropriate. There is no intention to teach computer programming; rather, the intent is to provide the opportunity to practice solving technical problems to those who have some knowledge of BASIC and have access to a suitable computer. A faculty member's diskette for use on an IBM PC will be available; it will include the 26 sample programs and 30 additional ones for departments adopting the text.

The third edition introduced SI metric units along with the conventional U.S. customary units in roughly 40–60 proportion so that either system or both could be used according to the instructor's emphasis. This ratio has been retained. However, the metric dimensions for unit stress have been changed to emphasize the units of newtons per square millimeter (N/mm^2), which shows the force-per-area nature of stress more clearly and conforms more closely to actual industrial practice. Thus, the pascal (Pa) has been deemphasized and the megapascal (MPa) has been retained, since it is interchangeable with the preferred unit; $1 \ N/mm^2 = 1$ MPa. Metric sample problems and end-of-chapter problems have been denoted with an asterisk (*) for easy identification.

New tables and problems have been introduced in Appendix B. They use a selection of metric fastener data, all of which conform to ISO and ANSI standards. This material will permit the user to become familiar with proper metric thread specifications.

A new section has been added to Chapter 14, Combined Stresses, to deal with the Mohr's circle method of representing and solving certain combined stress problems.

The latest available provisions of the American Institute of Steel Construction (AISC) Code have been incorporated to maintain pace with constantly changing design data for structures. In addition, several appendix tables have been revised to reflect the up-to-date AISC specifications for rolled shapes.

Typical allowable, yield, and ultimate stresses as well as other mechanical properties of selected engineering materials have also been brought up to date. A new table giving some properties for selected polymers of both thermoplastic and thermosetting types has been included in Appendix B, and representative plastics have been added to the table of linear expansion coefficients in Chapter 8.

The previous section on bolted joints related to the American Society of Mechanical Engineers (ASME) Boiler and Pressure Vessel Code has been deleted from Chapter 9, Bolted, Riveted, and Welded Joints and Thin-Walled Pressure Vessels. These changes follow industry trends whereby practically all pressure vessels are fabricated using welded construction or other techniques which do not require bolts. The treatment of bolted joints under the AISC code continues to be important and has been retained and updated.

The solutions manual has been expanded to incorporate some useful information and suggestions for utilization and further assignments regarding the sample computer programs in the text and on the faculty member's diskette. This manual contains the solutions to many problems and the answers to all of the problems for which answers are not given in the text. The text continues to provide answers at the back of the book to the first five and all odd-numbered end-of-chapter problems.

In addition to the changes mentioned above, many of the comments and suggestions of users have been incorporated. The authors would be grateful to receive suggestions, comments, and corrections from persons using the text.

We wish to acknowledge with thanks the many useful suggestions received from Professor Elliot Colchamiro, P.E., of the Construction Technology Department, and Professors Henry Ortiz and Richard T. Woytowich, P.E., of the Mechanical Engineering Technology Department, as well as others from both departments at New York City Technical College of the City University of New York. In addition, we wish to offer our thanks to Arlene A. Spadafino, P.E., Director of Codes and Standards, American Society of Mechanical Engineers, for her advice and assistance regarding materials properties and to Professor Josephine Accummano-Muth, of the Data Processing Department of New York City Technical College, for her recommendations regarding the sample computer programs.

Milton G. Bassin
Stanley M. Brodsky
Harold Wolkoff

STATICS AND STRENGTH OF MATERIALS

1

Fundamental Terms

1-1 INTRODUCTION TO MECHANICS

The importance of a thorough knowledge of fundamentals in any field cannot be overemphasized. Fundamentals have always been stressed in the learning of new skills, whether it be football or physics.

Similarly, the science of mechanics is founded on basic concepts and forms the groundwork for further study in the design and analysis of machines and structures.

Mechanics can be divided into two parts: (1) statics, which relates to bodies at rest, and (2) dynamics, which deals with bodies in motion. This text is concerned with statics.

In mechanics, the term *strength of materials* refers to the ability of the individual parts of a machine or structure to resist loads. It also permits the selection of materials and the determination of dimensions to ensure sufficient strength of the various parts.

1-2 BASIC TERMS

It is essential that the following 13 basic terms be understood, since they continually recur in all phases of this technical study.

Length. This term is applied to the linear dimension of a straight or curved line. For example, the diameter of a circle is the length of a straight line which divides the circle into two equal parts; the circumference is the length of the curved perimeter of the circle.

Area. The two-dimensional size of a shape or a surface is the area. The shape may be flat (lie in a plane) or curved; examples are the size of a plot of land, the surface of a fluorescent bulb, and the cross-sectional size of a shaft.

Volume. The three-dimensional or cubic measure of the space occupied by a substance is known as the volume.

Force. This term is applied to any action on a body which tends to make it move, change its motion, or change its size and shape. A force is usually thought of as a push or a pull, such as a hand pushing against a wall or the pull of a rope fastened to a body.

Pressure. The external force per unit area, or the total force divided by the total area on which it acts, is known as pressure. Water pressure against the face of a dam, steam pressure in a boiler, and earth pressure against a retaining wall are some examples.

Mass. The amount of matter in a body is called mass, and for most problems in mechanics, mass may be considered constant.

Weight. The force with which a body is attracted toward the center of the earth by the gravitational pull is called weight.

Density. This term may refer to either weight density or mass density. Weight density is the weight per unit volume of a body or substance; mass density is the mass per unit volume of a body or substance.

Load. This term is used to indicate that a body of some weight is applying a force against some supporting structure or part of a structure. For example, a desk weighing 100 lb is applying a load (and thus a force) of 100 lb against the floor on which it stands.

Moment. The *tendency* of a force to cause rotation about an axis through some point is known as a moment.

Torque. The action of a force which *causes* rotation to take place is known as torque. The action of a belt on a pulley causes the pulley to rotate because of torque. Also, if you grasp a piece of chalk near each end and twist your hands in opposite directions, it is the developed torque that causes the chalk to twist and, perhaps, snap.

Work. The energy developed by a force acting through a distance against a resistance is known as work. The distance may be along a straight line or along a curved path. When the distance is *linear,* the work can be found from *work = force × distance.* When the distance is along a *circular path,* the work can be found from *work = torque × angle.* Common forms of work include a weight lifted through a height, a pressure pushing a volume of a substance, and torque causing a shaft to rotate.

Power. The rate of doing work, or the work done per unit time, is called power. For example, a certain amount of work is required to raise an elevator to the top of its shaft. A 5-kW motor can raise the elevator, but a 20-kW motor can do the same job *four times faster.*

1-3 INTRODUCTION TO METRICS

Because industry has taken major steps toward a changeover to metrics, the technically trained individual should be familiar with metric usage.

Historically, a large number of different metric units have been in use throughout the world (countries have not necessarily used the same units as their neighbors or trading partners). In order to overcome the obvious disadvantages of dealing with a variety of systems, an international agreement was reached to limit and standardize the metric units to be used. The common system adopted is called the International System of Units, usually referred to as SI.[1] The United States has agreed to adopt this system of metrics. Thus, all reference and use of metric units in this book will be in the SI system.

The metric system offers major advantages relative to the U.S. customary system (our historical system of units). For example, the metric system uses only one basic unit for length, the meter, whereas the U.S. customary system uses many basic units for length: inch, foot, yard, mile, league, fathom, etc. Also, because the metric system is based on multiples of 10, it is easier to use and learn. For example, in the metric system there are 100 centimeters in 1 meter, 1000 meters in 1 kilometer, etc.; in the U.S. customary system there are 12 inches in 1 foot, 3 feet in 1 yard, 5280 feet in 1 mile, etc.

A major distinction exists between the SI system and the U.S. customary system. Whereas the U.S. system is a *gravitational* system with *force as a basic unit*—from which mass is derived—the SI system is an *absolute* system with *mass as a basic unit*—from which force is derived. Simply put, this means that in the U.S. customary system an object is referred to as having a *weight* of some number of pounds (lb), while in the SI system the same object would be referred to as having a *mass* of some number of kilograms (kg).

The following example will illustrate the difference in procedure between U.S. customary units and SI in order to determine the force exerted by an object resting on a table. In dealing with the U.S. customary system, the problem would state that the object has a weight of 100 lb, and we would then indicate that the force exerted by the object on the table is 100 lb, and, if needed, the mass would be calculated. In the SI system, the problem would state that the object has a mass of 45 kg. In order to establish the force the object exerts on the table, we would first calculate the weight of the object by using the formula $W = m \cdot g$ (weight = mass · gravitational constant). Thus, for $g = 9.81$ m/s² (gravitational constant for sea level on earth), $W = 45$ kg $\cdot$ 9.81 m/s² = 441 kg $\cdot$ m/s². Since 1 kg $\cdot$ m/s² is defined as a newton (N), $W = 441$ N. Since the force is equal to the weight, we can now indicate that $F = 441$ N.

The above example introduced you to SI units and symbols that you may not be familiar with—kg for kilograms, m for meters, s for seconds. As you continue into the book, you will be introduced to additional SI units and symbols. You will find that once you become familiar with the SI system, you will enjoy working with it, since the system employs powers of 10. For example, the basic unit for length is the meter (m). A kilometer (km) is equal to 10^3 m. A millimeter (mm) is equal to 10^{-3} m. As we just discovered, the basic unit for weight and force is the newton (N), which is equal to 1 kg $\cdot$ m/s². A meganewton (MN) is equal to 10^6 N. A micronewton (μN) is equal to 10^{-6} N. Thus we

---

[1] SI comes from the official name in French: *Le Système International d'Unités.*

see that the SI system employs prefixes (kilo, mega, milli, etc.) in front of basic units. Table 1-1 lists the most common prefixes used in engineering. Table 13, App. B, lists the SI units used in this text. Problems dealing with the SI system of metrics will be preceded by asterisks.

TABLE 1-1   COMMON SI METRIC PREFIXES

Factor	Prefix	SI Symbol
1 000 000 000 000 = 10^{12}	tera	T
1 000 000 000 = 10^9	giga	G
1 000 000 = 10^6	mega	M
1 000 = 10^3	kilo	k
0.001 = 10^{-3}	milli	m
0.000 001 = 10^{-6}	micro	μ
0.000 000 001 = 10^{-9}	nano	n
0.000 000 000 001 = 10^{-12}	pico	p

There are a number of specific rules that apply to the use of SI units. For example, a period is never used after a symbol, except when required to indicate the end of a sentence (e.g., kg not kg.). Commas are not used to separate groupings of three digits as is done in the U.S. customary system; instead, a space is used (e.g., 40,096 is written as 40 096).

1-4   VECTORS AND SCALARS

Any quantity that is specified by a *magnitude only* is a *scalar* term. That is, a scalar is not related to any definite direction in space. Examples of scalars are $10, 5 ft, 18°C, 16 hp, 6 volts, etc.

A *vector* quantity has *magnitude* and is *related to a definite direction in space.* Force is such a quantity. To properly specify a force, the magnitude and direction of the force must be known.

A vector quantity is represented by a line carrying an arrowhead at one end. The length of the line (to a convenient scale) equals the magnitude of the vector. The line, together with its arrowhead, defines the direction of the vector.

Suppose a man is using a rope to pull a box along the floor and, in order to exert as little effort as possible, he tilts the box onto one edge as shown in Fig. 1-1. He must exert a pull of 30 lb at an angle of 30° with the horizontal in order to keep the box moving. This force can be represented as a vector, as shown in Fig. 1-2, by drawing it to scale (each unit of length equals 10 lb of force) at the same angle and sense as the force. Sense refers to the placement of the arrow on the vector.

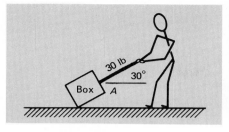

Fig. 1-1 Man pulling box with rope.   Fig. 1-2 Force represented as a vector.

1-5 NUMERICAL ACCURACY

Because the data used in engineering calculations vary in accuracy, it is important that we develop some means of indicating their accuracy. This is necessary so that, after performing a series of calculations, we arrive at an answer no more accurate than the original data.

We can establish the accuracy of a number by the use of *significant figures*. The larger the number of significant figures, the more accurate the data. For example, the number 2300.65 has six significant figures and is more accurate than the number 2301, which has four significant figures.

To establish the significant figures, we must keep in mind that the decimal point has nothing to do with the number of significant figures. Thus, each of the following has four significant figures: 27.61, 276.1, 0.2761, 0.002761, 2761, 276 100, and 2.761×10^6.

One of the ways you can identify well-trained technicians or engineers is that they carefully avoid giving numerical answers which have too many significant figures. They recognize that the *least* accurate number in the original data establishes the accuracy of the result. The following sample problem will illustrate the above principle.

Sample Problem 1   Determine the amount of work done (ft · lb) by a 55-lb force acting on a body for a distance of 1075 ft.

Solution: Work equals force times the distance through which the force acts.

$$\text{Work} = F \times d$$
$$\text{Work} = 55 \times 1075 = 59\ 125 \text{ ft} \cdot \text{lb}$$

This is the result which would be displayed by a calculator. But this answer has five significant figures. It is not proper to give such an accurate result from original data in which the least accurate number had only two significant figures (55 lb). Thus, the answer should be rounded off to two significant figures:

$$\text{Work} = 59\ 000 \text{ ft} \cdot \text{lb}$$

This means that with the information available in this Sample Problem we can only determine the work done to the nearest 1000 ft · lb. It should be noted that most engineering calculations do not require an accuracy greater than three significant figures.

1-6 ROUNDING OFF NUMBERS

If a number is given to the nearest hundredth and we wish to express it to the nearest tenth, what procedures do we follow? The following will apply to the rounding off of all numbers.

1. If the digit to be dropped is 5 or greater, increase the digit to the left by 1. Example: 36.48 becomes 36.5.
2. If the digit to be dropped is less than 5, simply drop it without changing the value of the digit to the left. Example: 36.42 becomes 36.4.

1-7 DIMENSIONAL ANALYSIS

In dealing with data, not only are we concerned with magnitude but we must also be concerned with the unit. For example, if we measured the length of a room and expressed it as 30, that would be meaningless. Do we mean 30 in or 30 ft or perhaps 30 yd? Thus, we see that the *dimension* of length consists of both *magnitude* and *unit*.

Very frequently we must change a dimension expressed in one unit to the same dimension expressed in another unit. For example, a length of 30 ft is equal to how many inches? To answer the question, we have to use a factor which will enable us to convert from feet to inches. The conversion factor will have to be used so that the value of the dimension will not be changed. This can be accomplished if we *multiply* the original form of the dimension by the conversion factor set up as a *fraction*. Thus,

$$30 \, \cancel{ft} \times \frac{12 \text{ in}}{1 \, \cancel{ft}} = 360 \text{ in}$$

Take note that the conversion factor is equal to 1, since the numerator is equal to the denominator. Take note also that units can be canceled. In fact, units, like numbers, can be divided, multiplied, squared, etc.

Example: express 30 mi/hr as ft/s.

$$30 \, \frac{\cancel{mi}}{\cancel{hr}} \times \frac{5280 \text{ ft}}{1 \, \cancel{mi}} \times \frac{1 \, \cancel{hr}}{3600 \text{ s}} = 44 \, \frac{\text{ft}}{\text{s}}$$

Example: express 60 km/h as m/s.

$$60 \, \frac{\cancel{km}}{\cancel{h}} \times \frac{1000 \text{ m}}{\cancel{km}} \times \frac{1 \, \cancel{h}}{3600 \text{ s}} = 16.7 \text{ m/s}$$

2

Resultant and Equilibrant of Forces

2-1 DEFINITIONS

External Force. When a force is applied to a body, it is called an external force.

Internal Force. The resistance to deformation, or change of shape, exerted by the material of a body is called an internal force.

Suppose a load F is suspended by a cable as in Fig. 2-1. The load tends to stretch or break the cable. The fibers of the cable resist the tendency. Thus, the load applies an *external force* to the cable, while the cable fibers exert an *internal force* to prevent being pulled apart.

Collinear Forces. These are forces whose vectors lie along the same straight line.

Concurrent Forces. Forces whose lines of action pass through a common point are called concurrent forces. In Fig. 2-2, forces F_1, F_2, and F_3 pass through the common point O and are concurrent.

Coplanar Forces. Forces whose lines of action lie in the same plane are called coplanar forces. Note that collinear forces must also be coplanar.

Rigid Body. When a body is acted upon by forces and the size and shape of the body are not changed, the body is said to be *rigid*. All real bodies are more or less elastic, and the application of a force would produce some change. In the study of mechanics, the bodies on which the forces act are often considered rigid.

Transmissibility of a Force. A force may be considered as acting at any point on its line of action as long as the direction and magnitude are unchanged.

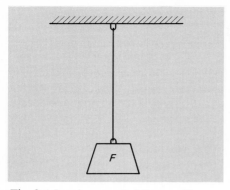

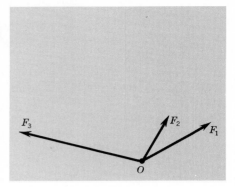

Fig. 2-1 Load suspended from cable.   Fig. 2-2 Concurrent forces.

Suppose a body (Fig. 2-3) is to be moved by a horizontal force F applied by hooking a rope to some point on the body. The force F will have the same effect if it is applied at points A, B, C (Fig. 2-4), or any point on its line of action. This property of a force is called *transmissibility.*

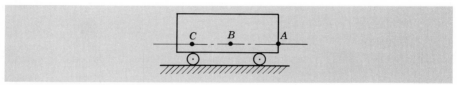

Fig. 2-3 Transmissibility of a force.

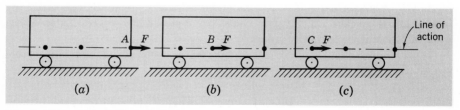

Fig. 2-4 Transmissibility of a force.

The principle of transmissibility can be used freely when overall effects on a body are being determined. It should be noted, however, that the local effects on individual members of the body will change when the force is moved. This can be seen in Fig. 2-4 by visualizing three eyebolts located at points A, B, and C. Only that eyebolt on which the rope is hooked will feel the effects of force F; the two unused eyebolts are unaffected by the force.

2-2 TYPES OF FORCE SYSTEMS

Most practical problems in statics involve several forces which act simultaneously. Such force systems may be classified as follows.

Concurrent-Coplanar. This system occurs when the lines of action of all forces lie in the same plane and pass through the common point. In Fig. 2-5, forces F_1, F_2, and W all lie in the same plane (the plane of the paper) and all their lines of action have point O in common. Collinear forces are the simplest type and are a special case of concurrent-coplanar forces. This chapter deals with concurrent-coplanar force systems.

Nonconcurrent-Coplanar. Such a system exists when the lines of action of *all* forces lie in the same plane but *do not* pass through a common point. Figure 2-6 shows a situation of this type. These systems will be treated in Chaps. 3 and

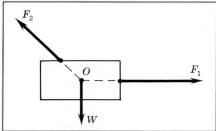

Fig. 2-5 Concurrent-coplanar forces.

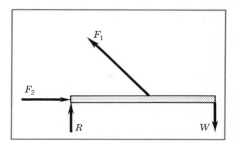

Fig. 2-6 Nonconcurrent-coplanar forces.

4. It should be noted that parallel-force systems, as in Fig. 2-7, are a special case of nonconcurrent-coplanar systems.

Concurrent-Noncoplanar. This system is evident when the lines of action of all forces *do not* lie in the same plane but *do* pass through a common point. An example of this force system is the forces in the legs of a tripod support for a camera (Fig. 2-8) or a transit. Chapter 5 is devoted to concurrent-noncoplanar systems.

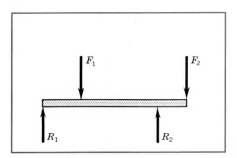

Fig. 2-7 Parallel forces.

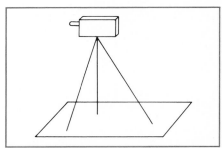

Fig. 2-8 Tripod support for camera.

Nonconcurrent-Noncoplanar. When the lines of action of all forces *do not* lie in the same plane and *do not* pass through a common point, a nonconcurrent-noncoplanar force system is present. This type will not be discussed in the text.[1]

2-3 RESULTANT OF CONCURRENT FORCE SYSTEMS

A *resultant* is a single force which can replace two or more concurrent forces and produce the same effect on the body as the concurrent forces.

It is a fundamental principle of mechanics, demonstrable by experiment, that when a force acts on a body which is free to move, the motion of the body is in the direction of the force and the distance traveled in a unit time depends on the magnitude of the force. Then, for a system of concurrent forces acting on a body, the body will move in the direction of the *resultant* of that system and the distance traveled in a unit time will depend on the magnitude of the *resultant.*

2-4 RESULTANT OF COLLINEAR FORCES

The simplest type of a concurrent force system involves collinear forces. When two or more forces act on a body along the same straight line, their resultant will also act along the same line.

If a body is acted upon by forces of 20 and 30 lb, as shown in Fig. 2-9, it is evident that a single force of $20 + 30 = 50$ lb will be the resultant.

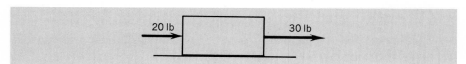

Fig. 2-9 Collinear forces.

If the forces act in opposite directions (Fig. 2-10), their resultant will be $30 - 20 = 10$ lb acting to the right.

Fig. 2-10 Collinear forces.

As a matter of convenience, when a force acts in a certain direction, say to the right, it is considered positive. If it acts to the left, it is negative. The force is

---

[1] For a discussion of nonconcurrent-noncoplanar force systems see A. Jensen and H. H. Chenoweth, *Applied Engineering Mechanics,* 4th ed., McGraw-Hill Book Company, New York, 1983.

read or expressed as so many pounds, with a plus or minus sign attached, depending on the direction in which the force acts. Thus, the forces shown in Fig. 2-10 are $+30$ and -20 lb. The purpose of the sign is to indicate direction of action only.

The rule for determining the resultant of collinear forces may be stated as follows: *The resultant of any number of forces acting along the same straight line is the algebraic sum of the forces.*

Although the choice of signs is a matter of convenience, the custom in mechanics is to observe the signs as in mathematics. Thus, for forces acting horizontally, a force acting to the right is positive and a force acting to the left is negative. When forces act along a vertical line, up forces are positive and down forces are negative.

Sample Problem 1   Find the resultant of the collinear force system shown in Fig. 2-11.

Fig. 2-11 Diagram for Sample Problem 1.

Solution: $R = +15 + 25 - 20 - 10 = +10$ lb
Thus, the four given forces may be replaced by a single force of 10 lb acting to the right.

*Sample Problem 2   Find the resultant of the collinear forces shown in Fig. 2-12.

Solution: $R = +8 + 12 - 10 = +10$ N
Therefore, the three given forces can be replaced by a single up force of 10 N.

Sample Problem 3   Find the resultant of the forces shown in Fig. 2-13.

Solution: There are two sets of collinear forces, which will be analyzed separately.
In the vertical direction, $R_y = +10 - 10 = 0$.
In the horizontal direction, $R_x = -6 - 4 + 8 = -2$ lb.
Then the five forces can be replaced by a single force of 2 lb acting horizontally to the left.

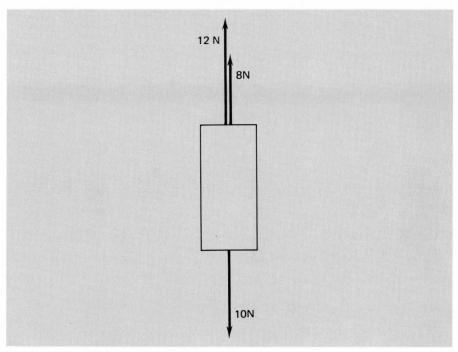

Fig. 2-12 Diagram for Sample Problem 2.

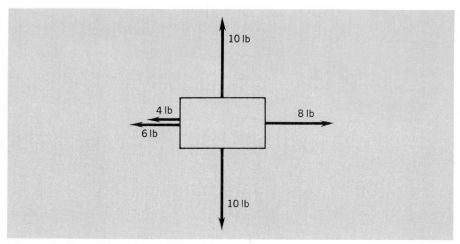

Fig. 2-13 Diagram for Sample Problem 3.

2-5 EQUILIBRIUM OF CONCURRENT FORCE SYSTEMS

When the resultant of any number of concurrent forces acting on a body is zero, the forces are said to be in *equilibrium.* The reader should note that when forces are in equilibrium, the body on which they act may be at rest or in uniform motion.

Statics is a study of forces in equilibrium.

2-6 EQUILIBRIUM OF COLLINEAR FORCES

If the resultant of forces acting on a body in the same straight line is zero, the forces are in *equilibrium.* The body on which the forces act will not move or, if in motion, will continue to move with the same speed. This condition of equilibrium might be stated in another way by saying that the sum of the forces in one direction must be equal to the sum of the forces in the opposite direction. Direction is not limited to either the horizontal or the vertical; it can make any angle with the horizontal.

2-7 ACTION AND REACTION

To every *action* there is an equal and opposite *reaction* (Newton's third law).

When one pushes against a wall, one exerts an action. The wall resists the push with an equal and opposite force called a reaction.

When a load is suspended from a hook, the load exerts an action downward. The hook exerts an equal, opposite force called the reaction.

A body weighing 10 lb rests on a table. The action force is the downward pull due to the earth's attraction. The table exerts an upward reaction N (Fig. 2-14). Since the body does not move, the forces are in equilibrium. Then $R = 0$. But $R = N - 10$; thus, $N - 10 = 0$ and $N = 10$ lb.

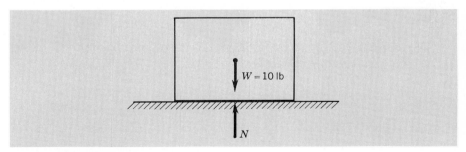

Fig. 2-14 Action and reaction.

2-8 TENSION AND COMPRESSION: TWO-FORCE MEMBERS

When two equal and opposite forces F act on a body, such as the bar shown in Fig. 2-15, the body is in *tension,* because the tendency is to stretch or pull the bar apart. A hoisting cable is a good illustration of this action.

If the forces act in the reverse directions, as in Fig. 2-16, the body is in *compression,* because the tendency of the forces *F* is to shorten, compress, or crush the body. This action occurs in a prop, brace, or column.

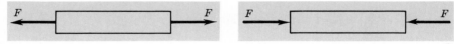

Fig. 2-15 Body in tension.   Fig. 2-16 Body in compression.

When a bar is acted upon by two opposite forces that are in equilibrium, the member is always either in tension or compression and the direction of the external forces is along the axis of the bar. These forces are termed *axial* forces.

It is important to distinguish between axial tension and axial compression forces. A long member subjected to axial tension force is not normally affected by its length. On the other hand, a long member which is subjected to axial compression force *is* affected by its length. If a member subjected to compression force is relatively *long* compared to its cross-sectional area, it will be subjected to both buckling and compression stresses. This type of member is discussed in Chap. 15. If a member subjected to compression force is relatively *short* compared to its cross-sectional area, it will be subjected to compression stress only. Unless otherwise indicated, it will be assumed in this text that a member subjected to compression force is a short compression member.

For a simple experiment in tension, a rope of any convenient length is chosen and a force of 50 lb is applied at each end, as shown in Fig. 2-17.

Fig. 2-17 Rope in tension.

The rope is cut at any point *C*, and two spring balances are fastened to the ends and hooked together (Fig. 2-18).

Fig. 2-18 Rope in tension.

Each balance then reads 50 lb, an indication that a force of 50 lb was required to hold the parts of the rope together. This is just the force that was exerted by the fibers of the rope before it was cut.

Members subjected solely to axial forces or to forces whose resultants are axial forces are called *two-force members.* The ability to recognize two-force members will be useful in analyzing structures and mechanisms.

2-9 RESULTANT OF TWO CONCURRENT FORCES[1]

Two forces that are concurrent can be added vectorially by means of the parallelogram law or the triangle law.

Figure 2-19 shows a body acted on by two concurrent forces, P and F. By

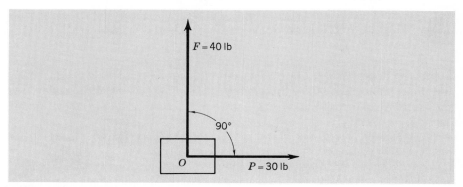

Fig. 2-19 Diagram for Sample Problem 4.

constructing a parallelogram whose sides are equal to P and F, respectively, the resultant of the forces can be determined. The diagonal of the parallelogram will be the resultant R (Fig. 2-20a).

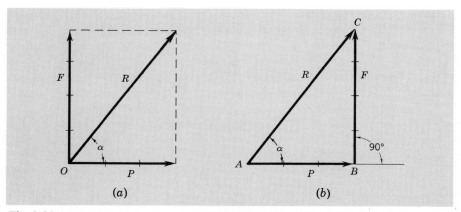

Fig. 2-20 (a) Parallelogram solution for Sample Problem 4. (b) Force triangle.

The resultant R can also be found by constructing one-half of the parallelogram. This figure will be a triangle. From any convenient point A, using an appropriate scale, draw a vector equal in magnitude and direction to one of the forces (Fig. 2-20b). From the tip of the arrowhead of the first vector, lay out the second vector to the same scale so that the vectors are connected head-to-tail. The closing side of the triangle, drawn from the starting point A of the first

[1] See also Sample Computer Programs 2 and 3, Chap. 17.

vector to the tip of the arrowhead of the second vector, is the resultant of the two forces (Fig. 2-20b). Note that the arrowhead of the resultant R is at the end away from the starting point. The magnitude of the resultant is determined by measuring its length with the scale used for drawing the other vectors.

Most cases of concurrent force systems are best solved by algebraic methods. Such a solution is demonstrated in the following examples, along with the graphical method described above.

Sample Problem 4 Two concurrent forces of 30 and 40 lb act on a body and make an angle of 90° with each other as in Fig. 2-19. Find the magnitude and direction of the resultant by (a) the graphical method and (b) the algebraic method.

Solution a (Graphical): From any point A, draw **AB** equal to 30 lb to any convenient scale and parallel to force P, as in Fig. 2-20b. From B, draw a line **BC** in the direction of force F and in length to represent 40 lb. Then the line **AC** is the resultant R in magnitude and direction. By measurement, R is found to be 50 lb. Angle α can be measured by use of a protractor.

Figure 2-20a indicates the graphical solution by means of the parallelogram law.

Solution b (Algebraic): Note that, since the resultant forms a triangle with the two given forces, an algebraic solution of the triangle will give R. Since ABC is a right triangle,

$$R = \sqrt{P^2 + F^2} = \sqrt{900 + 1600} = 50 \text{ lb}$$

The angle α that R makes with P can be found from

$$\tan \alpha = \frac{\textbf{BC}}{\textbf{AB}} = \frac{F}{P} = \frac{40}{30} = 1.333$$
$$\alpha = 53.13° \text{ or } 53°8'$$

*Sample Problem 5 Find the magnitude and direction of the resultant of two concurrent forces of 50 and 75 N acting on a body at an angle of 50° with each other as in Fig. 2-21. Use (a) the graphical method and (b) the algebraic method.

Solution a (Graphical): From any point A, draw **AB** equal and parallel to F_1 as in Fig. 2-22a. From B, draw line **BC** equal and parallel to F_2. Line **AC** is the resultant R. By measurement, R is found to be 114 N.

Figure 2-22b indicates the graphical solution by means of the parallelogram law.

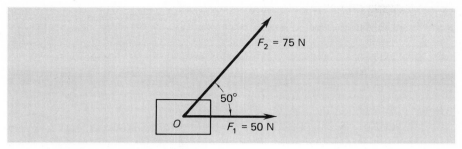

Fig. 2-21 Diagram for Sample Problem 5.

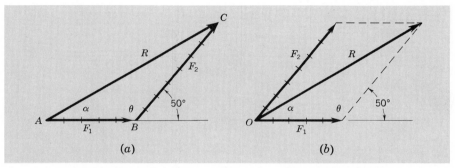

Fig. 2-22 (*a*) Force triangle for Sample Problem 5. (*b*) Parallelogram solution.

Solution b (Algebraic):   Triangle ABC (Fig. 2-22*a*) can be solved for R by using the law of cosines.

$$R^2 = F_1^2 + F_2^2 - 2F_1F_2 \cos \theta$$

But $\theta = 180 - 50 = 130°$.

$$\cos 130° = -\cos 50° = -0.6428$$
$$R^2 = 50^2 + 75^2 - 2(50)(75)(-0.6428)$$
$$= 2500 + 5625 + 4820 = 12\ 945$$
$$R = 114 \text{ N}$$

Angle α can be calculated by the law of sines.

$$\frac{F_2}{\sin \alpha} = \frac{R}{\sin \theta} \qquad \sin \alpha = \frac{F_2}{R} \sin \theta$$
$$\sin \theta = \sin 130° = \sin 50° = 0.7660$$
$$\sin \alpha = \frac{75}{114} (0.7660) = 0.505$$
$$\alpha = 30.3° \text{ or } 30°18'$$

The reader should appreciate that the degree of accuracy obtainable with the graphical method depends upon the care with which the drawing is made and the scale that is used.

2-10   EQUILIBRANT AND THE FORCE TRIANGLE

If two forces are acting on a body, the third force that will hold them in equilibrium is called the *equilibrant,* or the *balancing force.*

   The relations among three forces in equilibrium are best illustrated by an experiment. In Fig. 2-23, let *OA*, *OB*, and *OC* be fastened to a ring at *O*. Spring balances are fastened at *A*, *B*, and *C*. The springs are then stretched and the rings are fastened to pins *E*, *F*, and *G*. Since the forces acting on *OA*, *OB*, and *OC* are axial, they are two-force members.

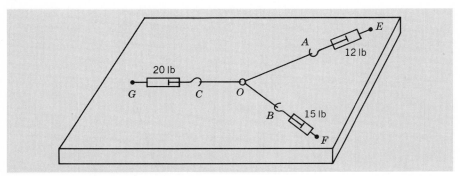

Fig. 2-23 Three forces in equilibrium.

   Balances *E*, *F*, and *G* show forces of 12, 15, and 20 lb, respectively. These forces are in equilibrium because the ring *O* is at rest.

   Now draw **MN** parallel to *OA* and to scale equal to 12 lb, as in Fig. 2-24. From *N*, draw **NP** equal to 15 lb and parallel to *OB*. Join *P* and *M*. **PM**, by careful measurement, will be found to scale 20 lb. If the angles of the triangle *MNP* are found with a protractor and compared with angles *AOC* and *BOC*, **PM** will be found to be parallel to *OC*.

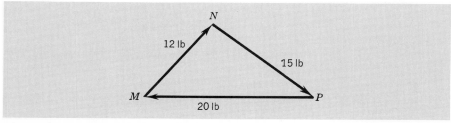

Fig. 2-24 Force triangle.

The equilibrant, or balancing force, of any two forces is the closing line of the triangle, with the arrowhead pointing in such a direction that the arrowheads of all the forces appear to follow each other around the triangle.

Triangle *MNP* is called the *force triangle*. Thus, three concurrent forces in equilibrium can be represented as sides of a force triangle.

2-11 PRINCIPLE OF CONCURRENCE

If three nonparallel forces acting on a body are in equilibrium, their lines of action must meet in a common point. As already shown, the resultant of any two concurrent forces is a single force acting through their point of intersection. When equilibrium exists, the resultant is balanced by a third force (equilibrant). However, in order to be in equilibrium, the resultant and the equilibrant must act along the same straight line. To meet that condition, the lines of action of the original forces must meet at a point. This principle of concurrence will be found useful in later problems in which the direction of some of the unknown forces is not at first evident.

2-12 METHODS OF SOLUTION

Since three concurrent forces in equilibrium can be represented as a triangle of forces, the solution of such a problem, in mechanics, is the solution of a triangle.

If the triangle of forces is constructed accurately, the unknown parts of the triangle can be found by measurement. This is called the *graphical solution.*

When the triangle is solved by means of algebra or trigonometry, the method is called *algebraic.*

The three theorems used in the algebraic method are as follows.

1. *Pythagorean theorem. In any right triangle, the square of the hypotenuse is equal to the sum of the squares of the two legs:*

$$c^2 = a^2 + b^2 \tag{2-1}$$

2. *Law of sines. In any triangle, the sides are to each other as the sines of the opposite angles:*

$$\frac{a}{\sin A} = \frac{b}{\sin B} = \frac{c}{\sin C} \tag{2-2}$$

3. *Law of cosines. In any triangle, the square of any side is equal to the sum of the squares of the other two sides minus twice the product of the sides and cosine of their included angle:*

$$c^2 = a^2 + b^2 - 2ab \cos C \tag{2-3}$$

Sample Problem 6 Two forces of 75 and 100 lb make an angle of 110° with each other as shown in Fig. 2-25. Find their resultant and the angle it makes with the 100-lb force.

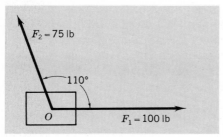

Fig. 2-25 Diagram for Sample Problem 6.

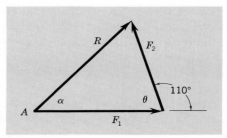

Fig. 2-26 Force triangle for Sample Problem 6.

Solution: Sketch a force-vector diagram as in Fig. 2-26.

$$\theta = 180 - 110 = 70°$$

By Eq. (2-3),

$$R^2 = F_1{}^2 + F_2{}^2 - 2F_1F_2 \cos \theta$$
$$= 100^2 + 75^2 - 2(100)(75)(0.3420)$$
$$= 10\,000 + 5625 - 5130 = 10\,495$$
$$R = 102 \text{ lb}$$

By Eq. (2-2),

$$\frac{F_2}{\sin \alpha} = \frac{R}{\sin \theta}$$

$$\sin \alpha = \frac{F_2}{R} \sin \theta = \frac{75}{102}(0.9397) = 0.691$$

$$\alpha = 43.7° \text{ or } 43°42'$$

*Sample Problem 7 Three concurrent forces of 45, 60, and 90 N are in equilibrium (Fig. 2-27). Find the angles that the forces must make with each other.

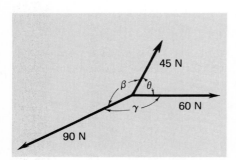

Fig. 2-27 Diagram for Sample Problem 7.

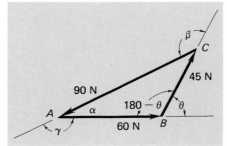

Fig. 2-28 Force triangle for Sample Problem 7.

Solution: Since the forces are in equilibrium, they can be represented as the three sides of a triangle (Fig. 2-28).

With A as a starting point, draw **AB** parallel and equal to the 60-N force. With B as a center and a line equal to 45 N as a radius, strike an arc. With A as a center and a radius equal to 90 N, draw an arc intersecting the first arc at C. Draw **AC** and **BC**. Then ABC is the required force triangle (Fig. 2-28) with sides parallel to the given forces. The required angles can be found by means of a protractor. A more accurate method is to apply the law of cosines.

By Eq. (2-3),

$$45^2 = 60^2 + 90^2 - 2(60)(90) \cos \alpha$$

$$\cos \alpha = \frac{60^2 + 90^2 - 45^2}{2(60)(90)} = \frac{9675}{10\ 800} = 0.896$$

$$\alpha = 26.4° \text{ or } 26°24'$$

By Eq. (2-2),

$$\frac{90}{\sin(180 - \theta)} = \frac{45}{\sin \alpha} \qquad \text{but} \qquad \sin(180 - \theta) = \sin \theta$$

$$\sin \theta = \frac{90}{45} \sin \alpha = \frac{90}{45}(0.4446) = 0.8892$$

$$\theta = 62.8° \text{ or } 62°48'$$

Referring to Fig. 2-28,

$$\gamma = 180 - \alpha = 180 - 26.4$$
$$= 153.6° \text{ or } 153°36'$$

Referring to Fig. 2-27,

$$\beta = 360 - \theta - \gamma = 360 - 62.8 - 153.6$$
$$= 143.6° \text{ or } 143°36'$$

The reader should note that the angles between the forces (Fig. 2-27) are supplements of the angles of the force triangle (Fig. 2-28).

2-13 FREE BODY

In solving problems of mechanics, it is often convenient to disregard bodies acting upon a single body and to replace those actions by force vectors that have their magnitude and direction the same as those of the actions produced by each of the acting bodies. A body thus represented is said to be a *free body*.

It is essential to designate the free body clearly and to account for all the forces that act on it. A sketch showing the body with the forces acting is an important aid in the investigation and solution of problems and should always be drawn.

Take, for example, a 10-ft ladder resting against a smooth wall. (A smooth surface is assumed to offer no resistance to sliding.) The ladder weighs 25 lb. A 155-lb man stands on it, as shown in Fig. 2-29, while his helper holds his foot against the bottom to prevent the ladder from slipping on the smooth floor.

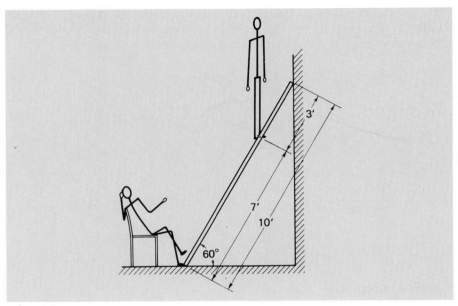

Fig. 2-29 Man on a ladder.

In order to analyze this problem, consider the ladder as a free body. To sketch a free-body diagram, draw a line to represent the ladder so that the direction and length are proper. Place force vectors at the points on the line to represent the external forces which act on the ladder. The free-body diagram for the ladder is shown in Fig. 2-30.

In this free-body diagram, the weight of the ladder and the man are shown acting downward. R_1 and R_2 are the reactions of the floor and wall, respectively. F is the force applied by the helper's foot. This particular problem involves nonconcurrent-coplanar forces and will be solved in Chap. 4 (Sample Problem 1). Free-body diagrams are useful for all types of equilibrium force systems.

*Sample Problem 8   A body with a mass of 300 kg is suspended from a beam by two ropes that make angles of 20° and 30° with the vertical, respectively. Find the forces in the ropes.

Solution: Let *AB* and *BC* be the ropes that support the body (Fig. 2-31). Both *AB* and *BC* are two-force members, and they are in tension.

Set point *B* out as a free body. There are three forces: the weight of the suspended body acting down ($W = m \cdot g = 300$ kg $\cdot$ 9.81 m/s² $= 2940$ N) and F_1 and F_2 acting up at angles 20° and 30°, as shown in Fig. 2-32.

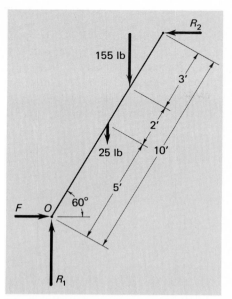

Fig. 2-30 Free-body diagram of ladder.

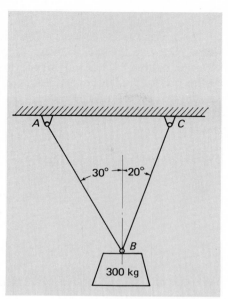

Fig. 2-31 Diagram for Sample Problem 8.

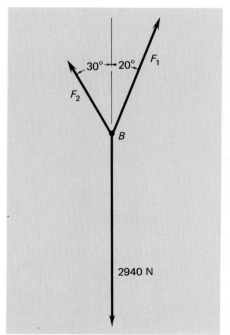

Fig. 2-32 Free-body diagram, at point B, for Sample Problem 8.

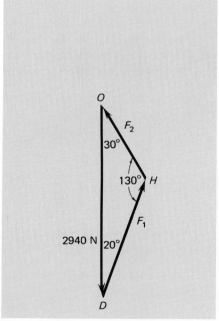

Fig. 2-33 Force triangle, at point B, for Sample Problem 8.

From any point O in Fig. 2-33, draw **OD** to represent 2940 N. From D, draw a line parallel to F_1 but indefinite in length. F_1 is known in direction but not in magnitude. Since the three forces must form a triangle, force F_2 must end at the point O. Therefore, draw a line through O, parallel to F_2, cutting the second line at H. Triangle ODH is the force triangle, and **DH** and **HO** to scale represent F_1 and F_2. These values can be obtained by direct measurement. By the law of sines,

$$\frac{F_1}{\sin 30°} = \frac{2940}{\sin 130°} \quad \text{and} \quad \frac{F_2}{\sin 20°} = \frac{2940}{\sin 130°}$$

$$F_1 = \frac{0.50(2940)}{0.7660} = 1919 \text{ N} \quad \text{and} \quad F_2 = \frac{0.3420(2940)}{0.7660} = 1313 \text{ N}$$

Say, $F_1 = 1920 \text{ N} = 1.92 \text{ kN (tension)}$

and $F_2 = 1310 \text{ N} = 1.31 \text{ kN (tension)}$

2-14 ANALYSIS OF A SIMPLE STRUCTURE (PINNED JOINTS)

A simple structure is a body formed of tension and compression members (two-force members) hinged or pinned together. The hinges or pins are assumed to be frictionless, and the weight of the parts is neglected. Most roof trusses, many bridge trusses, and certain mechanical devices are treated as simple structures.

Since the parts are in equilibrium, free-body diagrams can be drawn for any of the parts in a simple structure. However, it is usually more practical to use the pins at the joints as free bodies, since the pins are generally subjected to concurrent forces which are in equilibrium.

Consider a structure made up of a boom BC connected to a wall by pin C and a tie rod AB connected to the boom at pin B and to the wall at pin A. A load of W lb is suspended from the end of the boom, as shown in Fig. 2-34.

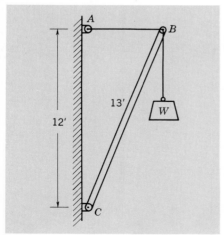

Fig. 2-34 A simple structure.

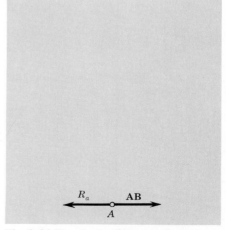

Fig. 2-35 Free-body diagram of pin A.

and
$$\frac{AB}{5} = \frac{4000}{12}$$
$$AB = 1670 \text{ lb (tension)}$$

Whenever the principle of similar triangles is applicable, it should be used.

2-15 COMPONENTS OF A FORCE

The resultant of two forces was previously defined as the single force that will produce the same effect as the two forces. Also, it was stated that the two forces and their resultant form a triangle.

The converse of this statement also is true: that a force can be replaced by any two forces which, with the given force, form a triangle. In Fig. 2-39, **AB** and **BC** are two forces that, if applied to a body at A, produce the same effect as **AC**. **AB** and **BC** are called the *components* of **AC**. Note the direction of the arrows.

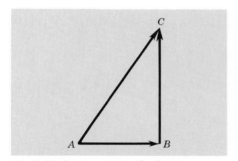

Fig. 2-39 Components of a force.

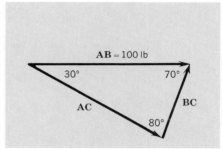

Fig. 2-40 Diagram for Sample Problem 10.

Sample Problem 10   Resolve a force of 100 lb into two components making angles of 30° and 70°, respectively, with the force.

Solution: Let **AB** be the given force (Fig. 2-40). Draw **AC** and **BC** making angles of 30° and 70°, respectively. Then **AC** and **BC** are the components. By the law of sines,

$$\frac{AC}{\sin 70°} = \frac{100}{\sin 80°} \qquad AC = \frac{0.9397(100)}{0.9848} = 95.4 \text{ lb}$$

$$\frac{BC}{\sin 30°} = \frac{100}{\sin 80°} \qquad BC = \frac{0.50(100)}{0.9848} = 50.8 \text{ lb}$$

Say, **AC** = 95 lb and **BC** = 51 lb.

Therefore, forces of 95 and 51 lb acting at A (as in Fig. 2-41) are the components of **AB**.

*Sample Problem 11   A force of 93 N has forces of 60 and 75 N for components. Find their angles to the given force (Fig. 2-42).

Solution: Triangle ABC can be solved for angles α and β by Eqs. (2-2) and (2-3). By using the law of cosines, we have

$$(BC)^2 = (AB)^2 + (AC)^2 - 2(AB)(AC) \cos \alpha$$

$$\cos \alpha = \frac{(AB)^2 + (AC)^2 - (BC)^2}{2(AB)(AC)}$$

$$= \frac{93^2 + 60^2 - 75^2}{2(93)(60)} = \frac{8650 + 3600 - 5625}{11\,160} = \frac{6625}{11\,160}$$

$$= 0.594$$

$$\alpha = 53.6° \text{ or } 0.935 \text{ rad}$$

(Refer to Table 14, App. B, for conversion from degrees to radians.) By the law of sines,

$$\frac{60}{\sin \beta} = \frac{75}{\sin 53.56°}$$

$$\sin \beta = \frac{60(0.8045)}{75} = 0.6436$$

$$\beta = 40.1° \text{ or } 0.699 \text{ rad}$$

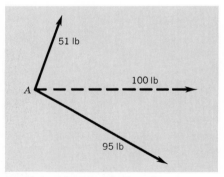

Fig. 2-41 Diagram for Sample Problem 10.

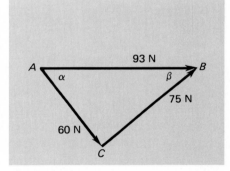

Fig. 2-42 Diagram for Sample Problem 11.

2-16   RECTANGULAR COMPONENTS OF A FORCE

When a force is resolved into two components that are at right angles to each other, the components are called *rectangular components*. With the given force as hypotenuse, they form a right triangle.

In the solution of many problems in mechanics, it is customary to choose an origin and x and y axes at right angles to each other. The force or forces acting are then resolved into their rectangular components along the axes. If the x axis is horizontal and the y axis is vertical, the components of a force along the axes are called *horizontal* and *vertical components*.

The axes are not always horizontal and vertical (an example is shown in Fig. 2-45), but the components will be referred to as x and y components and represented by F_x and F_y when the given force is F.

Let F be a force making an angle θ with the horizontal as shown in Fig. 2-43. From the point B, drop a perpendicular to the x axis. Then $F_x = \mathbf{OA}$, and $F_y = \mathbf{AB}$.

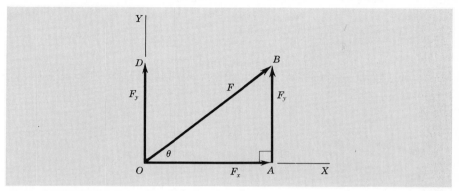

Fig. 2-43 Rectangular components of a force.

From trigonometry,

$$\cos \theta = F_x/F$$
$$\sin \theta = F_y/F$$
$$F_x = F \cos \theta$$
$$F_y = F \sin \theta$$

But F_x is the projection of F on the x axis; and since $\mathbf{AB} = \mathbf{OD}$, F_y is the projection of F on the y axis.

The following definitions can then be stated.

The x component of a force is the force times the cosine of the angle that the force makes with the x axis.

The y component of a force is the force times the cosine of the angle that the force makes with the y axis, or the force times the sine of the angle that the force makes with the x axis.

When a force is at right angles with a line, its component is zero along the line; that is, the force has no effect on a body in a direction at right angles to the force.

Sample Problem 12   A force of 50 lb makes an angle of 30° with the horizontal. Find F_x and F_y (Fig. 2-44).

Solution:
$$F_x = F \cos \theta = 50 \cos 30° = 43.3 \text{ lb} \qquad \text{say, 43 lb}$$
$$F_y = F \sin \theta = 50 \sin 30° = 25 \text{ lb}$$

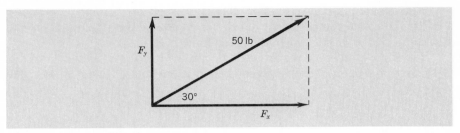

Fig. 2-44 Diagram for Sample Problem 12.

*Sample Problem 13   A body having a mass of 50 kg rests on an inclined plane making an angle of 20° with the horizontal. Find F_x and F_y of the body's weight (Fig. 2-45).

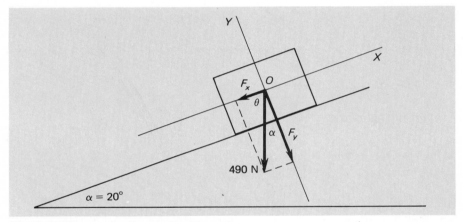

Fig. 2-45 Diagram for Sample Problem 13.

Solution: Take the x axis as parallel and the y axis as perpendicular to the plane. The weight of the body is a force of $(50 \text{ kg})(9.81 \text{ m/s}^2) = 490 \text{ N}$ acting down. But $\theta = 70°$ and $\alpha = 20°$. Then

$$F_x = 490 \cos 70° = 490(0.3420) = 167.6 \text{ N}$$
$$F_y = 490 \cos 20° = 490(0.9397) = 460.5 \text{ N}$$

Say, $F_x = 170 \text{ N}$ and $F_y = 460 \text{ N}$.

2-17   INCLINED PLANE

Let a body of weight W rest on an inclined plane making an angle α with the horizontal (Fig. 2-46).

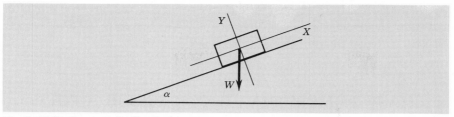

Fig. 2-46 Body on an inclined plane.

Choose the axes along and perpendicular to the face of the plane. Then

$$F_x = W \sin \alpha$$
$$F_y = W \cos \alpha$$

That is, the component of the weight of a body parallel to the plane is equal to the weight times the sine of the angle that the plane makes with the horizontal.

The component of the weight perpendicular to the plane is the weight times the cosine of the angle that the plane makes with the horizontal.

Since the y component is the one perpendicular to the plane, it produces no motion of the body.

The component of the weight parallel to the plane (F_x) is the force that acts to drag the body down the plane. If the body does not move, it is because of the reaction of some external force, such as friction, or of an applied force. (Friction is the force that acts between two surfaces in contact and tends to resist motion of one surface over the other. A more complete discussion of friction is to be found in Chap. 6.)

If there is no friction, then it can be seen that the applied force necessary to prevent the body from sliding down the plane will be equal to $W \sin \alpha$. A force parallel to the plane and slightly greater than $W \sin \alpha$ by the amount of the friction encountered will be needed to move the body up the plane. Frequent use is made of the inclined plane in loading heavy machinery on trucks, in jackscrews, and in ordinary screw threads. The general principle involved is that a relatively small force exerted parallel to the plane is capable of displacing the body along the plane and thereby raising the body to the desired height.

Sample Problem 14 The floor of a truck is 3 ft above the ground. An inclined plane is formed by using a 12-ft plank. It is desired to load a box of machine parts weighing 300 lb by pushing it up the 12-ft plane. How much force must be exerted parallel to the plane, neglecting friction?

Solution:

$$\sin \alpha = \tfrac{3}{12} = \tfrac{1}{4}$$
$$F_x = W \sin \alpha = 300(\tfrac{1}{4}) = 75 \text{ lb}$$

A force slightly in excess of 75 lb and applied to the box in a direction parallel to the plane will slide the box up the plane. It is seen that a vertical lift without using the inclined plane would require 300 lb.

2-18 RESULTANT OF MORE THAN TWO FORCES IN A PLANE

Suppose that three or more forces act through a common point O as shown in Fig. 2-47. Find the resultant.

Graphical Solution: From any convenient point A, as in Fig. 2-48, draw **AB** parallel and equal to F_1. From B, draw **BC** equal and parallel to F_2. **AC** is the resultant of F_1 and F_2. From C, draw **CD** equal and parallel to F_3. **AD** is the resultant of **AC** and **CD**. It is therefore the resultant of F_1, F_2, and F_3. From D, draw **DE** equal and parallel to F_4. **AE**, the resultant of **AD** and **DE**, is then the resultant of the four forces F_1, F_2, F_3, and F_4. *ABCDEA* is known as the *polygon of forces.*

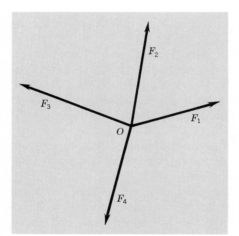

Fig. 2-47 Four concurrent forces.

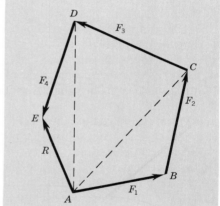

Fig. 2-48 Force polygon.

The method for determining the resultant of any number of concurrent forces is as follows. From any point, construct the force polygon by adding the vectors taken in order either clockwise or counterclockwise around the point. The closing line of the polygon drawn from the starting point of the first vector to the terminal point of the last vector is the resultant of the concurrent forces. (Note the direction of the arrowheads.)

Sample Problem 15 Given concurrent forces of 20, 30, 10, and 40 lb making angles of 10°, 30°, 120°, and 125°, respectively, with a horizontal line through the common point, find the resultant.

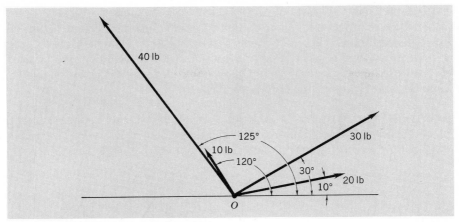

Fig. 2-49 Diagram for Sample Problem 15.

Solution: Figure 2-49 shows the given concurrent force vectors. A force poly-gon can now be constructed. From any point *A*, lay off the 20-lb vector **AB**. From point *B*, draw **BC** to represent the 30-lb force. Similarly, draw **CD** and **DE** for the 10- and 40-lb forces, respectively. Now, the resultant is the vector **AE** which closes the polygon, as in Fig. 2-50. By careful measurement, *R* is found to be 63 lb acting at an angle of 74° with the horizontal. These results may be verified by progressively solving triangles *ABC, ACD,* and *ADE* (Fig. 2-51) for unknown sides and angles by Eqs. (2-2) and (2-3). Note that θ is the sum of ∡*EAD,* ∡*DAC,* ∡*CAB,* and 10°.

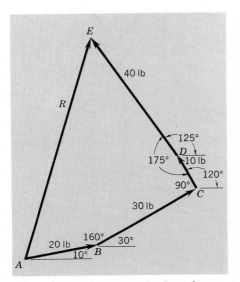

Fig. 2-50 Force polygon for Sample Problem 15.

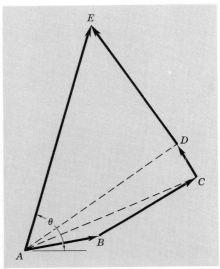

Fig. 2-51 Force polygon.

2-19 EQUILIBRIUM OF MORE THAN TWO FORCES

When three or more forces are in equilibrium, their resultant is zero. From Fig. 2-51, the resultant of the four forces shown is **AE**. The single force to be in equilibrium with **AE** is **EA**, a force equal in magnitude and opposite in direction. But **EA** closes the polygon. *Then when concurrent forces are in equilibrium, the force diagram must be a closed polygon.*

Sample Problem 16   A body weighing 50 lb rests on a rough horizontal plane. A force of 10 lb acts horizontally to the right. The body does not move. Find the force resisting motion and the reaction of the plane (Fig. 2-52).

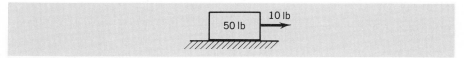

Fig. 2-52 Diagram for Sample Problem 16.

Solution: Let the body be considered as a particle (point O), and draw a free-body diagram (Fig. 2-53). At O, the weight of 50 lb acts down. The reaction of the plane is up and is represented by N. The force of 10 lb acts to the right. Since the body does not move, there must be a force acting between the two surfaces in contact to oppose the motion. Call it F. Construct the force polygon

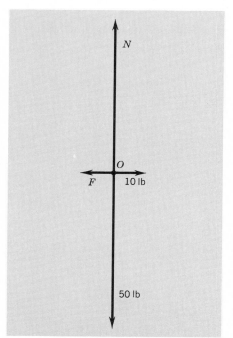

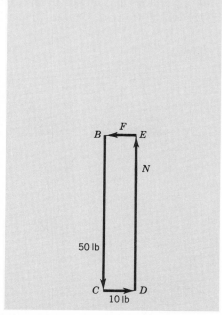

Fig. 2-53 Free-body diagram for Sample Problem 16.

Fig. 2-54 Force polygon for Sample Problem 16.

BCDEB (Fig. 2-54), taking the forces in order counterclockwise. Since *N* and *F* are known in direction only, first draw the forces completely known. **BC** is equal to 50 lb and **CD** to 10 lb. At *D*, erect a perpendicular indefinite in length. Since the four forces are in equilibrium, they must form a closed polygon. Then force *F* must end at *B*. From *B*, draw a line perpendicular to **BC** to meet the perpendicular from *D* at the point *E*. Then **DE** is *N* and **EB** is *F*. Since *BCDE* is a rectangle, lines **BC** and **DE** are equal in magnitude, or **DE** = 50 lb = *N*, and **EB** = 10 lb = *F*.

*Sample Problem 17   A bar *AC* (Fig. 2-55) is pinned to the wall at *A* and supports a mass of 1000 kg at *C*. It is held in a horizontal position by the strut, or compression member, *DB*, which is pinned at *D* and *B*. Find the force in *DB* and the pin reaction at *A*.

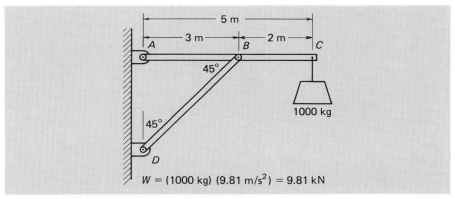

Fig. 2-55 Diagram for Sample Problem 17.

Solution: The bar *AC* is in equilibrium under the action of forces at *A*, *B*, and *C*. By the *principle of concurrence* (Sec. 2-11), *the lines of action of three (nonparallel) forces in equilibrium must pass through a common point.* The direction of the 9.81-kN load is vertical at point *C*. Since *DB* is a two-force member in compression, the direction of the thrust at point *B* must be along the axis of strut *DB*. Sketch *AC* as a free body and show the 9.81-kN load and the line of action of the force at *B*. Prolong these lines of action to meet at point *E* (Fig. 2-56). Then the reaction at pin *A* is a force that also must pass through point *E*, owing to concurrence. The force triangle can now be drawn, as in Fig. 2-57. The angle θ which R_a makes with the horizontal is found from Fig. 2-56. Since *BCE* is a right triangle and $\angle B = 45°$, length *EC* = length *BC* = 2 m. Now, from triangle *ACE*,

$$\tan \theta = \tfrac{2}{5} = 0.40$$
$$\theta = 21.8°$$

Thus, in Fig. 2-57,

$$\alpha = 90 + \theta = 111.8°$$

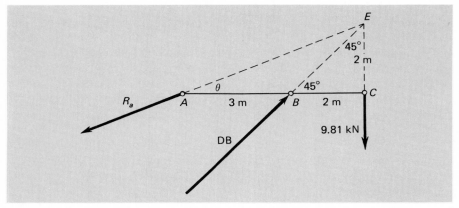

Fig. 2-56 Free-body diagram of member AC for Sample Problem 17.

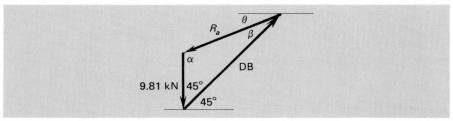

Fig. 2-57 Force triangle, at point E, for Sample Problem 17.

Hence, $\beta = 180 - 111.8 - 45 = 23.2°$

Now, by the law of sines,

$$\frac{R_a}{\sin 45°} = \frac{9.81}{\sin 23.2°} \qquad R_a = \frac{0.7071(9.81)}{0.3939} = 17.61 \text{ kN} \qquad \text{say, } 17.6 \text{ kN}$$

$$\frac{\text{DB}}{\sin 111.8°} = \frac{9.81}{\sin 23.2°}$$

But $\sin 111.8° = \sin (90 + \theta)$

And $\sin (90 + \theta) = \sin (90 - \theta)$

Then $\sin 111.8° = \sin 68.2°$

$$= 0.9285$$

$$\text{DB} = \frac{0.9285(9.81)}{0.3939}$$

$$\text{DB} = 23.12 \text{ kN (compression)} \qquad \text{say, } 23.1 \text{ kN}$$

Sample Problem 18   The beam AD is pinned to the wall at A and held in a horizontal position by the tension member BC. There is a load of 1000 lb at D. What are the force in BC and the reaction at A (Fig. 2-58)?

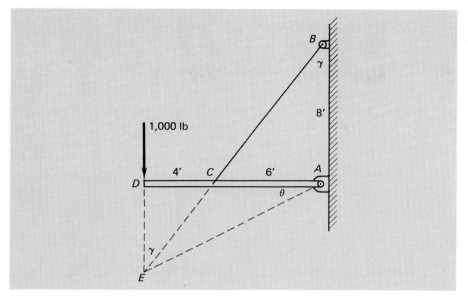

Fig. 2-58 Diagram for Sample Problem 18.

Solution: The beam AD is in equilibrium under the action of forces at A, C, and D. Since the member BC is acted on by forces at its ends only, it is a two-force member. The force at C is in the direction of BC. The load is vertical. Prolong line BC to intersect the vertical through D. Then E will be the point of concurrence of the forces. The reaction A is a force whose line of action must pass through E. The directions of all the forces are now known, and the force triangle can be constructed. Angles θ and γ can be determined from Fig. 2-58. From right triangle CAB,

$$\tan \gamma = \tfrac{6}{8} = 0.75$$
$$\gamma = 36.9° \text{ or } 36°54'$$

Note that angle CED also is γ.

Now triangle ABC is similar to triangle CED. Thus,

$$\frac{\text{length } DE}{\text{length } AB} = \frac{\text{length } DC}{\text{length } AC}$$

$$DE = \frac{8(4)}{6} = 5.33 \text{ ft}$$

In triangle AED,

$$\tan \theta = \frac{5.33}{10} = 0.533$$

$$\theta = 28.1° \text{ or } 28°6'$$

Figure 2-59 is the force triangle for the three concurrent forces acting on member AD

$$\alpha = 90 + \theta = 118.1° \text{ or } 118°6'$$
$$\beta = 180 - \alpha - \gamma = 180 - 118.1 - 36.9 = 25°$$

By the law of sines,

$$\frac{R_a}{\sin \gamma} = \frac{1000}{\sin \beta} \qquad R_a = \frac{0.6004(1000)}{0.4226} = 1420 \text{ lb}$$

$$\frac{BC}{\sin \alpha} = \frac{1000}{\sin \beta}$$

$$\sin \alpha = \sin (90 + \theta) = \sin (90 - \theta)$$

$$\sin (90 - \theta) = \sin 61.9° = 0.8821$$

$$BC = \frac{0.8821(1000)}{0.4226} = 2087 \text{ lb} \qquad \text{say, } 2100 \text{ lb (tension)}$$

Note that the point of concurrence need not lie on the body in equilibrium.

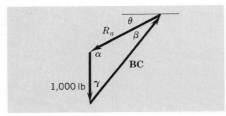

Fig. 2-59 Force triangle for Sample Problem 18.

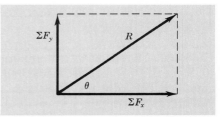

Fig. 2-60 Resultant by summation of components.

2-20   RESULTANT OF CONCURRENT FORCES BY SUMMATION

Any force can be resolved into two components at right angles to each other. If x and y axes are taken through the point common to all the forces and each force is resolved into its x and y components, the given forces are then replaced by two sets of forces acting along the axes of reference. The forces or component forces along the x axis can be combined by algebraic addition into a single force. This force is represented by ΣF_x. Similarly, there is found a force acting along the y axis and represented by ΣF_y (Fig. 2-60). The hypotenuse of the right triangle with legs equal to ΣF_x and ΣF_y is the resultant force in magnitude and direction of the given system of forces. That is,

$$F_x = F \cos \alpha \qquad (2\text{-}4)$$
$$F_y = F \sin \alpha \qquad (2\text{-}5)$$
$$R = \sqrt{(\Sigma F_x)^2 + (\Sigma F_y)^2} \qquad (2\text{-}6)$$
$$\tan \theta = \frac{\Sigma F_y}{\Sigma F_x} \qquad (2\text{-}7)$$

Sample Problem 19 Two 120-lb forces, one of which is horizontal and the other is at an angle of 60° with the horizontal, act on a body (point O). Find their resultant (Fig. 2-61).

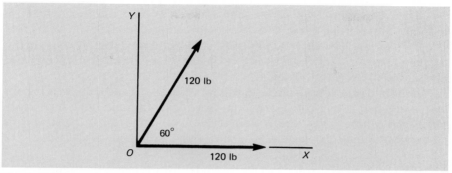

Fig. 2-61 Diagram for Sample Problem 19.

Solution: Establish x and y axes. The components can be tabulated as follows:

Force	Angle with x axis	F_x	F_y
120	0	120	0
120	60	60	104
		180	104

Then
$$\Sigma F_x = 180 \text{ lb}$$
$$\Sigma F_y = 104 \text{ lb}$$

In Fig. 2-62, ΣF_x and ΣF_y are shown at 90° with each other. Their resultant is the hypotenuse R. By Eq. (2-6),

$$R = \sqrt{180^2 + 104^2} = \sqrt{43\,200} = 208 \text{ lb}$$

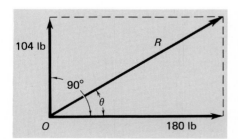

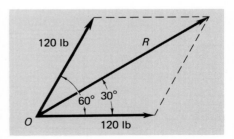

Fig. 2-62 Summation of components for Sample Problem 19.

Fig. 2-63 Parallelogram solution for Sample Problem 19.

By Eq. (2-7), $\quad \tan \theta = \dfrac{\Sigma F_y}{\Sigma F_x} = \dfrac{104}{180} = 0.5778$

$$\theta = 30°$$

Then a force of 208 lb acting at an angle of 30° with the horizontal is the resultant of the given forces.

That the angle of the resultant is 30° could have been seen by constructing the parallelogram of forces as shown in Fig. 2-63. Since the figure is equilateral, the diagonal bisects the 60° angle.

If the body at O were a pin fastened to a vertical wall and held rigid, the pin, by the law that action equals reaction, would exert a force equal to R in the opposite direction. Therefore, the pin reaction is a force of 208 lb acting from O, opposite R, and the angle is 30° below the negative side of the x axis.

Sample Problem 20   Find the resultant of the forces with magnitude and direction as shown in Fig. 2-64.

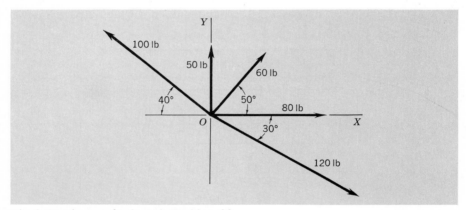

Fig. 2-64 Diagram for Sample Problem 20.

Solution: Resolve each force into its x and y components and arrange the data in a table.

Force	Angle with x axis	F_x	F_y
80	0	+ 80	0
60	50	+ 38.6	+ 46
50	90	0	+ 50
100	40	− 76.6	+ 64.3
120	30	+104	− 60
		+146	+100.3

Note that the forces in the left-hand column cannot be added algebraically because they do not act in the same straight line, whereas those in the

F_x and F_y columns can be added algebraically because they do act in the same line.

$$\Sigma F_x = 80 + 38.6 + 0 - 76.6 + 104 = +146 \text{ lb}$$
$$\Sigma F_y = 0 + 46 + 50 + 64.3 - 60 = +100.3 \text{ lb}$$

Then, from Fig. 2-65,

$$R = \sqrt{146^2 + 100.3^2} = \sqrt{31\ 380} = 177 \text{ lb}$$
$$\tan \theta = \frac{\Sigma F_y}{\Sigma F_x} = \frac{100.3}{146} = 0.687$$
$$\theta = 34.5° \text{ or } 34°30'$$

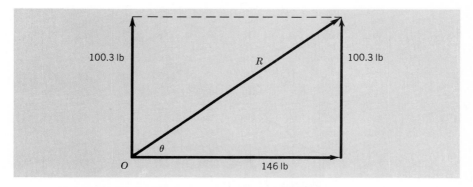

Fig. 2-65 Summation of components for Sample Problem 20.

Therefore, the resultant is a force of 177 lb acting at an angle of 34.5° with the horizontal.

2-21 EQUILIBRIUM OF CONCURRENT FORCES

In Sec. 2-20, it was found that, when concurrent forces act on a body, the resultant

$$R = \sqrt{(\Sigma F_x)^2 + (\Sigma F_y)^2}$$

Now, when forces are in equilibrium, their resultant is zero. Then

$$(\Sigma F_x)^2 + (\Sigma F_y)^2 = 0$$

But both expressions are plus, being squares of numbers. The sum of two positive numbers cannot be zero unless each one is zero. That is,

$$\Sigma F_x = 0 \quad \text{and} \quad \Sigma F_y = 0 \quad\quad\quad (2\text{-}8)$$

Then, when concurrent forces are in equilibrium, the sum of the components along each of two rectangular axes must be zero.

Conversely, if $\Sigma F_x = 0$ and $\Sigma F_y = 0$, the concurrent forces are in equilibrium and their resultant is zero.

The foregoing law can be illustrated in another way. A fundamental principle of mechanics is that a body cannot be in equilibrium when acted upon by a single force. Now, ΣF_x is the sum of all the components of acting forces along any arbitrarily chosen axis. If a body is in equilibrium, there can be no change in motion. Therefore, there cannot be a force acting in any direction, or the body would move or change its motion. Consequently,

$$\Sigma F_x = 0 \quad \text{and} \quad \Sigma F_y = 0$$

The meaning of Eq. (2-8) can also be illustrated by the force polygon shown in Fig. 2-66. Suppose $ABCDEA$ is the force polygon of five forces in equilibrium. Now, the x component of a force is its projection on the x axis as in Fig. 2-66.

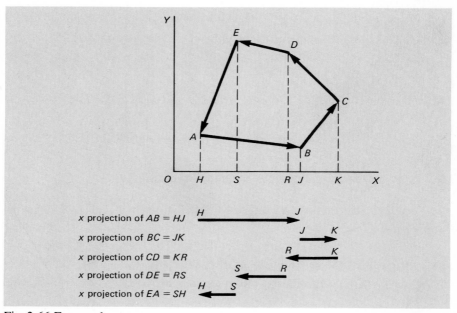

Fig. 2-66 Force polygon.

But $\Sigma F_x = \mathbf{HJ} + \mathbf{JK} + \mathbf{KR} + \mathbf{RS} + \mathbf{SH} = 0$. That is, the sum of the positive projections is equal and opposite to the sum of the negative projections.

A corollary to the preceding law is that, when the resultant effect of concurrent forces along any direction or axis is desired, simply project each of the forces onto the axis and find the algebraic sum of the projections.

PROBLEMS

*2-1. A body is acted on by an upward force of 12 N and a horizontal force of 20 N to the right. What are the amount and the direction of the resultant?

2-2. Two forces of 100 lb each act on a body at an angle of 120° with each other. Find the resultant, in magnitude and direction, by constructing the triangle and also by trigonometry.

2-3. Two equal forces P act on a body at an angle of 60° with each other. Find the resultant by trigonometry.

2-4. Two concurrent forces of 40 and 50 lb make an angle of 30° with each other. Find the direction and magnitude of their resultant.

2-5. Resolve a force of 60 lb into two components, one of which is 38 lb and makes an angle of 45° with the given force. *Hint:* Construct the triangle and apply cosine and sine laws.

*2-6. Three forces of 40, 50, and 75 N make angles of 10°, 30°, and 120°, respectively, with the x axis. What are the amount and the direction of the resultant?

*2-7. A body is supported by two cords, each of which makes an angle of 50° with the vertical. The tension in each cord is 65 N. What is the mass of the body?

2-8. A weight of 6000 lb is suspended from the point C shown in Fig. Prob. 2-8. Find the forces in AC and BC if $\alpha = 30°$ and $\beta = 60°$. Which member is a tie rod (tension), and which member is a brace (compression)?

*2-9. A mass of 10 metric tons (see Table 13, App. B) is carried by the simple derrick shown in Fig. Prob. 2-9. By making use of similar triangles, find the forces in AB and BC.

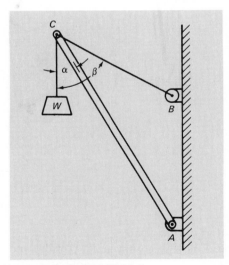 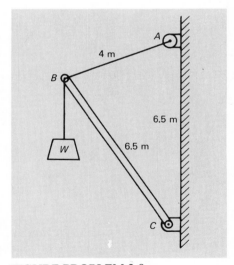

FIGURE PROBLEM 2-8   FIGURE PROBLEM 2-9

2-10. A body weighing 160 lb is suspended from a hook by a rope 12 ft long. If a horizontal force is applied to the body so as to swing the body to one side and cause the rope to make an angle of 30° with the vertical, what will be the tension in the cord and what is the amount of the horizontal force?

2-11. Find the rectangular components of a force of 200 lb making (*a*) an angle of 10°, (*b*) an angle of 35°, (*c*) an angle of 55°, (*d*) an angle of 80°, and (*e*) an angle of 90° with the *x* axis.

2-12. A simple truss supports a load of 8 kip (1 kip = 1000 lb) (Fig. Prob. 2-12). Find the forces (in kips) in members *AB*, *BC*, and *AC* and the reactions at *A* and *C*.

2-13. A wire 24 in long will stand a straight pull of 100 lb. The ends are fastened to two points 21 in apart on the same level. What weight suspended from the middle of the wire will break the wire?

*2-14. The lifting force of a balloon is 600 N. The anchor rope makes an angle of 65° with the vertical. Find the tension in the anchor rope and the horizontal force of the wind against the balloon.

2-15. A truck weighing 2000 lb rests on a slope of 30° inclination. Find the force that will be needed to start the stalled truck up the grade if the resistance to traction is 50 lb.

*2-16. A tractor is attached to a dump car by a cable which makes a horizontal angle of 10° with the track. A force of 1000 N is required to move the car. Find the force that the tractor exerts. What is the lateral force on the rails?

2-17. A boiler weighing 2 tons is supported by cables that make angles of 30° and 40° with the vertical while the boiler is being hoisted into place. Find the tension in each cable.

2-18. Find the forces in *AC* and *BC* of Fig. Prob. 2-18.

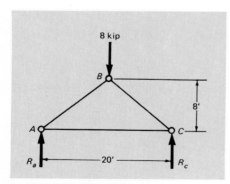

FIGURE PROBLEM 2-12

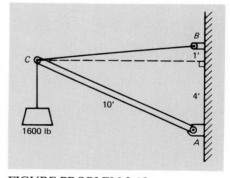

FIGURE PROBLEM 2-18

2-19. Find the forces in *BC* and *AB* of Fig. Prob. 2-19.

2-20. A 300-lb body rests on a plane. Find the components parallel and perpendicular to the plane for the following angles of inclination: (*a*) 20°, (*b*) 40°, (*c*) 70°, and (*d*) 85°.

*2-21. If a force of 200 N is sufficient to move a body on a horizontal plane, find the force necessary to move the body uniformly up a plane of slope 1 in 10, assuming the resistance is the same in both cases.

2-22. Find the thrusts in *AB* and *BC* in Fig. Prob. 2-22. Also, find the vertical and horizontal components of the reactions at *A* and *C*.

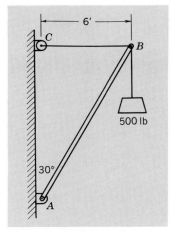

FIGURE PROBLEM 2-19

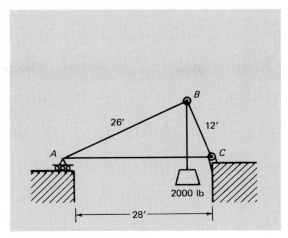

FIGURE PROBLEM 2-22

2-23. Figure Problem 2-23 shows a frame made of four pieces of equal length with pivots at points M, N, S, and R. With the forces acting as shown, find the forces in each of the arms and the horizontal reactions at S and N.

2-24. Given three concurrent forces of 50, 75, and 90 lb making angles of 0°, 70°, and 120°, respectively, with the x axis, find the resultant.

*2-25. In the toggle joint shown in Fig. Prob. 2-25, what vertical force is exerted against the plunger owing to the force of 4.0 kN? $AB = 0.4$ m. $BC = 0.4$ m. $BD = 0.3$ m. Assume joint C to be stationary.

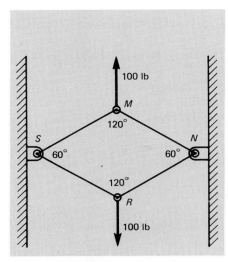

FIGURE PROBLEM 2-23

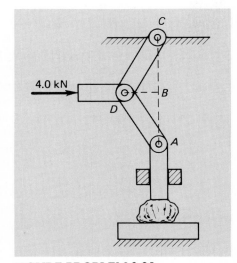

FIGURE PROBLEM 2-25

3

Moments

3-1 MOMENTS

A complete study of force systems and their applications requires the understanding of the concept of moments.

Figure 3-1 shows a person balancing a 100-lb load by means of a lever. A downward force of 40 lb applied at A is just sufficient to balance the load of 100 lb. The distance from the pivot to the end of the lever at which the force is applied is 5 ft. Then, 40 lb $\times$ 5 ft $= 200$ ft·lb is necessary to balance the turning effort of the load, which also is 200 ft·lb; that is, 100 lb $\times$ 2 ft $= 200$ ft·lb (Fig. 3-2). If the person were to try balancing the load by applying a force at B, which is 4 ft from the pivot, 50 lb would be required. That is, 50 lb $\times$ 4 ft $= 200$ ft·lb. At the 1-ft point, C, it would take 200 lb to do the job; 200 lb $\times$ 1 ft $= 200$ ft·lb. Thus, 200 ft·lb is a measure of the turning effort required to balance the load and is called the *moment* of the force about the pivot.

It should be noted that the moment of a force varies directly with the distance of the force from the pivot. The *moment is measured by the product of the force and the perpendicular distance from the pivot to the line of action of the force.*

For example, it is much easier to turn a revolving door by pushing at the outer edge of the door, as in Fig. 3-3, than by pushing at the center, as in Fig. 3-4.

The *moment of a force about an axis* is defined as the force multiplied by the perpendicular distance from the axis to the force. For simplicity, the moment of a force about a point in a plane is understood to be a moment about an axis perpendicular to the plane. A moment is expressed in foot-pounds or in

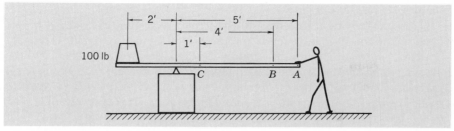

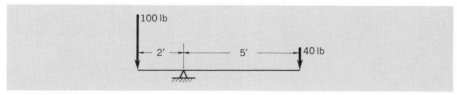

Fig. 3-1 Use of lever to balance weight.

Fig. 3-2 Lever.

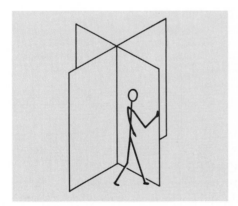

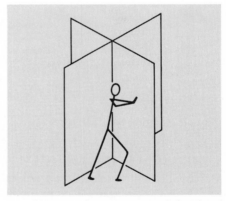

Fig. 3-3 Revolving door easily rotated.

Fig. 3-4 Revolving door rotated the "hard way."

inch-pounds, newton-meters, or other combinations of force and distance units.

As a further illustration, if a force is applied to the rim of a pulley mounted on a shaft, the pulley will rotate. Experience shows that it is easier to turn the pulley when the force is applied to the rim than when it is applied closer to the shaft.

It is quite apparent that any effort to turn a wheel will be applied not parallel to a spoke, but rather perpendicular to a spoke. Likewise, a force applied parallel to the axle will not rotate the wheel. From such observations, it is seen that no rotation or moment results when the line of action of the force either intersects the axis or is parallel to it. These several conditions will be found useful in deciding the axis about which moments may be taken.

3-2   SIGN OF MOMENTS

In Fig. 3-5, it is evident that the 100-lb force tends to rotate about AA in one direction and the 50-lb force tends to rotate about AA in the opposite direction. Some rule for the direction of rotation is then necessary. In this book, the following rotation rules will be used unless stated otherwise. *When a force tends to produce clockwise rotation, the moment is positive. When a force tends to produce counterclockwise rotation, the moment is negative.* For example:

Moment of 100-lb force $= +100 \times 4 = +400$ ft·lb

Moment of 50-lb force $= -50 \times 2 = -100$ ft·lb

Total turning effect $= +400 - 100 = +300$ ft·lb (clockwise rotation)

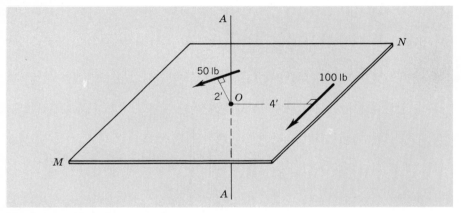

Fig. 3-5 Rotation about an axis.

3-3   EQUILIBRIUM OF PARALLEL FORCES

Forces whose lines of action are parallel are called *parallel forces.*

The principles involved in the study and application of parallel forces will be best understood by a simple experiment. Suspend a very light rod AC (Fig. 3-6) by means of spring balances at J and K. Now suspend 120 lb from point B. The scale at J will indicate 40 lb, and the one at K will show 80 lb. The forces at A, B, and C are parallel and in equilibrium (as shown in the free-body diagram, Fig. 3-7). Their sum is $40 + 80 - 120 = 0$. Take moments of all forces about point A as an axis. The algebraic sum of moments about A is represented by ΣM_a:

$$\Sigma M_a = (8 \times 120) - (12 \times 80) = 960 - 960 = 0$$

The algebraic sum of moments about point B is

$$\Sigma M_b = (8 \times 40) - (4 \times 80) = 320 - 320 = 0$$

Similarly, about C,

$$\Sigma M_c = (12 \times 40) - (4 \times 120) = 480 - 480 = 0$$

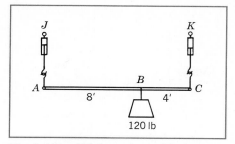

Fig. 3-6 Parallel forces in equilibrium.

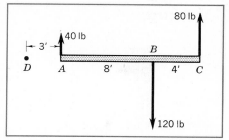

Fig. 3-7 Free-body diagram.

About D,

$$\Sigma M_d = -(3 \times 40) + (11 \times 120) - (15 \times 80)$$
$$= -120 + 1320 - 1200 = 0$$

These results illustrate the principle that *when parallel forces are in equilibrium, the sum of their moments about any axis through any point is zero.*

Parallel forces tend to translate and rotate the body on which they act. If such a body is in equilibrium, it must neither translate nor rotate. Then the *conditions for static equilibrium* are as follows:

1. The algebraic sum of the forces must be zero.

$$\Sigma F = 0 \tag{3-1}$$

2. The algebraic sum of the moments about any axis through any point must be zero.

$$\Sigma M = 0 \tag{3-2}$$

The resultant of the 40 lb at A and 80 lb at C must be the sum, $80 + 40 = 120$, and it must act upward at point B. Then

$$\frac{80}{40} = \frac{\text{length } AB}{\text{length } BC} = \frac{8}{4}$$

The resultant is applied at a point dividing the distance between the forces into two parts that are in an inverse ratio to the forces themselves.

In Fig. 3-8, the resultant of P_1 and P_2 is the sum, $P_1 + P_2$, and it acts at B such that

$$\frac{P_1}{P_2} = \frac{l_2}{l_1} \tag{3-3}$$

Clearing of fractions produces

$$l_1 P_1 = l_2 P_2 \tag{3-3a}$$

That is, the moments of the two forces about a point on the resultant are equal to each other.

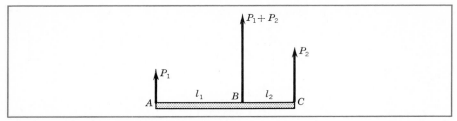

Fig. 3-8 Resultant of parallel forces.

Taking moments of the forces about A,

$$l_1(P_1 + P_2) = (l_1 + l_2)P_2$$

or
$$l_1 P_1 + l_1 P_2 = l_1 P_2 + l_2 P_2$$

The second equation could have been obtained by adding $l_1 P_2$ to both sides of Eq. (3-3a). The result shows that the moment of the resultant of two parallel forces about any point is equal to the sum of moments of the forces about the same point.

The preceding statement can be extended to include any number of co-planar parallel forces. Thus, the resultant of any number of parallel forces is parallel to and equal to the sum of the forces, and its moment about any point is equal to the sum of the moments of all the forces about the same point.

Sample Problem 1   A bar 10 ft long carries a load of 20 lb located 6 ft from the end. What force must be applied at each end to support the rod in equilibrium (Fig. 3-9)?

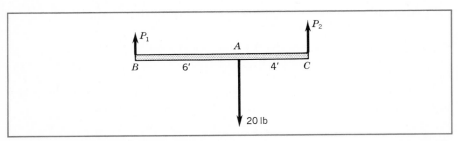

Fig. 3-9 Diagram for Sample Problem 1.

Solution a: The resultant of P_1 and P_2 must be 20 lb applied at A. By the inverse ratio,

$$\frac{P_1}{P_2} = \frac{4}{6}$$

That is, $\frac{4}{10}$ of the load is at B, or $\frac{4}{10}(20) = 8$ lb. Also, $\frac{6}{10}$ is at C, or $\frac{6}{10}(20) = 12$ lb.

Solution b: By the conditions of equilibrium,

$$\Sigma M_b = 6(20) - 10P_2 = 0$$
$$P_2 = 12 \text{ lb}$$

Since $\Sigma F_y = 0$ $P_1 + P_2 - 20 = 0$

$$P_1 = 20 - 12 = 8 \text{ lb}$$

Sample Problem 2   A beam resting on two end supports carries concentrated loads as shown in Fig. 3-10. Find the reactions of the supports.

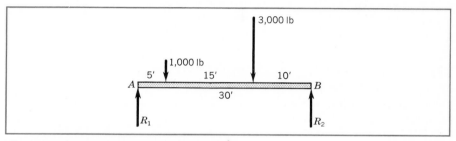

Fig. 3-10 Diagram for Sample Problem 2.

Solution a: Let R_1 and R_2 be the reactions. By the conditions of equilibrium, the algebraic sum of the moments about any point is zero. A convenient point to choose would be either the right or left support, since one of the unknowns would be eliminated. If point A is chosen,

$$\Sigma M_a = +1000(5) + 3000(20) - 30R_2 = 0$$
$$R_2 = 2170 \text{ lb}$$

But $\Sigma F_y = 0$, $R_1 + R_2 = 1000 + 3000 = 4000 \text{ lb}$

$$R_1 = 4000 - 2170 = 1830 \text{ lb}$$

Solution b: The problem can also be solved by the inverse ratio. The two components of 1000 lb acting down at A and B are found as follows:

$$\frac{\text{Component at } A}{\text{Component at } B} = \frac{25}{5} = \frac{5}{1}$$

Then   Component at $A = \frac{5}{6}(1000) = 833 \text{ lb}$

Component at $B = \frac{1}{6}(1000) = 167 \text{ lb}$

Similarly, $\dfrac{\text{Component of 3000 at } A}{\text{Component of 3000 at } B} = \dfrac{10}{20} = \dfrac{1}{2}$

Then   Component of 3000 at $A = \frac{1}{3}(3000) = 1000 \text{ lb}$

Component of 3000 at $B = \frac{2}{3}(3000) = 2000 \text{ lb}$

Total force at $A = 833 + 1000 = 1833 \text{ lb}$ say, 1830 lb

Total force at $B = 167 + 2000 = 2167 \text{ lb}$ say, 2170 lb

But R_1 and R_2 are the equilibrants, or balancing forces:

$$R_1 = 1830 \text{ lb} \qquad R_2 = 2170 \text{ lb}$$

3-4   UNIFORMLY DISTRIBUTED LOADS

Thus far, we have considered how to calculate the moment produced by a *concentrated* load. Since a concentrated load is assumed to be acting at a point, its moment arm is simply the perpendicular distance from the line of action of the load to the point about which the moment is assumed to be acting.

Another type of load frequently encountered is a *uniformly distributed load,* which is assumed to be acting over an area rather than at a point. Whereas a concentrated load is designated in terms of a force unit, such as 4000 lb, the

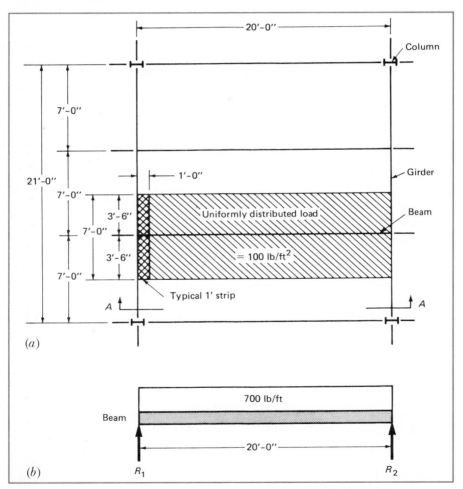

Fig. 3-11 (*a*) Partial plan view of structural steel floor framing. (*b*) Section *A-A* of Fig. 3-11*a*.

uniformly distributed load is usually designated in terms of a force unit per unit length, such as 750 lb/ft.

As an illustration of a uniformly distributed load, consider a multistory structural steel building whose partial typical floor is indicated in Fig. 3-11a. Assuming that the 21- × 20-ft bay will be subjected to a uniformly distributed load of 100 lb/ft², the beam indicated will be subjected to a uniformly distributed load of 700 lb/ft. This was arrived at as follows.

The floor beam in question is responsible for carrying the load halfway between itself and the beams on either side of it. Thus, this beam will be subjected to a 100-lb/ft² uniformly distributed load over an area measuring 20 × 7 ft (area shown shaded). Each *linear foot* of beam (typical 1-ft strip) will then be subjected to 100 lb/ft² × 7 ft = 700 lb/ft of uniformly distributed load. Figure 3-11b is a view of the beam supporting the load and in turn being supported by girders. (R_1 and R_2 indicate the reactions of the girders at the ends of the beam.)

In order to calculate the moment developed by a uniformly distributed load, the entire weight of the load is assumed to be concentrated at its centroid, which is the center of gravity of an area (see Sec. 10-2). Thus, the 700-lb/ft load produces a clockwise moment about R_1 equal to (700 lb/ft × 20 ft)(10 ft). The quantity 700 lb/ft × 20 ft represents the total weight of the load, and 10 ft represents the moment arm (the distance from the centroid of the area representing the load to R_1).

*Sample Problem 3   When a load is uniformly distributed over a beam, the total uniform load can be considered as acting at the center of the length of beam over which it is distributed if the moment of the load is desired. The weight of a beam is a uniformly distributed load. Find the reactions of the beam loaded as shown in Fig. 3-12. Ignore the load due to the beam itself.

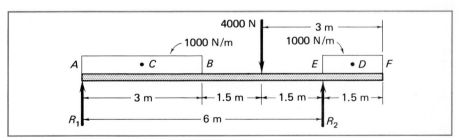

Fig. 3-12 Diagram for Sample Problem 3.

Solution: The beam is in equilibrium; thus, $\Sigma F = 0$.

$$R_1 + R_2 = 1000(3) + 4000 + 1000(1.5) = 3000 + 4000 + 1500 = 8500 \text{ N}$$

For the purpose of taking moments only, the uniform load AB may be assumed to be concentrated at point C, and load EF is assumed to be concentrated at point D. Taking moments about point A,

$$\Sigma M_a = 0$$
$$\Sigma M_a = 3000(1.5) + 4000(4.5) + 1500(6.75) - R_2(6) = 0$$
$$R_2 = 5440 \text{ N} = 5.44 \text{ kN}$$
$$R_1 = 8500 - 5440 = 3060 \text{ N} = 3.06 \text{ kN}$$

3-5   COUPLES

Figure 3-13 shows a sports car steering wheel. In order to turn the wheel clockwise, the driver must exert equal and opposite forces (20 lb) as shown. These two forces form a couple because they are *parallel, equal* in magnitude, and *opposite* in direction. Obviously, a couple can cause rotation only, since $\Sigma F_x = 0$ and $\Sigma F_y = 0$. The perpendicular distance between the two forces is called the *arm* of the couple.

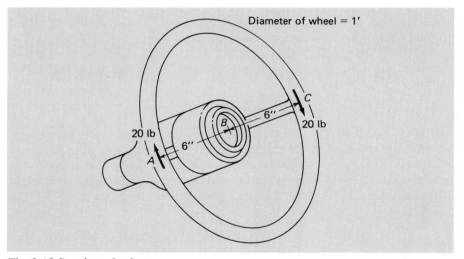

Fig. 3-13 Steering wheel.

Since the couple causes rotation, the moment, or torque, developed by the couple can be determined by taking the summation of the moments of the forces about some point. First we shall consider moments about the center of the wheel, point B,

$$\Sigma M_b = +20(\tfrac{1}{2}) + 20(\tfrac{1}{2}) = +20 \text{ ft} \cdot \text{lb}$$

And then moments about point A,

$$\Sigma M_a = +20(1) = +20 \text{ ft} \cdot \text{lb}$$

Choose a point D, 1 ft to the left of A, as a moment center. Then

$$\Sigma M_d = -20(1) + 20(2) = +20 \text{ ft} \cdot \text{lb}$$

The example shows that the moment of this couple is a constant and is equal to one of the forces times the arm, or distance between the forces. The statement is true in general; and if the forces are F and F with an arm a, the moment is $F.a$.

Since a single force produces translation and a couple produces rotation, a force cannot balance a couple.

In Fig. 3-14, a single force P acts at a point D. At A, b ft from D, draw two equal and opposite forces P. They neutralize each other, and the resultant effect remains the same. Now, forces along AB and DE form a couple with moment bP that tends to rotate the body clockwise. There remains the force P along AC, which tends to translate the body.

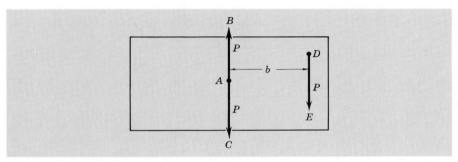

Fig. 3-14 Eccentric force = force + couple.

A single force acting at a point, not the center of the body, causes rotation around the center and a translation of the center.

A single force acting at one point is equivalent to the same force, acting at a different point, plus a couple.

Common illustrations of the couple are seen in the forces that are applied to the steering wheel of the automobile or to the arms of a lug wrench.

PROBLEMS

3-1. A beam 12 ft long is supported at its ends. A load of 600 lb is placed 5 ft from the left end. What are the reactions?

*3-2. In Fig. Prob. 3-2, what should be the values of the forces R and F to keep the member in equilibrium?

3-3. With loads as shown in Fig. Prob. 3-3, determine the values of R_1 and R_2 by moments and also by the inverse ratio principle.

3-4. Figure Problem 3-4 shows a beam weighing 40 lb per linear foot. This beam carries a load of 2000 lb located 4 ft from the right end and an additional load, uniformly distributed, of 300 lb/ft on the 6 ft adjacent to the left end. Find R_1 and R_2.

3-5. Find the position of the resultant load for a truck with a wheel base of 140 in if the distribution of the load is such that the reaction on the rear axle is 3.5 tons and that on the front axle is 2.5 tons.

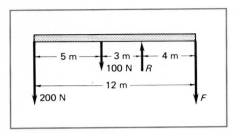

FIGURE PROBLEM 3-2

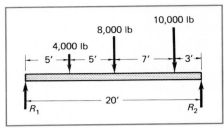

FIGURE PROBLEM 3-3

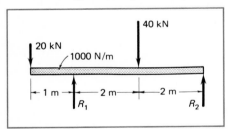

FIGURE PROBLEM 3-4

*3-6. In Fig. Prob. 3-6, the beam overhangs the left support and is supported at the right end. The beam weighs 1000 N/m. With the other loads as shown, what are the reactions?

3-7. What are the reactions of the beam shown in Fig. Prob. 3-7? The beam weighs 30 lb per linear foot.

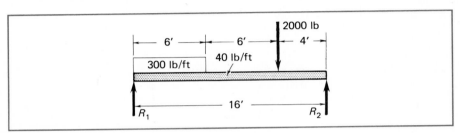

FIGURE PROBLEM 3-6   FIGURE PROBLEM 3-7

*3-8. A beam supported at its ends is 6 m long and weighs 700 N/m. It carries a concentrated load of 8 kN located 2.7 m from the right end and a total uniformly distributed load of 14 kN which extends the full length of the beam. What are the reactions?

3-9. In Fig. Prob. 3-9, a timber of uniform weight and cross section is shown in position on rollers while being moved. What are the roller reactions? With the rollers still 6 ft apart, what should be the position of the timber so that one roller will support twice as much as the other? The timber is 18 ft long.

3-10. What are the end reactions for simply supported floor beams 16 ft long

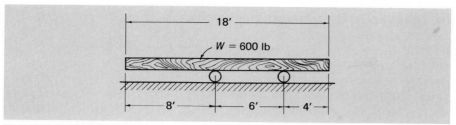

FIGURE PROBLEM 3-9

and spaced 4 ft from center to center? The total load on the floor, including the weight of the floor, is 140 psf.

3-11. A wheel 3 ft in diameter weighs 2500 lb with its load (Fig. Prob. 3-11). Find the horizontal force necessary to start the wheel over an obstruction 6 in high. (Forces act through the center.)

*3-12. Find the horizontal force F necessary to rotate the block about the point A shown in Fig. Prob. 3-12.

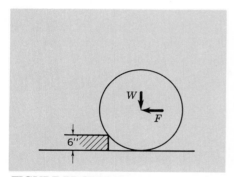

FIGURE PROBLEM 3-11   FIGURE PROBLEM 3-12

3-13. What force P is necessary to maintain equilibrium in the arrangement of the beams shown in Fig. Prob. 3-13? CD has a mass of 15.3 kg; AB has a mass of 10.2 kg and is pivoted at B.

3-14. The safety valve shown in Fig. Prob. 3-14 is 3 in in diameter and is just

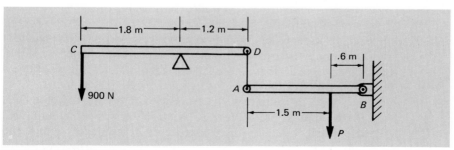

FIGURE PROBLEM 3-13

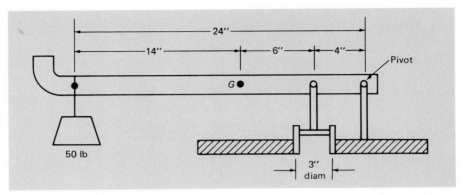

FIGURE PROBLEM 3-14

on the point of blowing off steam. The lever weighs 7 lb, and its center of weight is at G. The weight of the valve is 3 lb. If the weight on the end of the lever is 50 lb, find the pressure of the steam in the boiler. If the steam is to blow off at 80 psi, find the amount of the weight on the end of the lever.

*3-15. Figure Problem 3-15 shows an air cylinder and a system of levers. If the air pressure in the cylinder is 350 kPa, what force is exerted at F? If a force of 40 kN is desired, what should be the air pressure?

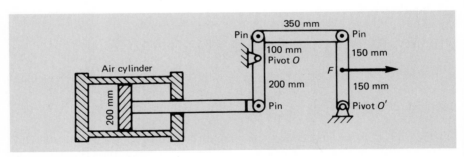

FIGURE PROBLEM 3-15

3-16. Figure Problem 3-16 represents a brake cylinder. The lever is bent at an angle of 120°. What force is produced in the brake rod by a pressure of 60 psi in the cylinder?

3-17. Forces of 10 lb each are applied on opposite sides of a steering wheel 18 in in diameter to form a couple. What is the moment of the couple? If a single force of 20 lb is applied to the rim of the steering wheel, how is a couple produced and what is its moment?

3-18. An iron pipe 1 in in diameter is embedded in a concrete base (Fig. Prob. 3-18). A horizontal force of 20 lb is applied perpendicular to BC. Describe the effect that this force has on each of the lengths of pipe and the elbow.

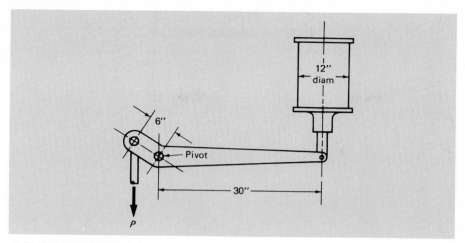

FIGURE PROBLEM 3-16

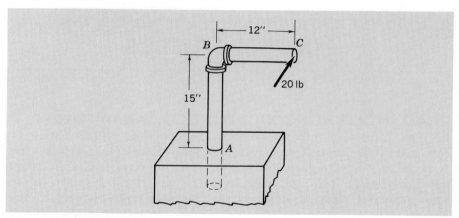

FIGURE PROBLEM 3-18

4

Nonconcurrent-Coplanar Forces; Trusses

4-1 RESULTANT OF NONCONCURRENT-COPLANAR FORCES

In Chap. 2, on concurrent forces, it was shown that the resultant of a set of concurrent forces was a single force determined by ΣF_x and ΣF_y. Also, in Chap. 3, on parallel forces, the resultant was found to be a single force and a couple determined by ΣF_x, ΣF_y, and ΣM. In this chapter, it will be shown that the resultant of nonconcurrent nonparallel forces is also a single force and a couple. The principle will be illustrated by an example before taking up the general proof.

Figure 4-1 shows three forces in action. Find the resultant in magnitude and position with reference to point O. At O, Fig. 4-2a, introduce two forces of 100 lb each that are opposite in direction and parallel to the original 100-lb force. This does not change conditions; for the forces introduced are in equilibrium. The two 100-lb forces, drawn as solid lines, form a couple with a moment arm of 2 ft that produces rotation around O in a counterclockwise direction. The other 100-lb force at O, shown by a broken line, causes translation exactly as if the original 100-lb force had been applied at that point. The other forces are treated in the same way. The 200-lb force, Fig. 4-2b, is equivalent to a couple of moment $+4(200) = +800$ ft·lb and a single force of 200 lb at O. Also, the 300-lb force, Fig. 4-2c, is equivalent to a couple of moment $-5(300) = -1500$ ft·lb and a single 300-lb force at O. If Figs. 4-2a, 4-2b, and 4-2c are superimposed, the combination is equivalent to the original system. The moment of the entire system is

$$\Sigma M = -200 + 800 - 1500 = -900 \text{ ft·lb}$$

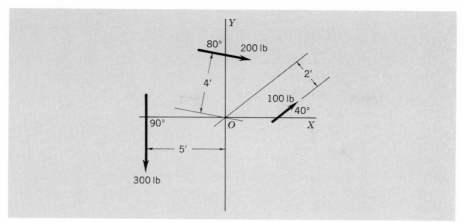

Fig. 4-1 Three nonconcurrent-coplanar forces.

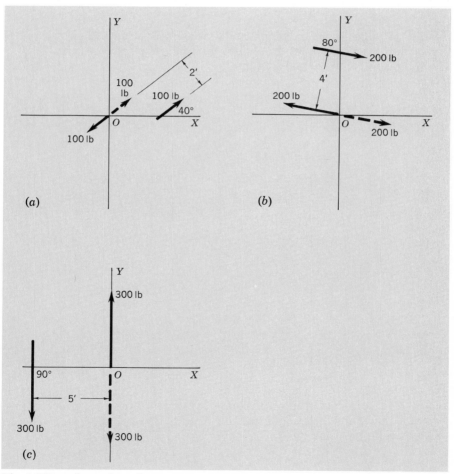

Fig. 4-2 Transformation of nonconcurrent forces into equivalent concurrent forces and couples.

The resultant of the forces (broken lines) at O is found as in Chap. 2.

$$\Sigma F_x = 100\cos 40° + 200\cos 10° + 300\cos 90° = +274 \text{ lb}$$
$$\Sigma F_y = 100\sin 40° - 200\sin 10° - 300\sin 90° = -270 \text{ lb}$$
$$R = \sqrt{274^2 + 270^2} = 385 \text{ lb}$$
$$\tan\theta = -\frac{270}{274} = -0.985$$
$$\theta = 315.42° \text{ or } 315°25'$$

The resultant is then a force of 385 lb making an angle of 315.42° with the x axis. It is directed to the right and down. It must have a moment of -900 ft·lb. Then

$$385r = 900$$
$$r = 2.34 \text{ ft}$$

Now, R must be drawn to the right and down. Since the moment is negative, R must cause counterclockwise rotation. Its location is shown in Fig. 4-3. The resultant is then a force and a couple.

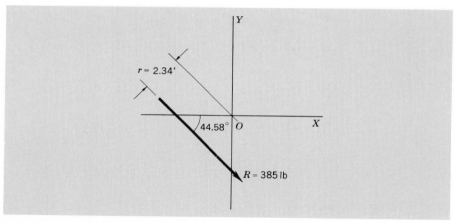

Fig. 4-3 Resultant of nonconcurrent force system.

4-2 GENERAL METHOD[1]

Now take a set of forces F_1, F_2, F_3 at the distances r_1, r_2, r_3, respectively, from the origin O in Fig. 4-4. At O, insert two forces F_1 parallel to F_1 and opposite in direction. Two of the forces form a couple of moment $F_1 r_1$, and a single force F_1 acts at O. By doing the same thing with F_2 and F_3, we obtain a set of couples tending to rotate around O, the resultant moment of which is

$$\Sigma M = F_1 r_1 + F_2 r_2 + F_3 r_3 = \Sigma(Fr) \tag{4-1}$$

[1] See also Sample Computer Program 4, Chap. 17.

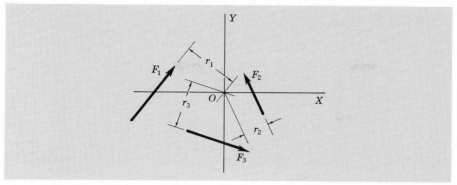

Fig. 4-4 Nonconcurrent-coplanar forces—general method.

There is also a set of concurrent forces at O, the resultant of which is

$$R = \sqrt{(\Sigma F_x)^2 + (\Sigma F_y)^2} \qquad (4\text{-}2)$$

$$\tan \theta = \frac{\Sigma F_y}{\Sigma F_x} \qquad (4\text{-}3)$$

$$r = \frac{\Sigma M}{R} \qquad (4\text{-}4)$$

The resultant R is located by drawing a vector that makes an angle θ with the x axis at a distance r from the origin. The resultant, therefore, is equivalent to a force and a couple.

If the system of forces is in equilibrium, $R = 0$. But that can be true only when $\Sigma F_x = 0$ and $\Sigma F_y = 0$. This condition shows that a body at O would not be translated by the system of forces. But, for equilibrium, there must be no rotation, a state that can exist when and only when $\Sigma M = 0$. There are then three equations of condition for the solution of forces in equilibrium. If the original forces F_1, F_2, and F_3 had been resolved into their F_x and F_y components in the position in which they are located, there would then have been two systems of forces, one parallel to the x axis and the other parallel to the y axis. But, from Chap. 3, the resultant of a set of parallel forces is the algebraic sum of the forces. The sums are ΣF_x and ΣF_y, just as if they had been resolved into their components at point O and their sum taken.

If $\Sigma F_x = 0$ and $\Sigma F_y = 0$ while ΣM does not equal 0, the resultant of the system of forces is a couple; also, if $\Sigma M = 0$ while ΣF_x and ΣF_y are not equal to 0, the resultant is a single force. When all three of the foregoing expressions equal zero, the system of forces is in equilibrium.

4-3   GRAPHICAL METHOD

Let F_1, F_2, and F_3 of Fig. 4-5 be three nonconcurrent-coplanar forces. To find their resultant graphically, extend the lines of F_1 and F_2 to intersect at A (Fig. 4-6a). Lay off F_1' and F_2' equal, respectively, to F_1 and F_2. Complete the

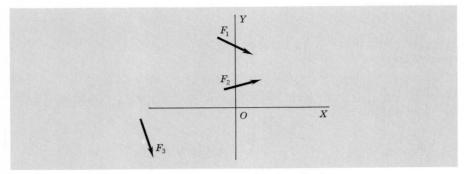

Fig. 4-5 Nonconcurrent-coplanar forces—graphical method.

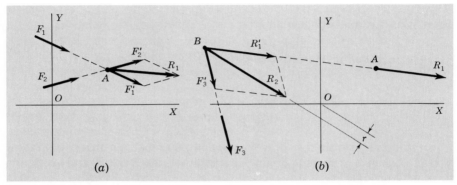

(a)   (b)

Fig. 4-6 (a) Resultant of F_1 and $F_2 = R_1$. (b) Resultant of F_3 and $R_1 = R_2$.

parallelogram. The diagonal R_1 is the resultant of F_1 and F_2. Now extend R_1 to intersect the line of action of F_3 at B (Fig. 4-6b). Lay off R_1' and F_3' equal, respectively, to R_1 and F_3. Complete the parallelogram. The diagonal R_2 is the resultant of R_1 and F_3, or of F_1, F_2, and F_3. With reference to a body at point O, not on R_2, the resultant is equivalent to a single force R_2 at point O and a couple of moment R_2r that causes rotation about point O.

If point O is on the line of action of R_2 (Fig. 4-7a), the resultant is a single force R_2. The moment of R_2 about point O is zero, since the moment arm is zero.

If F_3 should be equal and parallel to R_1 but opposite in direction (Fig. 4-7b), the resultant force is zero. However, there is a resultant couple of moment R_1r, where r is the distance between the parallel forces.

If F_3 is equal and opposite to R_1 and coincides with it, then the resultant force and couple are each equal to zero. The system of forces is then in equilibrium.

The preceding method of determining a resultant is general and can be extended to include any number of forces.

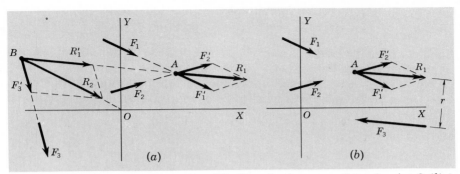

Fig. 4-7 (*a*) Resultant R_2 has no moment if its line of action passes through point O. (*b*) A moment $R_1 r$ exists and resultant R_2 is zero when F_3 is equal, parallel, and opposite to R_1.

4-4   APPLICATIONS

Many common nonconcurrent-coplanar force systems occur in pinned or hinged structures, machine linkages, and beams. The application of the foregoing principles will be demonstrated by a series of illustrative examples covering a variety of situations. Certain groups of structures will be subjected to more intensive study. For example, beams are discussed in Chaps. 11, 12, and 16, columns in Chap. 15, and trusses in this chapter.

Sample Problem 1   A 10-ft ladder rests against a smooth wall. The ladder weighs 25 lb. A 155-lb man stands on it, as shown in Fig. 4-8, while his helper

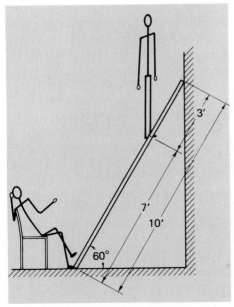

Fig. 4-8 Diagram for Sample Problem 1.

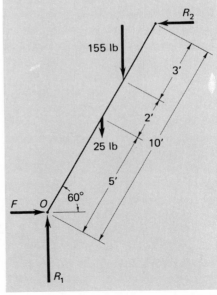

Fig. 4-9 Free-body diagram for Sample Problem 1.

holds his foot against the bottom to prevent the ladder from slipping on the smooth floor. Find the unknown force and reactions.

Solution: Use the ladder as a free body and indicate the various known and unknown forces, as in Fig. 4-9.

Since the system is in equilibrium, the sum of vertical forces must be zero, the sum of horizontal forces must be zero, and there can be no unbalanced moment.

$$\Sigma F_y = 0 \qquad\qquad R_1 - 25 - 155 = 0$$
$$R_1 = 25 + 155 = 180 \text{ lb}$$

$$\Sigma F_x = 0 \qquad\qquad F - R_2 = 0$$
$$F = R_2$$

$$\Sigma M_O = 0$$
$$25(5 \cos 60°) + 155 (7 \cos 60°) - R_2(10 \sin 60°) = 0$$
$$R_2 = 70 \text{ lb}$$
$$F = 70 \text{ lb}$$

*Sample Problem 2 A beam 6 m long and with a mass of 45.9 kg is pinned to the floor at A and rests against a smooth wall at D (Fig. 4-10). It carries a load of 900 N at C. Find the reactions at A and D.

Solution: Set AD out as a free body. Since the wall is perfectly smooth, the wall reaction at D will be normal to the wall. The reaction at A, being unknown in

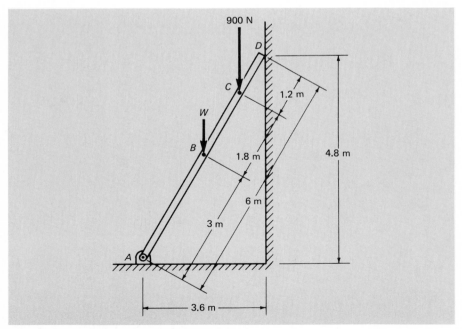

Fig. 4-10 Diagram for Sample Problem 2.

magnitude and direction, will be replaced by its two components F_x and F_y. The weight of the beam W is at B and is equal to $W = m \cdot g = (45.9 \text{ kg})(9.81 \text{ m/s}^2) = 450 \text{ N}$; the 900-N load is at C (see Fig. 4-11).

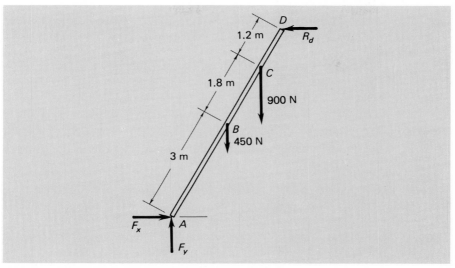

Fig. 4-11 Free-body diagram for Sample Problem 2.

$$\Sigma F_y = 0 \qquad\qquad F_y - 450 - 900 = 0$$
$$F_y = 450 + 900 = 1.35 \text{ kN}$$
$$\Sigma F_x = 0 \qquad\qquad F_x - R_d = 0$$
$$F_x = R_d$$
$$\Sigma M_a = 0 \qquad 450\left(3.6 \times \frac{3}{6}\right) + 900\left(3.6 \times \frac{4.8}{6}\right) - R_d(4.8) = 0$$

(Moment arm calculations are based on similar triangles.)

$$R_d = 709 \text{ N} = 0.709 \text{ kN}$$
$$F_x = 709 \text{ N} = 0.709 \text{ kN}$$

By Fig. 4-12,

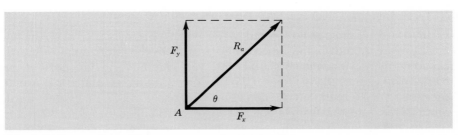

Fig. 4-12 Summation of force components at point A.

$$R_a = \sqrt{F_x^2 + F_y^2}$$
$$= \sqrt{0.709^2 + 1.35^2} = \sqrt{2.33}$$
$$= 1.53 \text{ kN}$$

$$\tan \theta = \frac{F_y}{F_x}$$
$$= \frac{1.35}{0.709}$$
$$= 1.90$$
$$\theta = 62.3°$$

Sample Problem 3   An A-frame carries a load of 1000 lb as shown in Fig. 4-13. Find the floor reactions at A and E and the pin reactions at B, C, and D.

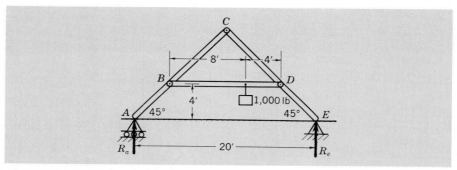

Fig. 4-13 Diagram for Sample Problem 3.

Solution: Set the entire A-frame out as a free body (Fig. 4-13). Since the load, or action, is vertical, the reactions at A and E also will be vertical, there being no tendency for the frame to move sidewise.

$$\Sigma F_y = 0 \qquad\qquad R_a + R_e - 1000 = 0$$
$$R_a + R_e = 1000$$
$$\Sigma M_a = 0 \qquad\qquad 1000(12) - R_e(20) = 0$$
$$R_e = 600 \text{ lb}$$
$$R_a = 400 \text{ lb}$$

Now set all three members out as free bodies (Fig. 4-14). Since all the pin reactions are unknown, they must be represented by their horizontal and vertical components. The reader should note very carefully the directions of the components. Since the member BD pulls downward on AC and also prevents the motion of AC to the left, F_{bx} acts to the right on AC, and F_{by} acts downward. Since AC prevents BD from falling and also exerts a pull to the left on BD in the free-body diagram of BD, F_{bx} acts to the left and F_{by} acts upward. A similar condition is shown at C and D. This result verifies the principle that action and

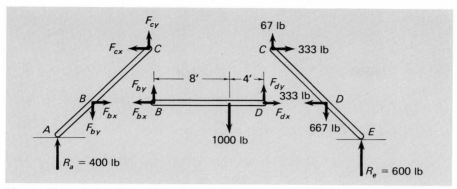

Fig. 4-14 Free-body diagrams for Sample Problem 3.

reaction are equal. When members of the A-frame are joined together, there must be equilibrium at each of points B, C, and D. That is, for each of the pins, $\Sigma F_x = 0$ and $\Sigma F_y = 0$. Referring again to the free-body drawing of AC, the reader will observe that there are four unknowns. Since only three equations are possible, the unknowns cannot be found without further information.

Now consider the free-body drawing of BD. From this sketch it is possible to determine the two vertical components by applying $\Sigma M = 0$ and $\Sigma F_y = 0$.

$$\Sigma F_y = 0 \qquad\qquad F_{by} + F_{dy} - 1000 = 0$$
$$F_{by} + F_{dy} = 1000$$
$$\Sigma M_b = 0 \qquad\qquad 1000(8) - F_{dy}(12) = 0$$
$$F_{dy} = 667 \text{ lb}$$
$$F_{by} = 333 \text{ lb}$$

When we insert this value of F_{by} in the free-body drawing of AC, we have only three unknowns remaining, and the solution can be completed. By taking moments around C, two of the three remaining unknowns are eliminated.

$$\Sigma M_c = 0 \qquad\qquad +400(10) - 333(6) - F_{bx}(6) = 0$$
$$F_{bx} = 333 \text{ lb}$$
$$\Sigma F_x = 0 \qquad\qquad 333 - F_{cx} = 0$$
$$F_{cx} = 333 \text{ lb}$$
$$\Sigma F_y = 0 \qquad\qquad 400 - 333 + F_{cy} = 0$$
$$F_{cy} = -67 \text{ lb}$$

The minus sign indicates that the wrong direction was assigned to F_{cy}, which acts downward. From the free-body diagram of BD, it is evident that $F_{dx} = F_{bx} = 333$ lb. As a check on the solution, insert the known values in the free-body drawing of CE (Fig. 4-14).

$$\Sigma F_x = 0 \qquad\qquad\qquad +333 - 333 = 0$$
$$\Sigma F_y = 0 \qquad\qquad\qquad +600 - 667 + 67 = 0$$
$$\Sigma M_e = 0 \qquad\qquad -4(667) - 4(333) + 333(10) + 10(67) = 0$$

The resultant pin reaction at B is found as follows:

$$F_b = \sqrt{F_{bx}^2 + F_{by}^2} = \sqrt{333^2 + 333^2} = 471 \text{ lb}$$

$$\tan \theta = \frac{333}{333} = 1.0$$

$$\theta = 45°$$

In the same way, the pin reactions at C and D can be found to be

$$F_c = 340 \text{ lb at } 11.3° \text{ from the horizontal}$$
$$F_d = 745 \text{ lb at } 63.5° \text{ from the horizontal}$$

Sample Problem 4   Figure 4-15 shows a derrick carrying a load of 2000 lb. The boom DG weighs 500 lb, and the mast weighs 400 lb. Find the reactions at E, D, C, B, and F. Neglect the weight of member CF.

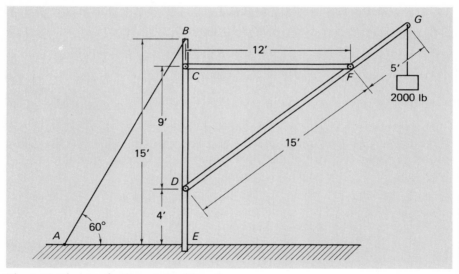

Fig. 4-15 Diagram for Sample Problem 4.

Solution: First, set the entire derrick out as a free body (Fig. 4-16). The derrick is acted upon by the load, the weights of the members, the cable AB, and the ground at E. Since AB is a two-force member, the tension F_b in cable AB must act along the cable. The reaction at E is unknown, and therefore we represent it by its components. By taking moments about point B, we have

$\Sigma M_b = 0$ $\qquad\qquad -15F_{ex} + 8(500) + 16(2000) = 0$

$$F_{ex} = 2400 \text{ lb}$$

$\Sigma F_x = 0$ $\qquad\qquad\qquad 2400 - F_b \sin 30° = 0$

$$F_b = 4800 \text{ lb}$$

$\Sigma F_y = 0$ $\qquad F_{ey} - 400 - 500 - 2000 - 4800 \cos 30° = 0$

$$F_{ey} = 7060 \text{ lb}$$

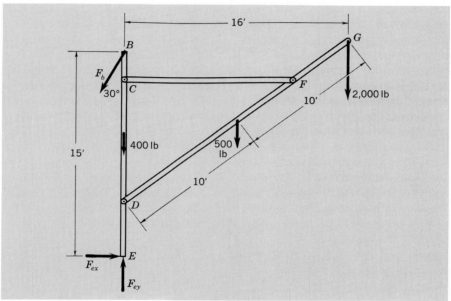

Fig. 4-16 Free-body diagram for Sample Problem 4.

Next, classify the members of the derrick. *AB* and *CF* are two-force members, whereas *EB* and *DG* are not. Since there is no point at which only two-force members meet, the next step in the solution will be to set out *DG* and *EB* as free bodies (Figs. 4-17 and 4-18). Applying the conditions of equilibrium to Fig. 4-17 gives us

$\Sigma M_d = 0$ $\qquad\qquad -9F_f + 8(500) + 16(2000) = 0$

$$F_f = 4000 \text{ lb} = F_c$$

$\Sigma F_x = 0$ $\qquad\qquad\quad F_{dx} - 4000 = 0$

$$F_{dx} = 4000 \text{ lb}$$

$\Sigma F_y = 0$ $\qquad\qquad\quad F_{dy} - 500 - 2000 = 0$

$$F_{dy} = 2500 \text{ lb}$$

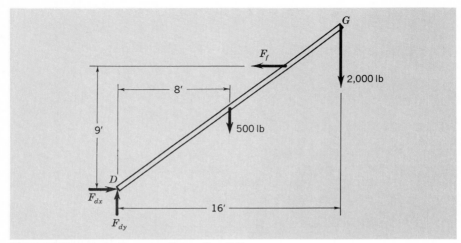

Fig. 4-17 Free-body diagram of member *DG*.

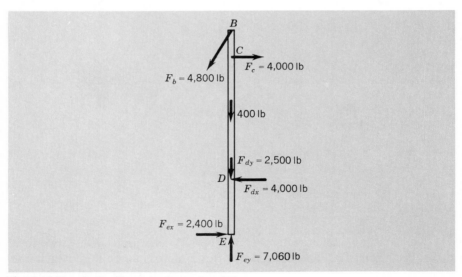

Fig. 4-18 Free-body diagram of member *EB*.

The reader can now find the resultant reactions by the method shown in the preceding problem. These results are

$$F_e = 7460 \text{ lb at } 71.2° \text{ from horizontal}$$
$$F_d = 4720 \text{ lb at } 32° \text{ from horizontal}$$
$$F_b = 4800 \text{ lb at } 60° \text{ from horizontal}$$
$$F_c = F_f = 4000 \text{ lb horizontal}$$

Check: Since all the forces are known, the free-body drawing of *EB* can now be used as a check on the solution (Fig. 4-18).

$$\Sigma F_x = 0 \qquad 2400 - 4000 + 4000 - 4800 \sin 30° = 0$$
$$\Sigma F_y = 0 \qquad 7060 - 2500 - 400 - 4800 \cos 30° = 0$$
$$\Sigma M_b = 0 \qquad -15(2400) + 11(4000) - 2(4000) = 0$$

4-5 TRUSSES

Trusses are structures whose members are connected to *form triangles.* The triangle is significant because it is stable and cannot collapse as long as a member does not break or deform. Thus, when designing a truss, it is necessary to find the force in each member of the truss and then select structural members which are adequate with regard to strength and resistance to buckling.

Such structures may exist as roof trusses, bridges, airplane wing trusses, booms of industrial cranes, and the like, as illustrated in Fig. 4-19. The method

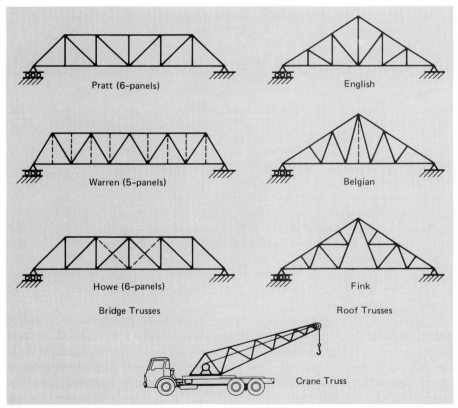

Fig. 4-19 Several common types of trusses (broken lines represent members occasionally added).

of connecting the members may be with pins, welds, or bolts as shown in Fig. 4-20. Rivets were widely used at one time, but they are seldom used today.

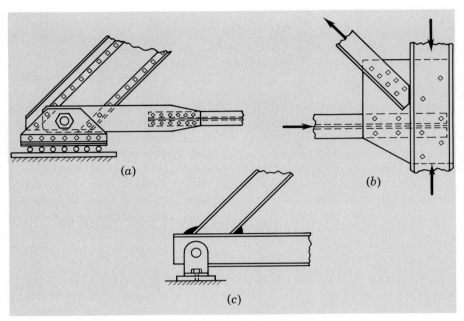

Fig. 4-20 Methods of connecting truss members.

Roof trusses are generally made of wood or steel depending upon the nature of the structure. Most other trusses are steel except for wing trusses, which are usually made of aluminum or magnesium alloys. Wood trusses are generally joined by bolts. Metal trusses are joined by either bolts or welds; pin connections are used only at the reaction points.

A structure consisting mainly of bolted or welded joints is complex to analyze with theoretical exactness. It is much simpler to assume that all the joints are connected by means of frictionless pins (which, of course, they are not). This assumption yields results which are suitable for many important types of structures. Even for structures which are unusually large or stiff it is possible to first assume the joints are pin-connected and then apply a correction factor.

There are two common methods for the analysis of trusses. One is called the *method of joints;* the other, the *method of sections.* Both are based on principles of static equilibrium which were established in Chap. 2, and, in both, the free-body method of analysis is important. The method of joints is most convenient when it is desired to determine the forces on all the members of the truss or those near the reactions only. The method of sections is most convenient when it is desired to find the forces on only one or a few members, particularly those which are not at or near the supports. Because each method has its advantages and areas of application, both methods will be described here.

In each method, the members of the truss are assumed to be connected by frictionless pins and to be two-force members.

The reader should recall that two-force compression members may be subjected to both buckling *and* compression owing to column action as mentioned in Sec. 2-8 and discussed in Chap. 15. In our treatment of trusses, we will be finding only the forces which the members must sustain and transmit. The next design step—that of selecting the actual sizes of compression members—would require the techniques shown in Chap. 15.

4-6 METHOD OF JOINTS

Since the members of a truss are two-force members connected by frictionless pins, each force acting at the end of the member is assumed to be acting axially along the member. Furthermore, since the conditions of static equilibrium apply, the end forces on a member must be equal and opposite.

The method of joints will be demonstrated in the following problem, where it is desired to find the internal forces in each member of the truss (Fig. 4-21).

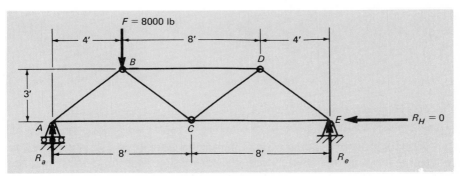

Fig. 4-21 Truss diagram to illustrate method of joints.

The reactions at the supports are found by setting out the entire truss as a free body and applying the equations for static equilibrium. First, we should note that, because support A is a roller-type support, no horizontal reaction can occur at A. However, because support E is a pinned-type support, it is possible for a horizontal reaction to occur at E. Since the truss is not subjected to any horizontal forces, the horizontal reaction at E (R_H) must be equal to zero and will not be considered in our calculations. Suppose we first take moments about an axis through support A and set the algebraic sum of those moments to zero ($\Sigma M_a = 0$). That gives us $F(4) - R_e(16) = 0$, from which R_e can be calculated because load F is known.

$$R_e(16) = 8000(4)$$
$$R_e = 2000 \text{ lb}$$

Now we sum the vertical forces algebraically and set the sum equal to zero

($\Sigma F_y = 0$). We obtain

$$R_a + R_e - F = 0$$
$$R_a = 8000 - 2000 = 6000 \text{ lb}$$

Now it is necessary to find a joint in which there are only two unknown forces. Then, by using $\Sigma F_x = 0$, $\Sigma F_y = 0$, and information obtained from the geometry of the situation, it is possible to solve for the two unknown forces.

By referring to Fig. 4-21, it can be seen that joints A and E have two unknown forces each, whereas all the other joints have more than two. (Joints A and E had three unknowns before the reactions were computed.) Therefore, it is necessary to start either with joint A or E.

We shall start with A and draw a free-body diagram of the joint (Fig. 4-22). In the free-body diagram, all the forces acting on joint A are shown. Included are the reaction R_a and the forces exerted by members AB and AC. Since AB and AC are two-force members, their forces are shown acting axially along the directions of the members. For simplicity, the symbols **AB** and **AC** are used to represent the forces exerted upon the pin by these members.

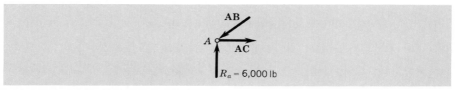

Fig. 4-22 Free-body diagram of joint A.

Although the lines along which the forces act are known, it is necessary to determine in which direction the arrows are pointing. The choice is easily made in this case; R_a is known to be acting upward. Since equilibrium is known to exist (joint A is at rest), R_a must be balanced by a downward force. Since **AC** cannot provide a downward component, such a component must come from **AB**, which then must be acting as shown in Fig. 4-22 in order that **AB**$_y$ is downward. This selection of direction for **AB** then automatically makes **AB**$_x$ act to the left, so that **AC** must act to the right in order to balance **AB**$_x$. This method of selecting the direction of forces will be found to work in most cases. In the remaining cases, it is necessary only to assume a direction for the force and solve the problem accordingly. Should the answer be negative, it is only necessary to change the direction of the force.

We can now set up equations for the forces on the joint. It is first desirable to resolve **AB** into its x and y components (Fig. 4-23).

$$\Sigma F_x = 0 \qquad \text{AC} - \text{AB}_x = 0 \qquad \text{AB}_x = \text{AC}$$
$$\Sigma F_y = 0 \qquad 6000 - \text{AB}_y = 0 \qquad \text{AB}_y = 6000 \text{ lb}$$

The third equation comes from the geometry of the situation, from which **AB** was resolved into **AB**$_x$ and **AB**$_y$. The basic information comes from the

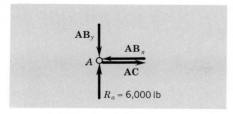

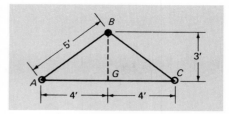

Fig. 4-23 Free-body diagram of joint A showing force **AB** resolved into its x and y components.

Fig. 4-24 Part of the truss showing the geometric relation of the members.

direction in which **AB** is known to act (line of action). Member AB is the hypotenuse of a 3-4-5 triangle, as can be seen in Fig. 4-24. Since force **AB** acts in the same direction as member AB, it must make the same angle with the horizontal and the vertical, and the two triangles ABG and $A'B'G'$ (Fig. 4-25) can be seen to be similar. Corresponding sides of similar triangles are proportional, so that

$$\frac{B'G'}{BG} = \frac{A'G'}{AG} = \frac{A'B'}{AB}$$

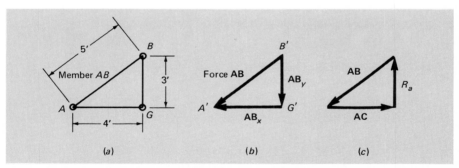

Fig. 4-25 (a) Part of the truss which forms a 3-4-5 right triangle. (b) Force **AB** and its components. (c) Force triangle for joint A.

But $B'G' = AB_y$

$A'G' = AB_x$

$A'B' = AB$ from the vector triangle Fig. 4-25b

and $AB = 5$ ft

$BG = 3$ ft

$AG = 4$ ft   from the truss-member triangle Fig. 4-25a

Therefore,

$$\frac{AB_y}{3} = \frac{AB_x}{4} = \frac{AB}{5}$$

Three equations result, since they mean

$$\frac{AB_y}{3} = \frac{AB_x}{4}$$

$$\frac{AB_x}{4} = \frac{AB}{5}$$

$$\frac{AB_y}{3} = \frac{AB}{5}$$

From these equations, we obtain

$$\frac{AB_y}{3} = \frac{AB}{5}$$

$$AB_y = \frac{3AB}{5}$$

and

$$\frac{AB_x}{4} = \frac{AB}{5}$$

$$AB_x = \frac{4AB}{5}$$

We had previously found that $AB_x = AC$ and $AB_y = 6000$ lb. By combining these equations with those above, we can solve for **AB** and **AC**.

$$AB_y = \frac{3AB}{5} = 6000 \text{ lb}$$

$$AB = \frac{5(6000)}{3} = 10\ 000 \text{ lb (compression)}$$

$$AB_x = \frac{4AB}{5} = AC$$

$$AC = \frac{4(10\ 000)}{5} = 8000 \text{ lb (tension)}$$

also, $$AB_x = AC = 8000 \text{ lb}$$

Since member AB is seen to be pushing on it, joint A is pushing back on member AB, and the member is in compression.

Since member AC is seen to be pulling on it, joint A is pulling back on member AC, and the member is in tension.

The unknown forces at joint A could have been determined by recognizing that the force triangle made by connecting **AB**, **AC**, and R_a (Fig. 4-25c) is similar to triangle ABG of Fig. 4-25a. From the proportionality of the triangles, we find that

$$\frac{AB}{5} = \frac{AC}{4} = \frac{R_a}{3}$$

Thus, $\quad$ $\mathbf{AB} = \dfrac{5R_a}{3} = \dfrac{5(6000)}{3} = 10\ 000$ lb (compression)

and $\quad$ $\mathbf{AC} = \dfrac{4R_a}{3} = \dfrac{4(6000)}{3} = 8000$ lb (tension)

The use of a force triangle and geometry gives a rapid solution for joints with three forces when one force is known.

A set of free-body diagrams showing joint A, member AB, and joint B is shown in Fig. 4-26. It is easy to see from the diagrams that if one end of a member acts with a certain force on one joint, it acts with an equal and opposite force on the other joint to which it is connected.

Knowing the value of force $\mathbf{AB}$ acting upon joint B is important, since it leaves us only two unknown forces acting on the joint: $\mathbf{BC}$ and $\mathbf{BD}$ (Fig. 4-27). These forces can be determined by using $\Sigma F_x = 0$ and $\Sigma F_y = 0$. Note that it is not necessary to calculate the components of $\mathbf{AB}$ again. As can be seen from Fig. 4-28, since the forces are equal and opposite, the components also are equal and opposite.

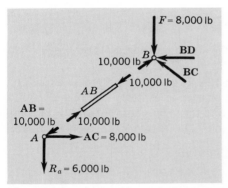

Fig. 4-26 Free-body diagrams of member AB and joints A and B.

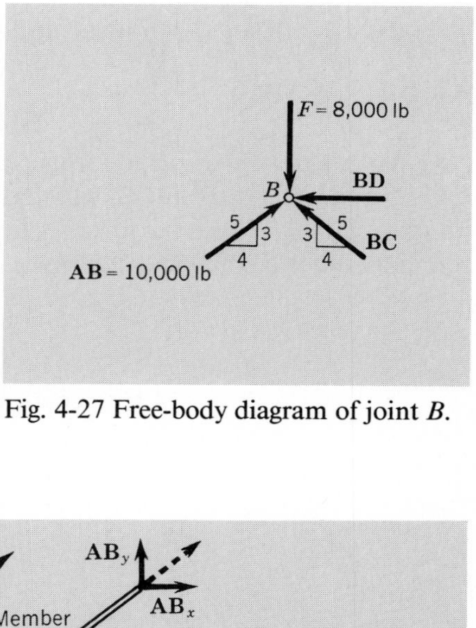

Fig. 4-27 Free-body diagram of joint B.

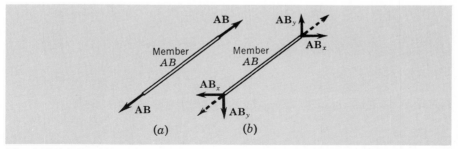

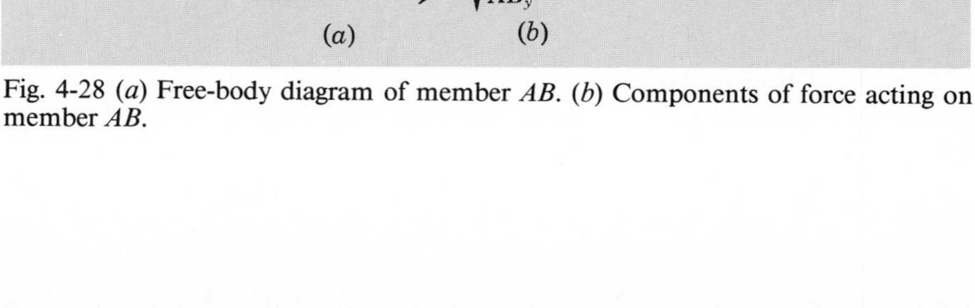

Fig. 4-28 (a) Free-body diagram of member AB. (b) Components of force acting on member AB.

With respect to joint B, $AB_x = 8000$ lb and $AB_y = 6000$ lb, as before. Then $\Sigma F_x = 0$ gives us

$$AB_x - BD - BC_x = 0$$
$$BC_x = -BD + 8000$$

and $\Sigma F_y = 0$ gives us

$$AB_y + BC_y - 8000 = 0$$
$$BC_y = 8000 - 6000 = 2000 \text{ lb}$$

From the geometry of the situation, using the similar triangle method as before,

$$\frac{BC_y}{3} = \frac{BC_x}{4} = \frac{BC}{5}$$

from which
$$BC_x = \frac{4BC_y}{3} = \frac{4(2000)}{3} = 2670 \text{ lb}$$

By substituting and solving for **BD**, we obtain

$$BC_x = 2670 = -BD + 8000$$
$$BD = 5330 \text{ lb (compression)}$$

Also,
$$BC = \frac{5BC_y}{3} = \frac{5(2000)}{3} = 3330 \text{ lb (compression)}$$

We can now proceed to joints C and D, by using the principles explained before, and finally to joint E, which can be used as a check. Figure 4-29 shows joints C, D, and E with the forces acting and the values computed thus far. The remainder of this solution is left to the reader. The results of the calculations are:

$$CD = 3330 \text{ lb (tension)}$$
$$DE = 3330 \text{ lb (compression)}$$
$$CE = 2670 \text{ lb (tension)}$$

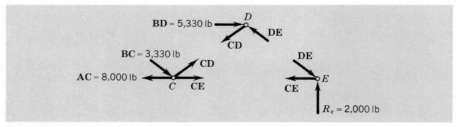

Fig. 4-29 Free-body diagrams of joints C, D, and E.

4-7　METHOD OF SECTIONS

Suppose we have a truss like the one loaded as shown in Fig. 4-30. It is desired to find the force in member FH.

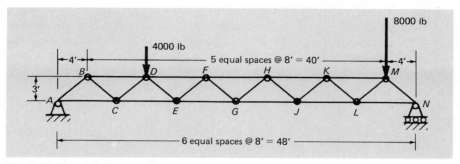

Fig. 4-30 Truss diagram to illustrate method of sections.

To do this by the method of joints, it is necessary to solve for the reaction at joint A and then solve for the forces involved at joints A, B, C, D, E, and F, which, of course, is quite involved.

A simpler solution is available through the method of sections. First solve for the reaction at support A by using $\Sigma M_n = 0$.

$\Sigma M_n = 0$ gives us

$$R_a(48) - 4000(36) - 8000(4) = 0$$
$$48R_a = 144\ 000 + 32\ 000$$
$$R_a = 3670\ \text{lb}$$

Note that it is not necessary to solve for both reactions in this method, although R_n can be calculated easily by using $\Sigma F_y = 0$.

Take a section through the truss as shown in Fig. 4-31. Note that the section passes through the unknown member FH and through two other members FG and EG. Now draw a free-body diagram of the truss to the left of the section, as in Fig. 4-32. The free-body diagram is not complete because there are forces acting upon the portion of members FH, FG, and EG which are part of the free body. The existence of those forces can be easily understood from the following

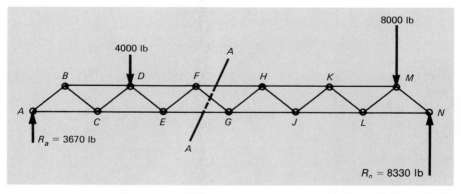

Fig. 4-31 Free-body diagram of truss showing section plane cutting unknown member FH.

reasoning. If the truss existed only as shown in Fig. 4-32 without forces acting at the "cut" members, it would fall down, as can be determined by inspection of the moment situation. Something must be helping to hold the truss up. Obviously, the rest of the truss is doing it, and it can do so only by means of forces acting at points 1, 2, and 3. The only things which can supply those forces are the remainders of members *FH*, *FG*, and *EG*, which are part of the rest of the truss. Members *FH*, *FG*, and *EG* are in either tension or compression. Take, for example, *FH* in compression. Since *FH* is in equilibrium, force *P* must also be acting at point 1 (see Fig. 4-33). A similar argument can be extended to the other members.

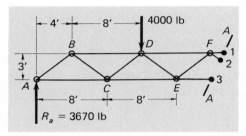

Fig. 4-32 Incomplete free-body diagram of portion of truss to left of section *A-A*.

Fig. 4-33 Equilibrium of a member cut by a section plane.

The correct free-body diagram then looks like Fig. 4-34. The directions of force **FH**, **FG**, and **EG** could have been selected arbitrarily. However, examination of the situation shows that we shall need an upward component to help balance the 4000-lb force at *D* (hence, **FG** in the direction shown). Since **FG** will probably be small (the amount required for $\Sigma F_y = 0$ is only 330 lb), we shall also need a counterclockwise moment to balance the clockwise moment caused by the 3670- and 4000-lb forces. Hence, **FH** and **EG** will be in the directions shown in Fig. 4-34.

Readers who desire to circumvent this reasoning exercise can select arbitrary directions for the vectors **FH**, **FG**, and **EG**. If an incorrect assumption is

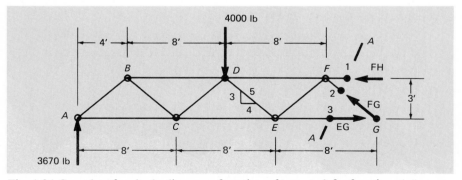

Fig. 4-34 Complete free-body diagram of portion of truss to left of section *A-A*.

made, the answer for that force will be negative and all that need be done is reverse the direction of the force.

The free-body diagram is now complete, and we can proceed to solve it. Note that we have three unknowns but we also have three equations available: $\Sigma F_x = 0$, $\Sigma F_y = 0$, and $\Sigma M = 0$.

Since the sum of the moments about any axis is zero, let us try to select an axis which will simplify our calculation to the greatest extent. Note that both forces **FG** and **EG** pass through joint G (located in Fig. 4-34). If we take moments about point G, forces **FG** and **EG** have zero moment about G, which leaves only the 4000-lb, 3670-lb, and **FH** forces with moments about G.

$\Sigma M_g = 0$ gives us

$$3670(24) - 4000(12) - 3\textbf{FH} = 0$$
$$88\,000 - 48\,000 - 3\textbf{FH} = 0$$
$$3\textbf{FH} = 40\,000$$
$$\textbf{FH} = 13\,300 \text{ lb (compression)}$$

Note that it was not necessary to use $\Sigma F_x = 0$ and $\Sigma F_y = 0$ to solve for **FH**.

This problem does not require a calculation for forces **FG** and **EG**; but if the forces are needed, several approaches are possible. For example, **EG** can be computed directly by taking moments about joint F and thereby eliminating the unknown force **FG**. Then **FG** can be obtained by taking moments about joint E, or, if preferred, the components of **FG** can be calculated by $\Sigma F_x = 0$ and $\Sigma F_y = 0$. Those calculations will give us:

$$\textbf{EG} = 13\,800 \text{ lb (tension)}$$
$$\textbf{FG} = 550 \text{ lb (compression)}$$

An extremely useful modification of the method of sections allows direct calculation of the forces in slanted internal truss members. The procedure employs $\Sigma F_y = 0$ and is often called the *method of shears.* Suppose it is necessary to calculate **FG** in Fig. 4-34 without finding **FH** or **EG**. It is again necessary to calculate the support reactions first and then choose a section through the truss which cuts no vertical members and no other slanted member except the one whose force is to be calculated. This requirement is met by the section taken in Fig. 4-34. Now simply apply the condition that $\Sigma F_y = 0$ and solve for the vertical component of **FG**.

$$3670 - 4000 + \textbf{FG}_y = 0$$
$$\textbf{FG}_y = 330 \text{ lb (upward)}$$

From the geometry of the truss,

$$\frac{\textbf{FG}_y}{\textbf{FG}} = \frac{3}{5}$$
$$\textbf{FG} = \frac{330(5)}{3}$$
$$= 550 \text{ lb (compression)}$$

This method can be applied conveniently to members AB, BC, CD, DE, EF, and any other slanted member of the truss.

Sample Problem 5[1]   Find the forces in all members of the truss shown in Fig. 4-35. Use the method of joints.

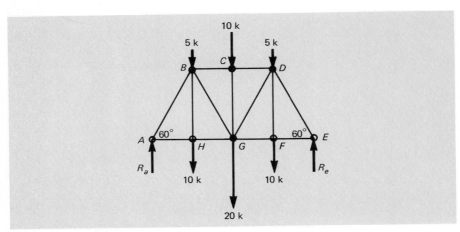

Fig. 4-35 Diagram for Sample Problem 5.

Solution: First determine the reactions at the supports. Since the loads are symmetrically placed, the support reactions are each one-half of the total load, or

$$\Sigma F_y = 0 \qquad R_a + R_e = 5 + 10 + 5 + 10 + 20 + 10 = 60$$
$$R_a = R_e = 30 \text{ kip}$$

At joint A (Fig. 4-36),

$$\Sigma F_y = 0 \qquad R_a - \mathbf{AB} \sin 60° = 0$$

$$\mathbf{AB} = \frac{30}{\sin 60°} = \frac{30}{0.866} = 34.6 \text{ kip (compression)}$$

$$\Sigma F_x = 0 \qquad \mathbf{AH} - \mathbf{AB} \cos 60° = 0$$
$$\mathbf{AH} = 34.6(0.50) = 17.3 \text{ kip (tension)}$$

Now, we cannot go to joint B next because there are three unknown forces: **BH**, **BG**, and **BC**. At joint H, however, there are only two unknowns; therefore, we choose joint H next. At joint H (Fig. 4-37),

$$\Sigma F_y = 0 \qquad \mathbf{BH} - 10 = 0$$
$$\mathbf{BH} = 10 \text{ kip (tension)}$$

$$\Sigma F_x = 0 \qquad \mathbf{HG} - 17.3 = 0$$
$$\mathbf{HG} = 17.3 \text{ kip (tension)}$$

[1] See also Sample Computer Program 5, Chap. 17.

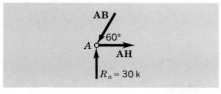

Fig. 4-36 Free-body diagram of joint A.

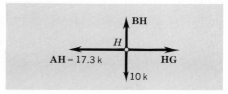

Fig. 4-37 Free-body diagram of joint H.

At joint B (Fig. 4-38),

$\Sigma F_y = 0$

$34.6 \cos 30° - 5 - 10 - \mathbf{BG} \cos 30° = 0$

$$\mathbf{BG} = \frac{15}{0.866} = 17.3 \text{ kip (tension)}$$

$\Sigma F_x = 0$

$34.6 \sin 30° + \mathbf{BG} \sin 30° - \mathbf{BC} = 0$

$\mathbf{BC} = 17.3 + 8.7$

$\mathbf{BC} = 26 \text{ kip (compression)}$

At joint C (Fig. 4-39),

$\Sigma F_y = 0 \qquad \mathbf{CG} - 10 = 0$

$\mathbf{CG} = 10 \text{ kip (compression)}$

$\Sigma F_x = 0 \qquad 26 - \mathbf{CD} = 0$

$\mathbf{CD} = 26 \text{ kip (compression)}$

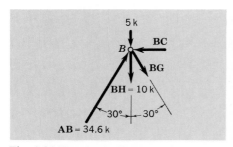

Fig. 4-38 Free-body diagram of joint B.

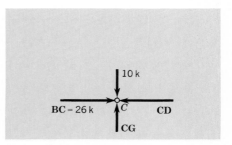

Fig. 4-39 Free-body diagram of joint C.

At joint G (Fig. 4-40),

$\Sigma F_y = 0$

$17.3 \sin 60° + \mathbf{GD} \sin 60° - 10 - 20 = 0$

$\mathbf{GD}(0.866) = 30 - 15 = 15$

$$\mathbf{GD} = \frac{15}{0.866} = 17.3 \text{ kip (tension)}$$

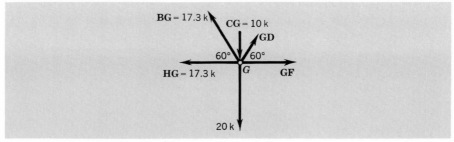

Fig. 4-40 Free-body diagram of joint G.

$$\Sigma F_x = 0$$
$$\mathbf{GF} + 17.3 \cos 60° - 17.3 \cos 60° - 17.3 = 0$$
$$\mathbf{GF} = 17.3 \text{ kip (tension)}$$

The reader will find that the remaining forces can be predicted from the forces already calculated because the loading is symmetrical, as are the truss members. Thus, **AB = DE, AH = FE, BH = DF, BG = GD, HG = GF**, and **BC = CD**. If the loading were not symmetrical, it would be necessary to proceed from joint to joint until all the forces were evaluated. There will always be one joint that has not been used. Summing forces about this "leftover" joint provides a check on the calculations.

The results for this sample problem are tabulated below:

Member	Force, kip	Type
AB	34.6	*C*
BC	26.0	*C*
CD	26.0	*C*
DE	34.6	*C*
EF	17.3	*T*
FG	17.3	*T*
GH	17.3	*T*
HA	17.3	*T*
HB	10.0	*T*
BG	17.3	*T*
CG	10.0	*C*
GD	17.3	*T*
DF	10.0	*T*

*Sample Problem 6   Find the forces in members *CD* and *DL* in the truss shown in Fig. 4-41.

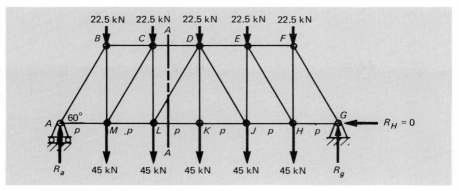

Fig. 4-41 Diagram for Sample Problem 6.

Solution: The method of sections applies conveniently to this problem. The support reactions share the total load equally, owing to the symmetrical arrangement.

$$\Sigma F_y = 0 \qquad R_a + R_g - 112.5 - 225 = 0$$
$$R_a = R_g = 168.8 \text{ kN}$$

Take a section through the truss which cuts the member to be solved. In this case, both CD and DL can be cut by the same section. Set out the left portion of the sectioned truss as a free body (Fig. 4-42). Take moments about joint L to obtain a direct solution for **CD**. Note that the widths of all panels are equal (but unknown, p).

$$\Sigma M_l = 0$$
$$168.8(2p) - 22.5(p) - 45(p) - \textbf{CD}(\sqrt{3}p) = 0$$
$$\textbf{CD}(\sqrt{3})p = 337.6p - 67.5p = 270.1p$$
$$\textbf{CD} = \frac{270.1p}{\sqrt{3}p} = 156 \text{ kN (compression)}$$

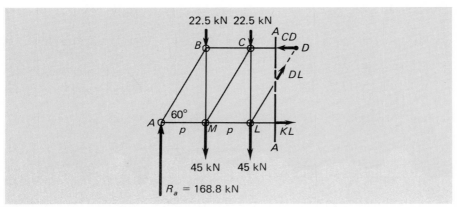

Fig. 4-42 Free-body diagram of portion of truss to left of section A-A.

Now, sum forces vertically to find DL_y.

$$\Sigma F_y = 0 \quad 168.8 + DL_y - 22.5 - 22.5 - 45 - 45 = 0$$

$$DL_y = -168.8 + 135 = -33.8$$

The direction of **DL** in Fig. 4-42 was incorrectly chosen. **DL** should act downward and put the member in compression instead of tension. Thus,

$$DL_y = 33.8 \text{ kN (downward)}$$

But from the direction of **DL** (60° with the horizontal), we note that

$$\frac{DL}{2} = \frac{DL_y}{\sqrt{3}} = \frac{DL_x}{1}$$

Then
$$DL = \frac{2DL_y}{\sqrt{3}} = \frac{2(33.8)}{\sqrt{3}} = 39 \text{ kN (compression)}$$

Incidentally, the direction of **KL** in Fig. 4-42 is correct, which can be verified by inspecting the moment situation about joint D.

PROBLEMS

4-1. Three smooth cylinders, each 18 in in diameter and weighing 30 lb, are placed in a box as shown in Fig. Prob. 4-1. What are the forces at all points of contact?

4-2. An ocean liner has an arrangement for supporting lifeboats and for lowering them over the side as shown in Fig. Prob. 4-2. There is a socket at A, and there is a smooth hole through the deck rail at B. If the boat and its load weigh 2000 lb, what are the reactions at A and B? Two identical davits support each lifeboat.

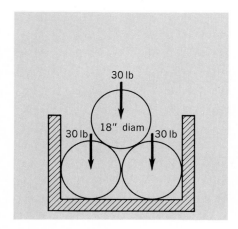

FIGURE PROBLEM 4-1

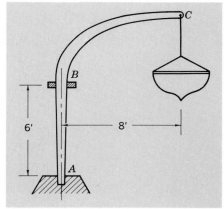

FIGURE PROBLEM 4-2

*4-3.   A member of uniform cross section having a mass of 68.8 kg is hinged at its upper end to a beam and is held in the position shown in Fig. Prob. 4-3 by means of the horizontal brace *BD*. What will be the reaction at *C* and the force in *BD*?

4-4.   The jointed frame in Fig. Prob. 4-4 carries loads as shown. What are the vertical and horizontal components of the pin reactions?

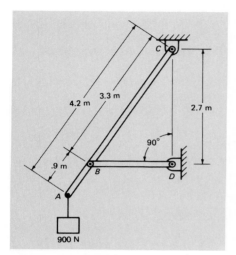

FIGURE PROBLEM 4-3

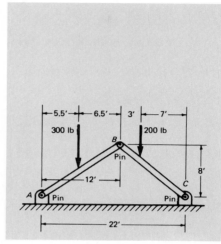

FIGURE PROBLEM 4-4

*4-5.   A brace is hinged at one end to a vertical wall and at the other end to a beam 3.6 m long. The beam has a mass of 183.5 kg and is also hinged to a vertical wall as shown in Fig. Prob. 4-5. The beam carries a load of 2.7 kN at the free end. What will be the compressive force in the brace, and what will be the values of the vertical and horizontal components of the reaction at hinge *A*?

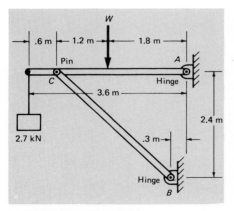

FIGURE PROBLEM 4-5

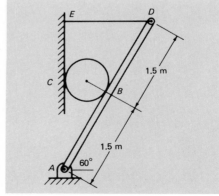

FIGURE PROBLEM 4-6

*4-6. A timber of uniform cross section with a mass of 90 kg is hinged at its lower end and held at an angle of 60° with the horizontal by a rod attached as shown in Fig. Prob. 4-6. A cylinder which has a mass of 30 kg is placed between the timber and the wall. What are the horizontal and vertical components of the reaction at A? What is the force between the cylinder and the timber?

4-7. A cherry-picker has a bucket at the end of the boom which is hinged to the boom as shown in Fig. Prob. 4-7. The diameter of the bucket is 3 ft; its depth is 3 ft; and its weight is 300 lb. A man weighing 180 lb is standing in the bucket. What will be the reactions at the hinges?

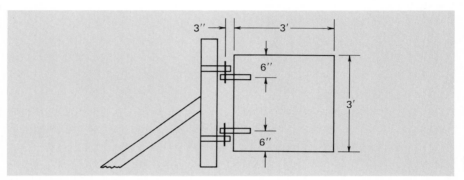

FIGURE PROBLEM 4-7

4-8. In an irrigation project it was found necessary to cross low ground or else swing the canal to the left by cutting into solid rock. It was decided to run the canal as a flume and support it on a number of frames, as shown in Fig. Prob. 4-8. The two members rest in sockets in solid rock

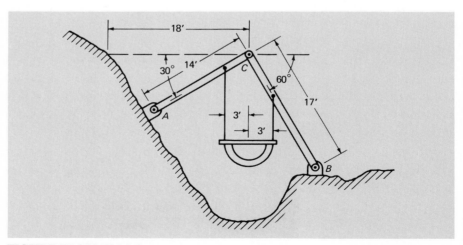

FIGURE PROBLEM 4-8

at points A and B. The sockets can be considered as hinges. What will be the vertical and horizontal components of the reactions at A and B? The weight of the water in the flume supported by each frame is estimated as 18 200 lb.

4-9. The A-frame in Fig. 4-13 has an additional load of 1200 lb at the center of BD. All members weigh 90 lb per foot of length. What are the pin reactions at B, C, and D?

4-10. Figure Problem 4-10 shows a wooden brace hinged to a floor at A and held by means of the rod BC. The weight of the brace is 300 lb and can be assumed to be applied as shown. What will be the horizontal and vertical components of the force at A? What is the force in BC?

4-11. In the crane shown in Fig. Prob. 4-11, the member BG weighs 600 lb, which is considered to act midway between B and F. If other weights of members are neglected, what will be the forces in the two tension rods? What will be the horizontal and vertical components of the reactions at A and B?

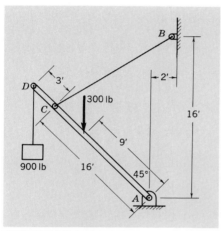

FIGURE PROBLEM 4-10   FIGURE PROBLEM 4-11

4-12. In a shipyard, a jib crane, shown in Fig. Prob. 4-12, was used to handle heavy machinery. With the moving load in the position shown, what will be the forces in members A, B, and C? What is the horizontal component of the reaction at the upper end of the vertical member?

4-13. For the Pratt truss shown in Fig. Prob. 4-13, find the magnitudes and types of forces in members AB, AH, BH, BC, HG, BG, and CG.

4-14. A cantilever truss, shown in Fig. Prob. 4-14, carries a load of 1200 lb. Find the reactions at A, B, C, and E.

4-15. Figure Problem 4-15 shows a truss that is supported at A and B and carries a load of 4000 lb. Find the external reactions and the forces in all the members, assuming that all joints are pin-connected.

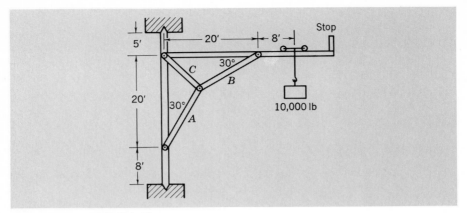

FIGURE PROBLEM 4-12

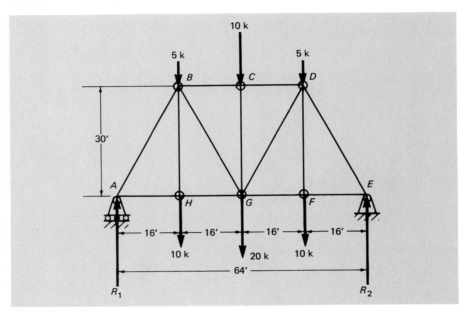

FIGURE PROBLEM 4-13

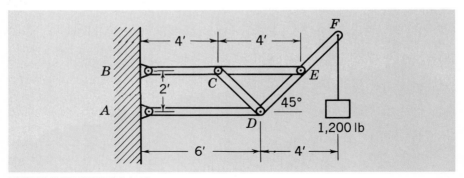

FIGURE PROBLEM 4-14

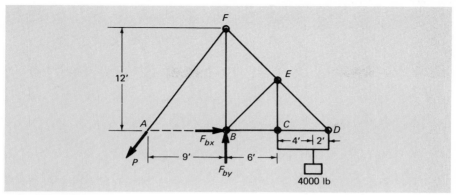

FIGURE PROBLEM 4-15

*4-16.   Determine the magnitude and direction of the reactions at roller A and pin B in Fig. Prob. 4-16.

 4-17.   Find the pin reaction for the link shown in Fig. Prob. 4-17.

*4-18.   Find the forces in the members of the truss shown in Fig. Prob. 4-18.

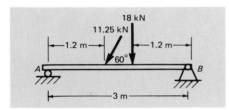

FIGURE PROBLEM 4-16

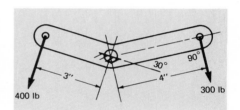

FIGURE PROBLEM 4-17

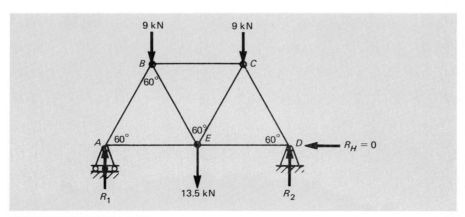

FIGURE PROBLEM 4-18

4-19. Find all the forces in the Warren truss shown in Fig. Prob. 4-19. All angles are 60°.

4-20. The members in one plane of a four-cornered tank tower are shown in Fig. Prob. 4-20. The weights given are due to the tank and water above. The horizontal forces are due to the wind. Find the forces in the members.

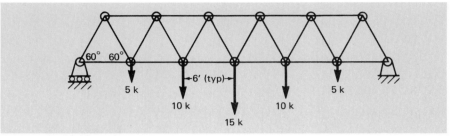

FIGURE PROBLEM 4-19

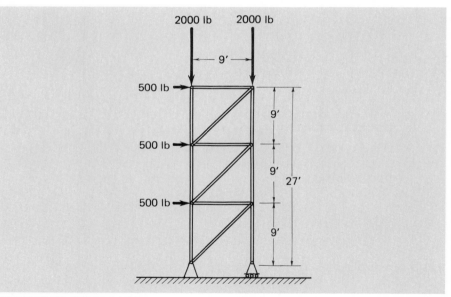

FIGURE PROBLEM 4-20

CHAPTER

5

Concurrent-Noncoplanar Forces

5-1 RESULTANT OF CONCURRENT-NONCOPLANAR FORCES

When concurrent forces are not in the same plane, they are called *noncoplanar*. To find the resultant of such forces, it is best to resolve each force into components along three axes that make angles of 90° with each other.

Let x, y, and z be the axes and O be the point through which the concurrent forces pass. Take **OE** as one of the forces, making the angles α, β, and γ with the axes (Fig. 5-1). Then the components, or projections, of **OE** $= F$ are $F_x = F \cos \alpha$, $F_y = F \cos \beta$, and $F_z = F \cos \gamma$. But the three projections **OG**, **OA**, and **OC** are the three edges of a rectangular parallelepiped of which the force F is the diagonal. From geometry, the square of the diagonal is equal to the sum of the squares of the three edges. Then

$$F^2 = F_x^2 + F_y^2 + F_z^2 = F^2(\cos^2 \alpha + \cos^2 \beta + \cos^2 \gamma)$$

But that cannot be true unless

$$\cos^2 \alpha + \cos^2 \beta + \cos^2 \gamma = 1 \tag{5-1}$$

Angles α, β, and γ are called the *direction angles of the force F*, and Eq. (5-1) gives the condition that their cosines must satisfy. For example, if $\alpha = 60°$ and $\beta = 45°$,

$$\cos^2 60° + \cos^2 45° + \cos^2 \gamma = 1$$
$$0.25 + 0.5 + \cos^2 \gamma = 1$$
$$\cos^2 \gamma = 0.25$$
$$\cos \gamma = 0.5$$
$$\gamma = 60°$$

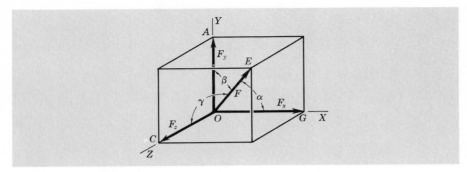

Fig. 5-1 x, y, and z components of a force.

If each of the concurrent systems of forces is resolved into its x, y, and z components, the original system is replaced by three sets of forces acting along the x, y, and z axes, respectively. Since all the forces along the x axis are in the same straight line, they can be added algebraically and their sum can be represented by ΣF_x. Similarly, those along the y axis and the z axis can be added algebraically and their sums can be represented by ΣF_y and ΣF_z, respectively. Their resultant, the diagonal of the parallelepiped formed by ΣF_x, ΣF_y, and ΣF_z, is equal to

$$R = \sqrt{(\Sigma F_x)^2 + (\Sigma F_y)^2 + (\Sigma F_z)^2} \qquad (5\text{-}2)$$

Sample Problem 1   The force F in Fig. 5-1 is 100 lb. What will be the components along the three axes if $\alpha = 60°$, $\beta = 45°$, and $\gamma = 60°$?

Solution:

$$F_x = 100 \cos \alpha = 100 \cos 60° = 100 \times 0.5 = 50 \text{ lb}$$
$$F_y = 100 \cos \beta = 100 \cos 45° = 100 \times 0.707 = 70.7 \text{ lb}$$
$$F_z = 100 \cos \gamma = 100 \cos 60° = 100 \times 0.5 = 50 \text{ lb}$$

*Sample Problem 2[1]   A system of concurrent forces has been reduced to the following components:

$$\Sigma F_x = 45 \text{ N} \qquad \Sigma F_y = 90 \text{ N} \qquad \text{and} \qquad \Sigma F_z = 67.5 \text{ N}$$

Find the magnitude of the resultant and its direction angles.

Solution: The resultant of the three forces can be represented as the diagonal of a parallelepiped whose sides are the three given components. Thus, by Eq. (5-2),

---

[1] See also Sample Computer Program 6, Chap. 17.

$$R = \sqrt{45^2 + 90^2 + 67.5^2} = 121 \text{ N}$$

$$\cos \alpha = \frac{\Sigma F_x}{R} = \frac{45}{121} = 0.372 \qquad\qquad \alpha = 68.2°$$

$$\cos \beta = \frac{\Sigma F_y}{R} = \frac{90}{121} = 0.744 \qquad\qquad \beta = 42.0°$$

$$\cos \gamma = \frac{\Sigma F_z}{R} = \frac{67.5}{121} = 0.558 \qquad\qquad \gamma = 56.1°$$

Of course, once α and β are known, γ can be found by using Eq. (5-1).

5-2 CONDITIONS FOR EQUILIBRIUM

When forces are in equilibrium, their resultant is zero; that is,

$$\sqrt{(\Sigma F_x)^2 + (\Sigma F_y)^2 + (\Sigma F_z)^2} = 0$$

But each quantity under the radical sign, being a square, is positive. The sum of positive numbers cannot be zero unless each number is equal to zero. Then

$$\Sigma F_x = 0 \qquad \Sigma F_y = 0 \qquad \Sigma F_z = 0 \qquad\qquad (5\text{-}3)$$

This result gives three equations of condition, from which three unknowns can be determined.

   If the force polygon were drawn for a set of concurrent-noncoplanar forces that were in equilibrium, the polygon would close. Then the projection of this polygon on any plane would be a closed-plane polygon. But since it is a closed figure, the forces that form it are in equilibrium. We then have coplanar forces in equilibrium, and the principles of Chap. 2 can be applied.

   $\Sigma F_x = 0$ and $\Sigma F_y = 0$ are the equations of condition that must be satisfied when concurrent forces in equilibrium are projected on the xy plane. Also, $\Sigma F_z = 0$ is the equation that the projection of such forces on the z axis must satisfy.

   The forces might also be projected on the xz plane and the y axis or the yz plane and the x axis. The conclusion can be summarized as follows: When a set of concurrent-noncoplanar forces is in equilibrium, project all the forces on a convenient plane and also on an axis perpendicular to the plane. Then form $\Sigma F_x = 0$, $\Sigma F_y = 0$, and $\Sigma F_z = 0$ and solve simultaneously.

*Sample Problem 3   A body having a mass of 46 kg is suspended from the ceiling by three cords of equal length. The cords make angles of 30° with the vertical, and they are attached to the ceiling at points equally spaced on the arc of a circle. Find the tensions in the cords (Fig. 5-2).

Solution: Let the tensions in OA, OB, and OC be F_1, F_2, and F_3. Set point O out as a free body (Fig. 5-3). Now project all the forces on a horizontal plane xz and a vertical axis y (Figs. 5-4 and 5-5) after calculating the weight of the body. $W = m \cdot g = (46 \text{ kg})(9.81 \text{ m/s}^2) = 451 \text{ N}$.

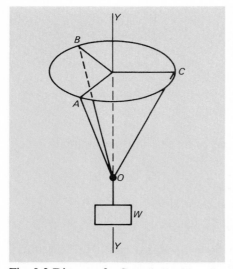

Fig. 5-2 Diagram for Sample Problem 3.

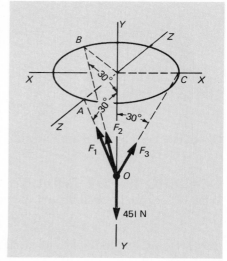

Fig. 5-3 Free-body diagram of point O for Sample Problem 3.

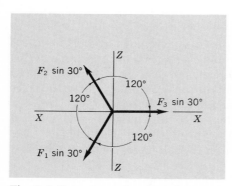

Fig. 5-4 Forces at point O projected on horizontal (xz) plane.

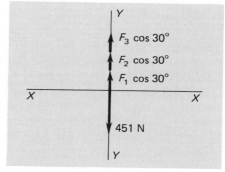

Fig. 5-5 Forces at point O projected on vertical (y) axis.

$\Sigma F_x = 0 \qquad -F_1 \sin 30° \cos 60° - F_2 \sin 30° \cos 60° + F_3 \sin 30° = 0$

$\Sigma F_y = 0 \qquad (F_1 + F_2 + F_3) \cos 30° - 451 = 0$

$\Sigma F_z = 0 \qquad -F_1 \sin 30° \sin 60° + F_2 \sin 30° \sin 60° = 0$

From the equation for ΣF_z,

$$F_1 = F_2$$

From the equation for ΣF_x,

$$F_1 = F_2 = F_3$$

From the equation for ΣF_y,

$$3F(0.866) = 451$$

$$F = \frac{451}{2.6} = 173 \text{ N (tension)}$$

Sample Problem 4[1]   A 200-lb weight is suspended from a wall by means of a bracket as shown in Fig. 5-6. Find the forces in OA, OB, and OD.

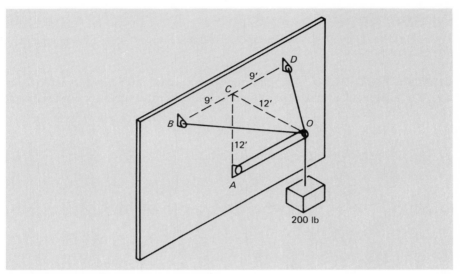

Fig. 5-6 Diagram for Sample Problem 4.

Solution: Let the forces be represented by F_1, F_2, and F_3 (Fig. 5-7a). Take point O as a free body. First project the forces onto a vertical axis through O. Since F_2 and F_3 are perpendicular to this axis, their projections are zero. Then

$$F_1 \sin \measuredangle COA - 200 = 0$$

But

$$\measuredangle COA = 45°$$

$$F_1 = \frac{200}{0.707} = 283 \text{ lb (compression)}$$

Now project the forces on a horizontal plane through O (Fig. 5-7b). Since F_2 and F_3 make equal angles with F_1, they are equal to each other.

$\Sigma F_x = 0$ $\qquad -2F_2 \cos \measuredangle COD + 283 \cos 45° = 0$

But $\qquad DO = \sqrt{9^2 + 12^2} = 15 \text{ ft} \qquad$ and $\qquad \cos \measuredangle COD = \frac{12}{15} = \frac{4}{5}$

Then $\qquad -2(\frac{4}{5})F_2 + 200 = 0$

$$F_2 = F_3 = 125 \text{ lb (tension)}$$

[1] See also Sample Computer Program 7, Chap. 17.

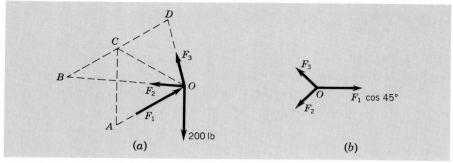

Fig. 5-7 (*a*) Free-body diagram of point *O* for Sample Problem 4. (*b*) Forces at point *O* projected on horizontal plane.

*Sample Problem 5   An object having a mass of 230 kg is suspended from a wall by means of a bracket as shown in Fig. 5-8. Find the forces in members *OA*, *OB*, and *OD*.

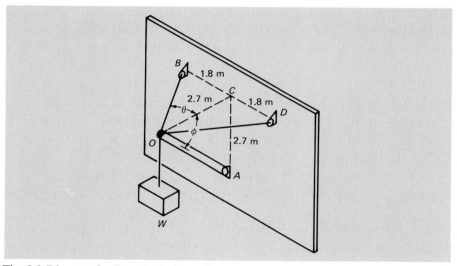

Fig. 5-8 Diagram for Sample Problem 5.

Solution: $W = m \cdot g = (230 \text{ kg})(9.81 \text{ m/s}^2) = 2260 \text{ N} = 2.26 \text{ kN}$. Represent the forces by P_1, P_2, P_3. Take *BOD* as a free body. The forces acting on *BOD* are P_1, 2.26 kN, and the reactions at *B* and *D*. Taking moments about line *BD*, we have

$$P_1 \times 2.7 \sin \phi - 2.26 \times 2.7 = 0$$

But $\phi = 45°$.

$$P_1 = \frac{2.26 \times 2.7}{2.7 \sin 45°} = \frac{2.26}{0.707} = 3.20 \text{ kN (compression)}$$

Now project the three forces on a horizontal plane with point O as the free body (Fig. 5-9).

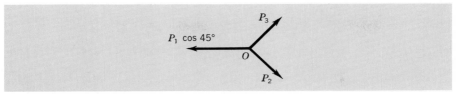

Fig. 5-9 Forces at point O projected on horizontal plane.

$$\angle BOC = \angle COD = \theta$$

$$\tan \theta = \frac{1.8}{2.7} = 0.67$$

$$\theta = 33.7°$$

$$P_3 = P_2$$

$\Sigma F_x = 0$ $\qquad 2P_2 \cos \theta = P_1 \cos 45°$

$$2P_2(0.832) = 3.20(0.707)$$

$$P_2 = P_3 = 1.36 \text{ kN (tension)}$$

Sample Problem 6 The boom of the derrick shown in Fig. 5-10 can swing from the horizontal to 25° from the vertical and through the larger angle CDA. With the boom horizontal and bisecting the angle CDA, find the forces in all members of the derrick.

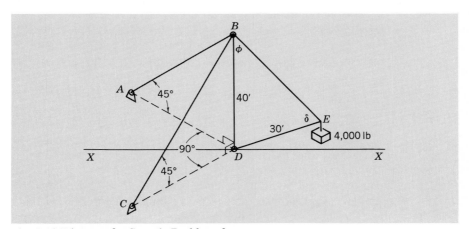

Fig. 5-10 Diagram for Sample Problem 6.

Solution: Since all the members are two-force members, the lines of action of all forces are known. At point B there are four forces in equilibrium. Since they are noncoplanar, there are only three equations of condition that can be formed, and the solution cannot start at this point.

At point E, there are three coplanar forces, one of which is known. Since for this case two equations exist, the solution is possible. Set point E out as a free body (Fig. 5-11a) and draw the force triangle LMN (Fig. 5-11b). Since ∡ BDE is 90°, side $BE = 50$ ft because DBE is a 3-4-5 right triangle. Hence,

$$\sin \delta = \tfrac{40}{50} = 0.80$$
$$\delta = 53.13° \text{ or } 53°8'$$

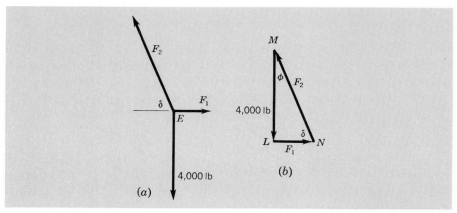

Fig. 5-11 (a) Free-body diagram of point E for Sample Problem 6. (b) Force triangle for point E.

and $\qquad \phi = 90° - \delta = 36.87° \text{ or } 36°52'$

From Fig. 5-11b, $\quad \sin \delta = \dfrac{4000}{F_2}$

$$F_2 = \frac{4000}{0.80} = 5000 \text{ lb (tension)}$$

$$\sin \phi = \frac{F_1}{F_2}$$

$$F_1 = 5000(0.60) = 3000 \text{ lb (compression)}$$

The results for F_1 and F_2 could have been obtained by use of similar triangles DBE and LMN.

Point B can now be set out as a free body, since the force in BE is known and there are only three unknowns left (Fig. 5-12). First, project the forces on a horizontal plane through B as shown in Fig. 5-13a. The projection of F_5 is zero, since F_5 is perpendicular to the plane of projection. Now draw the force triangle (Fig. 5-13b). Since it is an isosceles right triangle, F_3 equals F_4.

$$\cos 45° = \frac{F_4 \cos 45°}{3000}$$

$$F_4 = 3000 \text{ lb} = F_3 \text{ (tension)}$$

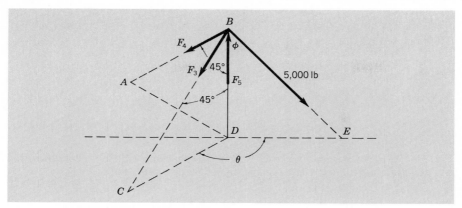

Fig. 5-12 Free-body diagram of point B.

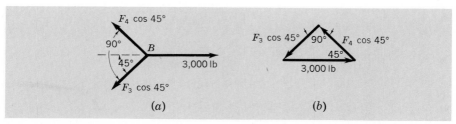

Fig. 5-13 (a) Forces at point B projected on horizontal plane with boom horizontal and central. (b) Force triangle for point B.

Now project all forces at point B onto a vertical line through B.

$\Sigma F_z = 0$ $F_3 \cos 45° + F_4 \cos 45° + 5000 \cos \phi - F_5 = 0$

But $F_3 = F_4$ and $\cos \phi = 0.80$

Then $2(3000)(0.707) + 4000 = F_5$

$$F_5 = 8240 \text{ lb (compression)}$$

When a derrick is to be designed, the maximum forces that exist in the members are to be found for a given load while the boom moves through its entire range. Since the force triangle LMN is similar to DBE, two of the sides of which are of fixed lengths, the force in BE will be a maximum when BDE is a right angle, that is, when the boom is horizontal (assuming BDE cannot exceed 90°).

Now the maximum forces will occur in AB, CB, and BD when the force in BE is a maximum. It remains to find what angle boom DE, when it is horizontal, must make with CD to produce maximum forces in the mast and legs. Let $\angle CDE = \theta$. Set point B out as a free body. Now project all the forces onto a horizontal plane through point B (Fig. 5-14a). Next, draw the force triangle (Fig. 5-14b), from which

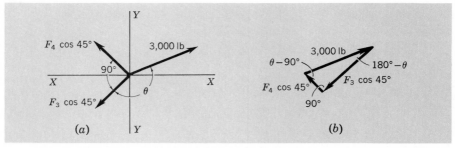

Fig. 5-14 (*a*) Forces at point *B* projected on horizontal plane with boom horizontal and positioned for maximum tension in leg *CB*. (*b*) Force triangle for point *B*.

$$\sin (\theta - 90°) = \frac{F_3 \cos 45°}{3000}$$

$$F_3 = \frac{3000 \sin (\theta - 90°)}{\cos 45°}$$

Now, F_3 is greatest when $\sin (\theta - 90°)$ is greatest.

$$\sin (\theta - 90°) = 1$$
$$\theta - 90° = 90°$$
$$\theta = 180°$$

This makes the plane of the boom perpendicular to the vertical plane through the leg *AB*. Although this case is special, in that the planes of the legs are perpendicular to each other, the same method of procedure for any angle *CDA* will lead to the same conclusion.

The maximum tension is found in one leg of a derrick when the plane of the boom is perpendicular to the plane of the other leg. By substituting $\theta = 180°$ in the equation for F_3, we have

$$F_3 = \frac{3000 \sin 90°}{0.707} = 4240 \text{ lb (tension)}$$

$$F_4 = \frac{3000 \sin 0°}{0.707} = 0$$

The analysis for maximum tension could be applied as well to maximum compression if it were possible to swing the plane of the boom into the smaller angle *CDA*. However, the maximum compression will be developed in either leg when the plane of the boom coincides with the plane of that leg and is on the same side of the mast. Again, set point *B* out as a free body, with the boom horizontal and in the position for maximum compression in leg *CB*, and project the forces onto a horizontal plane (Fig. 5-15). By inspection,

$$F_4 = 0$$

and
$$F_3 \cos 45° = 3000$$

$$F_3 = 4240 \text{ lb (compression)}$$

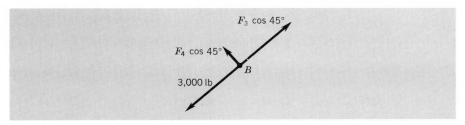

Fig. 5-15 Forces at point B projected on horizontal plane with boom horizontal and positioned for maximum compression in leg CB.

To obtain the force in DB for each of the foregoing cases, project onto the vertical axis DB. Then

$$5000 \cos \phi + 4240 \cos 45° = F_5$$
$$4000 + 3000 = F_5$$
$$F_5 = 7000 \text{ lb (compression)}$$

This result is less than the value obtained when the plane of the boom bisected angle CDA. To prove this, let $\angle CDE = \theta$ (Fig. 5-14a). Then, by Fig. 5-14b,

$$F_3 = \frac{3000 \sin (\theta - 90°)}{\cos 45°} = -\frac{3000 \cos \theta}{0.707}$$
$$F_4 = \frac{3000 \sin (180° - \theta)}{\cos 45°} = \frac{3000 \sin \theta}{0.707}$$

From the vertical projection on DB,

$$F_5 = 5000 \cos \phi + F_3 \cos 45° + F_4 \cos 45°$$
$$= 4000 + 3000(-\cos \theta + \sin \theta)$$

Now, F_5 is a maximum when $\sin \theta - \cos \theta$ is greatest. This condition occurs when both functions are numerically equal and $\cos \theta$ is negative; i.e., when $\theta = 135°$. When angle CDA is not $90°$, the same line of reasoning will show maximum force in the mast when the plane of the boom bisects the angle between the legs.

Sample Problem 7 In Sample Problem 6, in order to lift the weight vertically with the boom stationary and horizontal, it is connected to E by a block and tackle. Four lengths of cable run from E to W. One end is fastened at E, and the other runs along DE over a pulley at D and then to a winch. Between B and E, there are five strands of another cable, the free end of which runs down BD to a pulley at D and then to a second winch (see Fig. 5-16). Find the maximum forces in all members and the reaction at D.

Solution: Set W out as a free body (Fig. 5-17). If we neglect friction on the pulleys, the tension in the cable is the same at all points, and each of the upward forces is T_1. *For vertical equilibrium,*

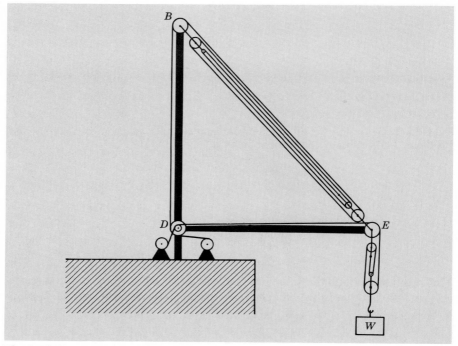

Fig. 5-16 Diagram for Sample Problem 7.

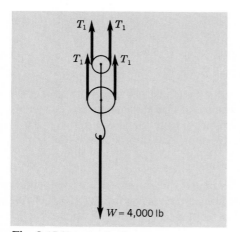

Fig. 5-17 Free-body diagram at W.

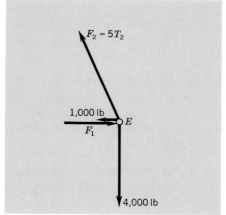

Fig. 5-18 Free-body diagram at point E.

$$4T_1 = 4000$$
$$T_1 = 1000 \text{ lb (tension)}$$

Now set point E out as a free body (Fig. 5-18). By taking $\Sigma F_y = 0$, $F_2 = 5000$ lb, as in Sample Problem 6; but this is carried by five cables. Thus

$$5T_2 = 5000$$
$$T_2 = 1000 \text{ lb (tension)}$$

From $\Sigma F_x = 0$,

$$F_1 = 1000 + 3000 = 4000 \text{ lb (compression)}$$

where 3000 lb is the horizontal component of F_2.

Now set point B out as a free body (Fig. 5-19). When we project the forces onto a horizontal plane, we see that F_3 and F_4 have the same values as in Sample Problem 6. By summing all vertical components along BD and setting them equal to zero, we obtain $F_5 = 9300$ lb (compression), or 1060 lb greater than in the preceding example.

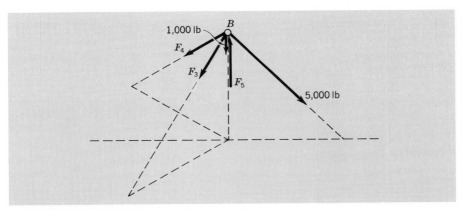

Fig. 5-19 Free-body diagram at point B.

For determining the reaction at D, set out D as a free body (Fig. 5-20). The reaction not being known, the components are used. By $\Sigma F_y = 0$ and $\Sigma F_x = 0$,

$$F_{dy} = 9300 + 1000 - 1000 = 9300 \text{ lb}$$
$$F_{dx} = 4000 - 1000 - 1000 = 2000 \text{ lb}$$
$$F_d = \sqrt{9300^2 + 2000^2} = 9510 \text{ lb}$$
$$\tan \theta = \frac{9300}{2000} = 4.65$$
$$\theta = 77.85° \text{ or } 77°50'$$

Sample Problem 8 Figure 5-21 shows a shear leg crane lifting a 40 000-lb load. The legs are 60 ft long and are 30 ft apart at their lower ends. The backstay is 70 ft long. All members are pin-connected, and A, E, and C are in the same horizontal plane. Find the forces in the members.

Solution: All members are two-force members concurrent at B. Set B out as a free body (Fig. 5-22). Project all forces onto a vertical plane ABF. Since the two

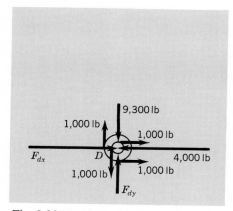

Fig. 5-20 Free-body diagram at point D.

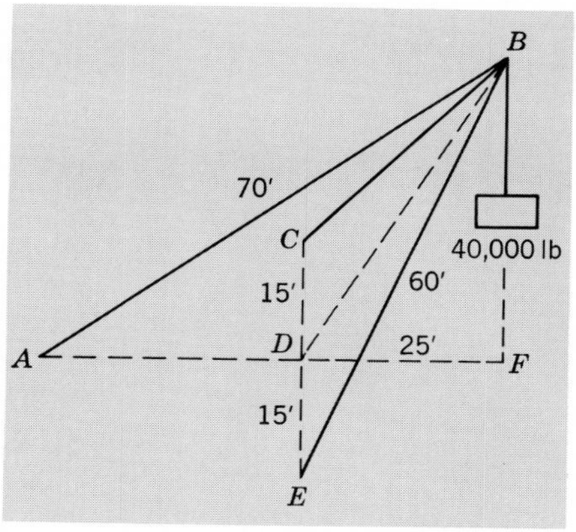

Fig. 5-21 Diagram for Sample Problem 8.

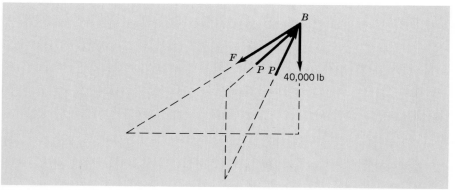

Fig. 5-22 Free-body diagram at point B.

forces P are in a plane BCE, which is perpendicular to the plane of projection, they project into P', their resultant (Fig. 5-23a).

Next, form the force triangle LMN (Fig. 5-23b). Angles M, L, and N must now be found.

$$BD^2 = 60^2 - 15^2 = 3375$$
$$BD = 58.1 \text{ ft}$$
$$BF^2 = BD^2 - 25^2 = 2750$$
$$BF = 52.4 \text{ ft}$$
$$\cos \angle ABF = \frac{52.4}{70} = 0.7490$$
$$\angle ABF = 41.57° \text{ or } 41°34'$$
$$\angle L = 180° - 41°34' = 138°26'$$
$$\tan \angle DBF = \tan \angle M = \frac{25}{52.4} = 0.477$$

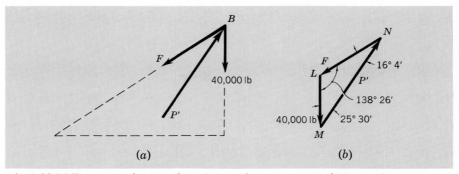

Fig. 5-23 (*a*) Forces at point *B* projected on vertical plane *ABF*. (*b*) Force triangle at point *B*.

$$\angle M = 25.5° \text{ or } 25°30'$$
$$\angle N = 180° - (138°26' + 25°30') = 16°4'$$

From triangle *LMN*,

$$\frac{F}{\sin 25°30'} = \frac{40\ 000}{\sin 16°4'}$$
$$F = 62\ 200 \text{ lb (tension)}$$
$$\frac{P'}{\sin 138°26'} = \frac{40\ 000}{\sin 16°4'}$$

Since P' is the resultant of the thrusts P, it will equal the sum of the components along BD, or

$$P' = 95\ 900 \text{ lb}$$
$$2P \cos \angle DBC = 95\ 900$$
$$P = \frac{95\ 900}{2(58.1/60)} = 49\ 500 \text{ lb (compression)}$$

*Sample Problem 9   A tripod has legs 3.6 m long. The lower ends of the legs are placed at the corners of an equilateral triangle 1.8 m on a side. What are the forces in the legs when a load of 4.5 kN is placed on top?

Solution: Figure 5-24 shows an elevation of the tripod and a bottom view of the base. Point D' is the projection of the vertex of the tripod on the plane of the base. Since the sides of the base are equal in length, this projection falls at the geometric center of the triangle. The angle that each leg makes with the vertical can be determined from the triangle ADD'. From the geometry of the triangles,

$$AD' = \tfrac{2}{3}AE = \tfrac{2}{3}\sqrt{1.8^2 - 0.9^2}$$
$$= \tfrac{2}{3}(1.56) = 1.04 \text{ m}$$
$$DD' = \sqrt{(AD)^2 - (AD')^2} = 3.45 \text{ m}$$

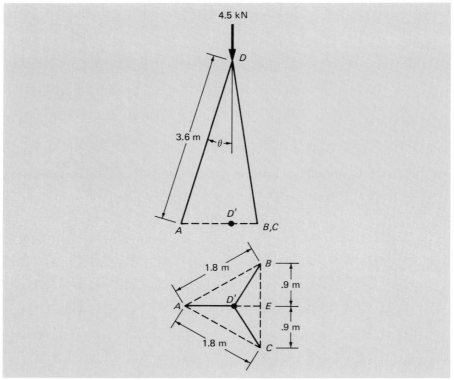

Fig. 5-24 Diagram for Sample Problem 9.

Since the legs are symmetrically arranged and make equal angles with the vertical, we have, from $\Sigma F_y = 0$,

$$3F \cos \theta = 4.5$$

$$3F \left(\frac{3.45}{3.6} \right) = 4.5$$

$$F = \frac{3.6(4.5)}{3(3.45)} = 1.57 \text{ kN} \qquad \text{say, } F = 1.6 \text{ kN (compression)}$$

PROBLEMS

5-1. Figure Problem 5-1 shows a tripod used in unloading heavy machinery. What is the force in each of the legs if the load is 4 tons?

*5-2. Find the forces in all members in the bracket shown in Fig. Prob. 5-2. Members AD and CD are horizontal.

5-3. In the derrick shown in Fig. Prob. 5-3, what is the force in member BD when boom AC is in the position shown?

5-4. In the derrick shown in Fig. Prob. 5-3, what would be the force in BD if boom AC were rotated into plane ABD?

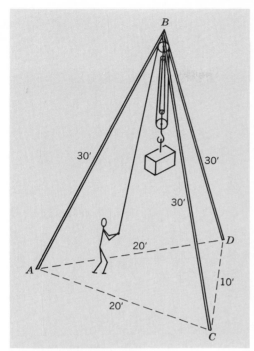

FIGURE PROBLEM 5-1

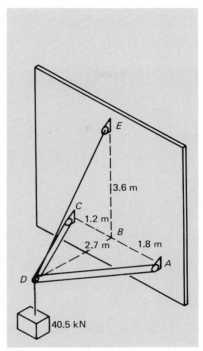

FIGURE PROBLEM 5-2

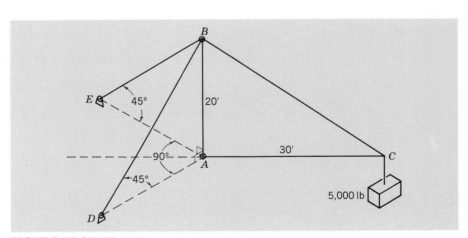

FIGURE PROBLEM 5-3

5-5. What are the forces in all parts of the shear leg crane shown in Fig. Prob. 5-5 if a naval gun weighing 50 tons is lifted?

*5-6. Figure Problem 5-6 shows a derrick used in a quarry. The mast and boom are pivoted at the top and the bottom so they can be rotated through an angle α of 360°. The angle θ that the boom makes with the

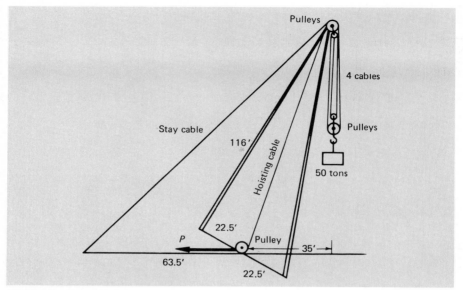

FIGURE PROBLEM 5-5

vertical also is adjustable. If a load having a mass of 9.20 metric tons is raised, what is the force in the boom when $\theta = 45°$? The length of the mast DF is 7.2 m, and the boom EF is 5.4 m (1 metric ton = 1000 kg).

5-7. The derrick shown in Fig. Prob. 5-6 is stabilized by guy wires AD, BD, and CD. In erecting the derrick, wires AD and BD are each pulled taut

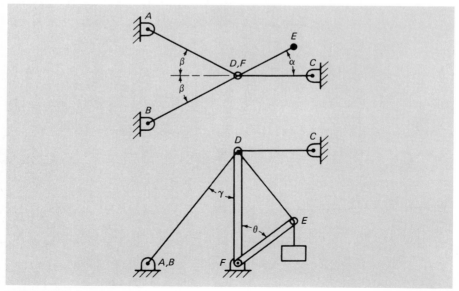

FIGURE PROBLEM 5-6

to an initial tensile force of 2500 lb by means of turnbuckles. What should be the initial tensile force in wire CD so that forces in a horizontal plane through point D are fully balanced if $\alpha = 0°$, $\beta = 60°$, and $\gamma = 75°$? (The derrick is not lifting a load.)

5-8. What force do the guy wires in Prob. 5-7 impart to member DF?

5-9. The derrick shown in Fig. Prob. 5-6 lifts a load of 18 tons. Members DF and EF are 30 and 20 ft long, respectively. The various angles are $\alpha = 15°$, $\beta = 60°$, $\gamma = 70°$, and $\theta = 60°$. Wires AD and BD were initially tightened to balance the horizontal initial pull of CD. The tension in guy wire CD is 5000 lb after the derrick is loaded. Find (a) the force in DE, (b) the resultant force in DF, and (c) the resultant forces in the guy wires.

*5-10. A load having a mass of 9.20 metric tons is to be lifted by means of a contractor's derrick as shown in Fig. Prob. 5-10. If the vertical plane

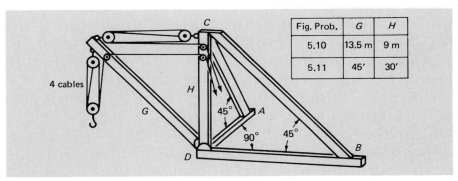

Fig. Prob.	G	H
5.10	13.5 m	9 m
5.11	45'	30'

FIGURE PROBLEM 5-10

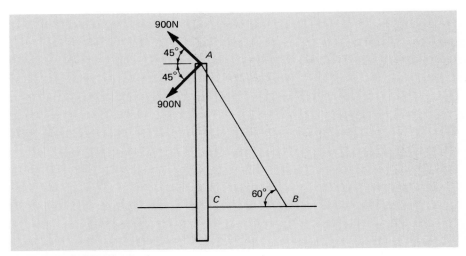

FIGURE PROBLEM 5-12

containing the boom bisects the angle *ADB*, what reactions must be provided at points *A* and *B* to keep the derrick from tipping? The boom has a mass of 1.84 metric tons, and it makes an angle of 45° with the horizontal. (1 metric ton = 1000 kg.)

5-11. When the boom of the derrick shown in Fig. Prob. 5-10 swings into the plane *BCD*, what will be the force in leg *BC*? The load is 7 tons, and the boom weighs 3000 lb and makes an angle of 60° with the horizontal.

*5-12. Both of the 900-N forces in Fig. Prob. 5-12 are acting in a horizontal plane and are applied to the top of pole *AC* at point *A*. To eliminate bending of the pole, guy wire *AB* is used. Determine the force acting in the pole and the guy wire.

6

Static and Kinetic Friction

6-1 MAXIMUM STATIC FRICTION

Friction was defined in Sec. 2-17 as a force between two surfaces in contact acting to resist the motion of one surface over the other. Friction is a passive force. It acts only when an effort is made to move one body over the other.

In Fig. 6-1, let a 100-lb body rest on a rough plane and be acted on by a horizontal force P. If the body does not move, the forces are in equilibrium. Then $P = F$ and $N = 100$ lb; that is, when $P = 10$ lb, $F = 10$ lb, and when $P = 20$ lb, $F = 20$ lb, etc.

But if $P = 30$ lb and the body is just on the point of moving, $F = 30$ lb is its greatest, or maximum, static friction. Any value of P, say 31 lb or more, will cause motion.

Experiments have shown that, within certain limits, the maximum value of friction depends on the normal force on the plane and the materials in the two rubbing surfaces. So long as the normal force per square inch does not exceed the elastic limit of the bearing surface, the friction is independent of the area of contact between the two surfaces; but it does depend on the normal force between the surfaces.

Experiments have also shown that, for practical purposes, the ratio of the maximum friction to the normal force on the plane is a constant for any given material. This ratio is defined as the *coefficient of static friction.* Let f represent the coefficient of friction and F_m the maximum friction force. Then

$$f = \frac{F_m}{N} \tag{6-1}$$

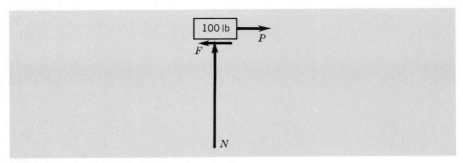

Fig. 6-1 Forces acting on body on rough plane.

Thus, if $F_m = 30$ lb, for Fig. 6-1,

$$f = \tfrac{30}{100} = 0.3$$

The resultant of F_m and N when motion is impending is a single force R_m, as shown in Fig. 6-2. Since P, W, and R are in equilibrium, they form a triangle such as OCB (Fig. 6-3a).

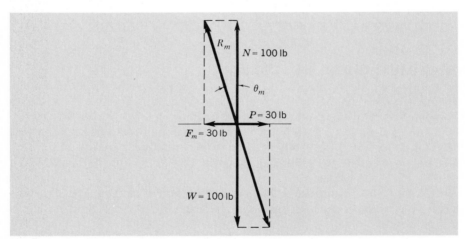

Fig. 6-2 Resultant of normal force N and maximum friction force F_m.

Now let θ_m be the angle that R_m makes with the normal to the plane. Then $\tan \theta_m = \tfrac{30}{100} = 0.3$; that is, when motion is impending, the resultant R_m makes an angle whose tangent is the coefficient of friction f. R is always drawn on the opposite side of the normal from the direction of motion. That is, if motion would take place to the right, R bears to the left. This principle is evident from the fact that F is always resisting motion.

In Fig. 6-3b, $\angle AOB = \theta_m$, the maximum angle of friction. Suppose P is less than F_m; then the friction F is less than F_m, say **BC** (Fig. 6-3a). Then **OC** is the resultant making an angle COB with the normal. Motion cannot take place until P is greater than **BA** (Fig. 6-3b).

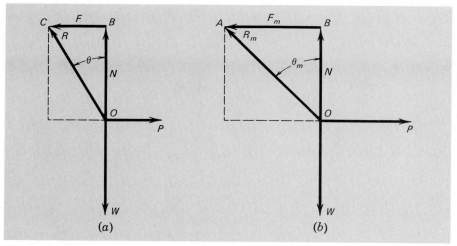

Fig. 6-3 (*a*) Resultant of normal force N and a static friction force F. Note that R acts at angle θ, opposing force P and supporting the weight W. (*b*) Impending motion. Resultant R_m at angle θ_m due to normal force N and maximum static friction force F_m. Note that $\tan \theta_m = F_m/N = f$, the coefficient of friction.

From Eq. (6-1), $F_m = fN$. If $f = 0.25$ and $N = 50$ lb, then $F_m = 12.5$ lb. If $f = 0.2$ and $N = 75$ lb, then $F_m = 15$ lb.

Every surface has depressions on its face. These depressions vary with the material and are more evident in a surface that is rough than in one that is smooth. When one body rests on another, the elevations on one tend to fill the depressions on the other. Before there can be motion, one body must be lifted out of the depressions or the surface must be leveled or smoothed. Although this illustration does not explain all the action of friction forces, it at least furnishes an explanation of why a body moves more easily over ice than over a brick sidewalk and also why it is easier to move a weight by pulling at an angle to give a small vertical or lifting component.

The force necessary to continue uniform motion is usually less than the force necessary to start it. The friction of motion in rotating shafts, drums on brake shoes, etc., is called *kinetic friction*.

Sample Problem 1
A 75-lb body rests on a horizontal plane for which $f = 0.4$. The body is acted on by a force of 20 lb at $30°$ with the horizontal as shown in Fig. 6-4. Will the body move?

Solution: Set the object out as a free body (Fig. 6-5). The forces are in equilibrium vertically.

$$\Sigma F_y = 0 \qquad\qquad N - 75 + 20 \sin 30° = 0$$

$$N = 65 \text{ lb}$$

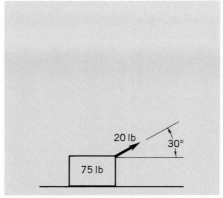

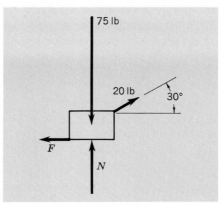

Fig. 6-4 Diagram for Sample Problem 1.

Fig. 6-5 Free-body diagram for Sample Problem 1.

By Eq. (6-1), maximum friction

$$F_m = 0.4N = 0.4(65) = 26 \text{ lb}$$

For equilibrium horizontally,

$$\Sigma F_x = 0 \qquad\qquad -F + 20 \cos 30° = 0$$
$$F = 17.32 \text{ lb}$$

But the body will not move until the maximum friction of 26 lb is overcome. Since only a force of 17.32 lb is necessary to prevent motion, the body remains at rest. Friction equals 17.32 lb.

˙Sample Problem 2 A body having a mass of 46 kg rests on a horizontal plane for which $f = 0.4$. A force P acts on the body at an angle of 20° with the horizontal. Find its magnitude for impending motion (Fig. 6-6).

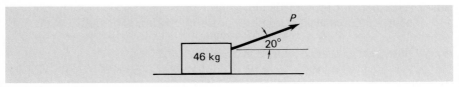

Fig. 6-6 Diagram for Sample Problem 2.

Solution: When motion is impending, it will often be found convenient to use the resultant R_m of N and F_m, which must make the $\angle\theta_m$ with the normal (tan $\theta_m = f$). By setting out the free body (Fig. 6-7), we find there are three forces in equilibrium: R_m, P, and W.

$$\tan \theta_m = f = 0.4$$
$$\theta_m = 21.8° \text{ or } 21°48'$$

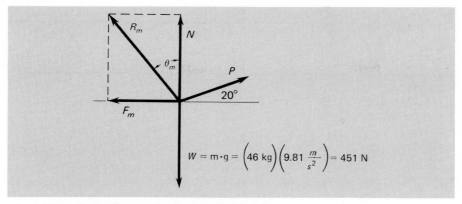

Fig. 6-7 Free-body diagram for Sample Problem 2.

This problem can be solved by using the summation method. Since the components of the forces in the horizontal direction are in equilibrium at the instant of impending motion,

$\Sigma F_x = 0$ $\qquad P \cos 20° - R_m \sin \theta_m = 0$

$$P \cos 20° = R_m \sin \theta_m$$

$\Sigma F_y = 0$ $\qquad R_m \cos \theta_m + P \sin 20° - 451 = 0$

$$R_m = \frac{451 - P \sin 20°}{\cos \theta_m}$$

$$P \cos 20° = \frac{\sin \theta_m}{\cos \theta_m} (451 - P \sin 20°)$$

$$= \tan \theta_m (451 - P \sin 20°)$$

$$P(\cos 20° + 0.4 \sin 20°) = 0.4(451) = 180$$

$$P = \frac{180}{0.9397 + 0.1368} = 167 \text{ N}$$

It can be seen that the force P exerts a lifting component and reduces the amount of the normal pressure; thus, $N = 394$ N.

Table 6-1 gives several values of coefficient of friction. The degree of smoothness of the surfaces affects the value of f very materially, and that accounts for the range of values. The kinetic coefficients are usually from 20 to 40 percent lower than the values of static f.

6-2 FRICTION ON AN INCLINED PLANE

Let a body weighing W lb rest on a rough inclined plane, Fig. 6-8a. Suppose the weight is just on the point of slipping down the plane. Then friction is a maximum F_m. From Sec. 2-17, the component of W down the plane is $W \sin \alpha$ and that normal to the plane is $W \cos \alpha$, Fig. 6-8b.

TABLE 6-1 COEFFICIENTS OF FRICTION (*f*)*

Material	Static		Sliding	
	Dry	Lubricated	Dry	Lubricated
Steel on steel	0.78	0.23	0.42	0.08
On babbitt	0.42	0.17	0.35	0.14
On Teflon	0.04			0.04
On lead	0.95	0.50	0.95	0.30
On aluminum	0.61		0.47	
On copper	0.53		0.36	0.18
On brass	0.51		0.44	
Cast iron on cast iron	1.10		0.15	0.07
On copper	1.05		0.29	
On zinc	0.85		0.21	
Aluminum on aluminum	1.05		1.40	
Nickel on nickel	1.10		0.53	0.12
Glass on glass	0.94	0.01	0.40	0.09
Leather on wood	0.61		0.52	
On cast iron			0.56	0.13
Teflon on Teflon	0.04			0.04
Wood on wood	0.62		0.48	0.16
Laminated plastic on steel			0.35	0.05
Nylon 6/6 on nylon 6/6	0.09			
Polyester (reinf.) on polyester (reinf.)	0.17			
On steel	0.13			

* Theodore Baumeister, Eugene A. Avalone, and Theodore Baumeister III: *Marks' Standard Handbook for Mechanical Engineers,* 8th ed., p. 3–26, McGraw-Hill Book Company, New York, 1978, except for polyester and nylon.

$$F_m = W_x = W \sin \alpha$$
$$N = W_y = W \cos \alpha$$

Dividing, $$\frac{F_m}{N} = \frac{W \sin \alpha}{W \cos \alpha} = \tan \alpha$$

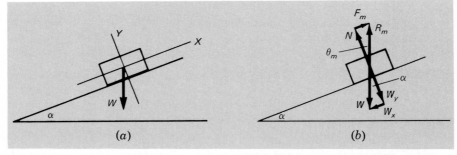

Fig. 6-8 (*a*) Body on an inclined plane with axes parallel and perpendicular to the plane. (*b*) Impending motion due to weight of the body. Note that $R_m = W$, $N = W_y$, $F_m = W_x$, and $\theta_m = \alpha$.

But $F_m/N = f = \tan \theta_m$, where θ_m is the maximum angle of friction, or the angle that the resultant makes with the normal. Then $\tan \theta_m = \tan \alpha$ and $\theta_m = \alpha$.

When a body rests on an inclined plane, if the maximum angle of friction θ_m is less than α, the angle of the plane, the body will slip. If $\theta_m > \alpha$, the body will not slide down the plane.

Sample Problem 3 A body weighing 50 lb rests on a 20° plane. Let $f = 0.2$. Will the body slide down the plane?

Solution:

$$\tan \theta_m = 0.2$$
$$\theta_m = 11°20'$$

Since $\theta_m < \alpha$, the body will slide. Alternatively,

$$N = W \cos \alpha = 50 \cos 20° = 47 \text{ lb}$$
$$F_m = 0.2N = 0.2(47) = 9.4 \text{ lb}$$

The component of W down the plane

$$W_x = W \sin \alpha = 50 \sin 20° = 17.1 \text{ lb}$$

Since 17.1 lb > 9.4 lb, the body must move.

Sample Problem 4 In the preceding example let $f = 0.3$ and $\alpha = 15°$. Will the body move? How large is the frictional force?

Solution:

$$\tan \theta_m = 0.3$$
$$\theta_m = 16°42' = 16.7°$$

Since $\theta_m > \alpha$, the body will not move. Now,

$$N = W \cos \alpha = 50 \cos 15° = 48.3 \text{ lb}$$
$$F_m = 0.3(48.3) = 14.49 \text{ lb}$$

But the force trying to move the body down the plane is

$$W_x = W \sin \alpha = 50 \sin 15° = 12.9 \text{ lb}$$

Since 12.9 lb < 14.49 lb, it is evident that the body will not move. But friction is a passive force. Only enough resistance is offered to prevent motion. Therefore, the frictional force actually developed is 12.9 lb.

6-3 FORCE NECESSARY TO MOVE A BODY UP A ROUGH PLANE

Figure 6-9 shows a body on a plane with a force P acting on the body parallel to the plane. If motion is impending up the plane, the resultant R_m of F_m and N

makes an angle θ_m(tan $\theta_m = f$) with the normal to the plane and is drawn on the left-hand side. There are then three concurrent forces in equilibrium. Construct the triangle of forces by drawing **AB** equal to W, the only known force. From B, draw a line parallel to P. Since the figure closes, R_m must end at A. Then from A draw a line parallel to R_m to form triangle ABC (Fig. 6-10). Since the normal makes an angle α with the vertical, angle A of the triangle is $\theta_m + \alpha$.

$$\angle B = 90° - \alpha$$
$$\angle C = 180° - (90° - \alpha + \theta_m + \alpha) = 90° - \theta_m$$

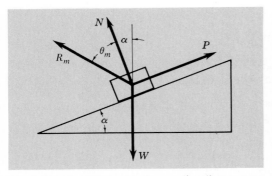

Fig. 6-9 Body with motion impending (but not yet moving) up a rough plane.

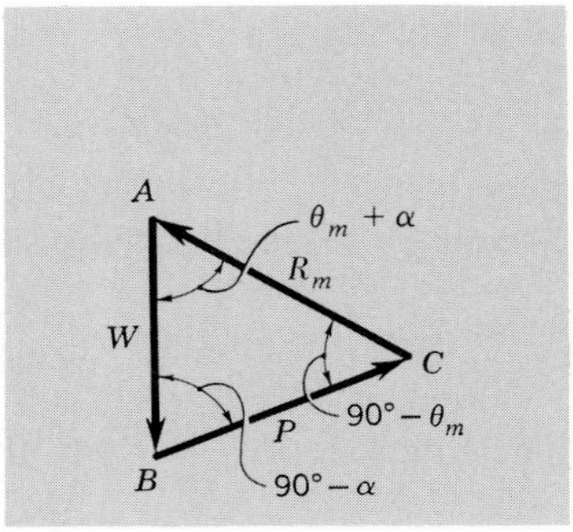

Fig. 6-10 Force triangle for impending motion up a rough plane.

This is easily seen from Figs. 6-9 and 6-10.
By the law of sines, the force P necessary to just start the body up the plane is

$$\frac{P}{\sin(\theta_m + \alpha)} = \frac{W}{\sin(90° - \theta_m)} = \frac{W}{\cos\theta_m}$$
$$P = \frac{W\sin(\theta_m + \alpha)}{\cos\theta_m}$$

*Sample Problem 5 Find the force required to move a body having a mass of 46 kg up a 20° inclined plane if $f = 0.3$.

$$W = m \cdot g = (46 \text{ kg})(9.81 \text{ m/s}^2) = 451 \text{ N}$$

Solution a: From Sample Problem 4, $\theta_m = 16°42' = 16.7°$.

$$P = \frac{451\sin(20° + 16.7°)}{\cos 16.7°} = \frac{451(0.5976)}{0.9580} = 281 \text{ N}$$

Solution b: Choose axes as in Fig. 6-11.

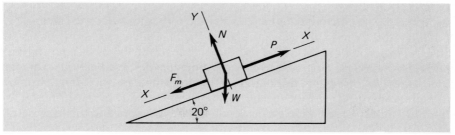

Fig. 6-11 Diagram for Sample Problem 5.

$\Sigma F_x = 0 \qquad P - F_m - W \sin 20° = 0$

$\Sigma F_y = 0 \qquad N - W \cos 20° = 0$

$$N = 451(0.94) = 424 \text{ N}$$

$$F_m = 0.3(424) = 127 \text{ N}$$

$$P = 127 + 451 \sin 20°$$

$$= 127 + 154 = 281 \text{ N}$$

Sample Problem 6 Find the force P parallel to the plane (Fig. 6-12) needed to just start the body down the plane.

Solution: From Fig. 6-13 and the law of sines,

$$\frac{P}{\sin (\theta_m - \alpha)} = \frac{W}{\sin (90° - \theta_m)} = \frac{W}{\cos \theta_m}$$

$$P = \frac{W \sin (\theta_m - \alpha)}{\cos \theta_m}$$

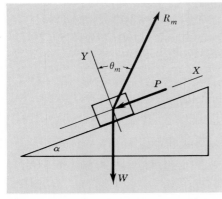

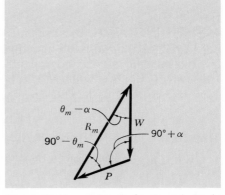

Fig. 6-12 Diagram for Sample Problem 6.   Fig. 6-13 Force triangle for Sample Problem 6.

6-4   LEAST FORCE

Let a body W rest on a horizontal plane (Fig. 6-14). A force P acts at an angle γ. Find the least force to just move the body. R, the resultant reaction, makes an angle θ with the normal. There are three concurrent forces in equilibrium. Draw **AB** equal to W (Fig. 6-15). Since the direction of P is unknown, we draw a line parallel to R from A. The diagram must close, the P must be drawn from B to some point on **AD** and is to be the least force, represented by the shortest line. But the shortest line from B is a perpendicular to **AD**, which in this case is line **BC**.

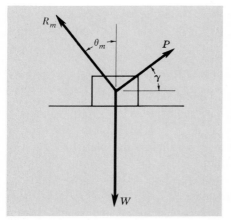

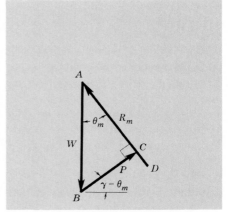

Fig. 6-14 Free-body diagram for finding least force to move a body.

Fig. 6-15 Force triangle for finding least force.

Since P is perpendicular to R, P makes an angle θ with the horizontal; $\gamma = \theta$. But ABC of Fig. 6-15 is a right triangle.

$$P = W \sin \theta$$

Sample Problem 7[1]   Let $W = 50$ lb and $f = 0.3$. Find the least horizontal force needed to move the body. Find the least force for any direction (Fig. 6-14).

Solution: If P is horizontal, $P = F$.

$$N = W = 50 \text{ lb} \qquad f = 0.3$$
$$F_m = Nf = 50(0.3) = 15 \text{ lb}$$

Therefore, $P = 15$ lb (least horizontal force)

But, for any direction the least force is (note $\theta_m = 16°42'$ from Sample Problem 4)

$$P = W \sin \theta_m = 50 \sin 16°42' = 14.37 \text{ lb (at } 16°42' \text{ or } 16.7°)$$

[1] See also Sample Computer Program 8, Chap. 17.

6-5 CONE OF FRICTION

If line **OA** of Fig. 6-3b, which makes the maximum angle of friction θ_m with the normal, is revolved about **OB** as an axis, the cone generated is called the *cone of friction* (Fig. 6-16).

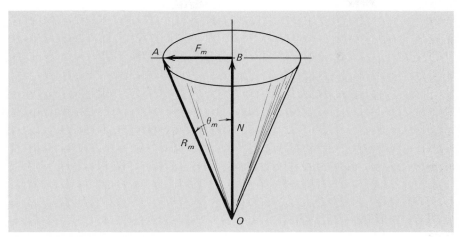

Fig. 6-16 Cone of friction.

If the resultant R_m of the normal reaction and the friction falls within the cone of friction, the forces acting on the body are not great enough to cause motion.

This principle is used in self-locking mechanisms and also in taper pins. For example, if the angle of the taper pin is less than the angle of friction, no force at a right angle to the axis of the pin could cause it to move in the direction of its axis. This principle will be illustrated by applying it in the solution of a problem.

Sample Problem 8   A lift slides on a vertical shaft 3 in in diameter. Find the greatest distance from the shaft at which any load W can be placed and still permit the lift to slide on the shaft. Let $f = 0.3$. Disregard the weight of the lift (Fig. 6-17).

Solution a: When the load is placed on the lift, force is exerted perpendicular to the shaft on the right side at the bottom and on the left side at the top. Let A and B be the points of application of the resultant horizontal forces F_{ax} and F_{bx}. Friction at A and B is called F_m in the direction indicated by vectors. Since motion is taking place, or impending, maximum friction is developed and the resultant reaction at A is along AE at angle $\theta_m = \tan^{-1} 0.3$ with the normal AH. At B, it is along BE at the same angle with the horizontal (Fig. 6-18).

For impending motion, the lift is in equilibrium under the action of three forces. By the principle of concurrence, the forces must meet at the point at

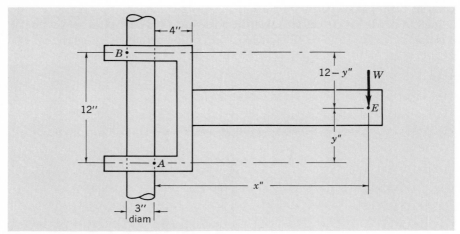

Fig. 6-17 Diagram for Sample Problem 8.

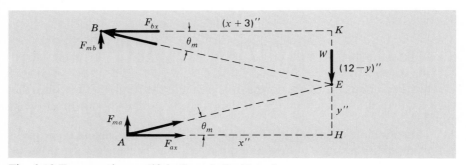

Fig. 6-18 Forces acting on lift in Sample Problem 8.

which AE and BE meet (point E). From triangle HAE,

$$\tan \theta_m = \frac{y}{x} = 0.3$$

$$y = 0.3x$$

From the dimensions in Fig. 6-17 and triangle KBE,

$$\frac{12 - y}{x + 3} = 0.3$$

$$12 - y = 0.3x + 0.9$$

Substituting for y, $\qquad 12 - 0.3x = 0.3x + 0.9$

$$0.6x = 11.1$$

$$x = 18.5 \text{ in}$$

That is, with W less than 18.5 in from the edge of the shaft, the lift will slide.

Solution b (Summation Method): The conditions for equilibrium give

$$\Sigma F_y = 0 \qquad\qquad F_{ma} + F_{mb} - W = 0$$

$$\Sigma F_x = 0 \qquad\qquad F_{ax} - F_{bx} = 0 \qquad F_{ax} = F_{bx}$$

$$\Sigma M_a = 0 \qquad F_{mb}(3) + Wx - F_{bx}(12) = 0$$

Since $F_{ax} = F_{bx}$ and triangles KBE and HAE are similar,

$$F_{ma} = F_{mb} = F_m$$

From ΣF_y above,

$$2F_m = W \qquad \text{or} \qquad F_m = \frac{W}{2}$$

Since $F_m = fN$ and $N = F_{ax} = F_{bx}$,

$$N = \frac{F_m}{f} = \frac{W}{2f} = F_{bx}$$

By substituting in the expression for $\Sigma M_a = 0$, we have

$$\frac{W}{2}(3) + Wx - \frac{W}{2f}(12) = 0$$

Dividing by W produces

$$\frac{3}{2} + x - \frac{12}{2(0.3)} = 0$$

$$x = \frac{6}{0.3} - 1.5 = 20 - 1.5 = 18.5 \text{ in}$$

6-6   WEDGE ACTION

A wedge is a piece of wood or metal used in lifting heavy loads, transmitting power, etc. It may be made with one or two inclined faces. Figure 6-19 shows a load of 3000 lb to be lifted by means of two wedges with faces inclined at an angle of 15° with the horizontal. The coefficient of friction for all rubbing surfaces is 0.3. Find the forces P needed to just lift the load.

First, set the load out as a free body (Fig. 6-20a). It is in equilibrium under the action of the 3000-lb weight and the two resultant reactions R_1 of the two

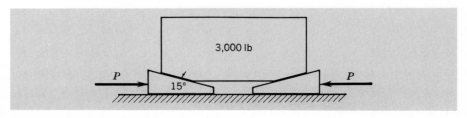

Fig. 6-19 Lifting a load by means of wedges.

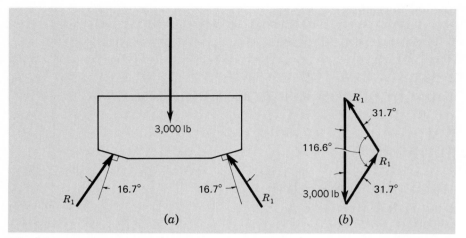

Fig. 6-20 (*a*) Free-body diagram of load for impending upward motion. (*b*) Force triangle for load.

rubbing surfaces. The direction of R_1 is determined by first drawing a line normal to each of the inclined surfaces at their centers. The motion of the load relative to the wedge is upward. Therefore, the force R_1 must act inclined to the normal so that it will resist this motion as shown in the figure. The angle that R_1 makes with the normal is $\theta_m = \tan^{-1} 0.3 = 16.7°$. Now, draw the force triangle, Fig. 6-20*b*. By the law of sines,

$$\frac{R_1}{3000} = \frac{\sin 31.7°}{\sin 116.6°}$$

Note: $\sin 116.6° = \sin 63.4°$

$$R_1 = 1760 \text{ lb}$$

Now, set the wedge out as a free body (Fig. 6-21*a*). It is in equilibrium under the action of force P and the resultant reactions of the two rubbing

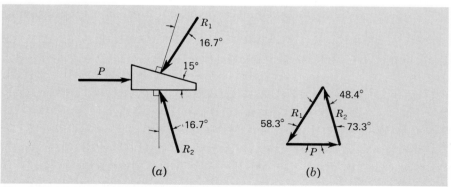

Fig. 6-21 (*a*) Free-body diagram of wedge for impending upward motion. (*b*) Force triangle for wedge.

surfaces. Since motion is impending, R_1 and R_2 must make the angle of friction with the normals at the center of the rubbing surfaces. The motion of the wedge is to the right. This motion is resisted by the two surfaces in contact with the wedge. Therefore, R_1 and R_2 are drawn to the right of and at an angle of 16.7° with their respective normals. Now draw the force triangle (Fig. 6-21b). By the law of sines,

$$\frac{P}{R_1} = \frac{\sin 48.4°}{\sin 73.3°}$$

$$P = 1370 \text{ lb}$$

The two free-body diagrams show that R_1 in one figure is equal, parallel, and opposite to R_1 in the other. This fact is easily understood because action is equal and opposite to reaction.

For this type of problem, the solution by means of the force triangle is probably the simplest. However, for purposes of comparison, the summation method will now be applied. For the load (Fig. 6-20a),

$$\Sigma F_y = 0 \qquad 2R_1 \cos 31.7° - 3000 = 0$$

$$R_1 = \frac{1500}{\cos 31.7°} = 1760 \text{ lb}$$

For the wedge (Fig. 6-21a),

$$\Sigma F_y = 0 \qquad R_2 \cos 16.7° - R_1 \cos 31.7° = 0$$

$$R_2 = \frac{1760(0.8508)}{0.9578} = 1560 \text{ lb}$$

$$\Sigma F_x = 0 \qquad P - 1760 \sin 31.7° - 1560 \sin 16.7° = 0$$

$$P = 1760(0.5255) + 1560(0.2873) = 1370 \text{ lb}$$

With the same conditions in Fig. 6-22a, suppose the problem is to find

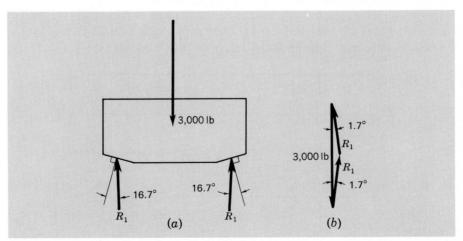

Fig. 6-22 (*a*) Free-body diagram of load for impending downward motion. (*b*) Force triangle for load.

what force applied to the wedge will cause impending downward motion of the load. First, set the wedge (Fig. 6-23a) and load (Fig. 6-22a) out as free bodies. Since the motion of the load is downward, the resultant reaction of the surfaces of the wedges must act on the side of the normal opposite to that acted on in the preceding problem. Also, since the wedge tends to move to the left, the reactions must act on the left side of the normals. The force triangles are now drawn. The one for the wedge shows P acting to the left. The mechanism is self-locking, and it requires a force to pull the wedge from under the load.

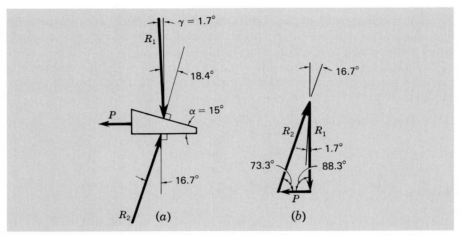

Fig. 6-23 (a) Free-body diagram of wedge for impending downward motion. (b) Force triangle for wedge.

If the wedge angle α is continually increased, the self-locking effect will be reduced and eventually disappear. In the preceding example of downward impending motion, with R_2 constant as the angle α is increased, the load P required to overcome the self-locking force is reduced. A tabulation of results for several values of α shows that P becomes zero when $\alpha = 33.4°$.

α, deg	γ, deg	R_1, lb	∡ between R_1 and R_2 deg	P, lb
15	1.7	1500	18.4	495
20	3.3	1503	13.4	364
25	8.3	1516	8.4	231
30	13.3	1541	3.4	95.4
33.4	16.7	1566	0	0

6-7 JOURNAL FRICTION

When a shaft or axle rotates in its bearings, a suitable amount of lubrication is usually applied in order to reduce the effect of friction. However, since there is sliding between surfaces in contact, a force of friction is present and must be recognized. Owing to the lubrication, the value of f is substantially reduced compared to the value of f for a similar unlubricated surface. The effect of the friction is to cause wear on the bearings as well as to increase the turning moment needed to continue rotation. Figure 6-24 shows a shaft and bearing. The resultant R_m of the friction F_m and the normal reaction N will be equal and opposite to the resultant of the weight W and the force P. The friction will therefore be at a slight angle with the horizontal in order to satisfy the requirements for equilibrium. In other words, the shaft will at first climb up on the bearing until a point at which the forces are in equilibrium is reached.

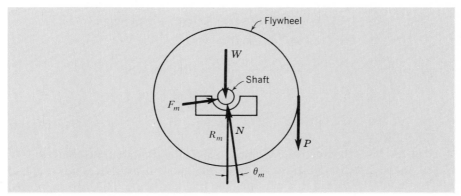

Fig. 6-24 Force acting in a journal bearing.

*Sample Problem 9   A flywheel having a mass of 920 kg is 1.8 m in diameter. The coefficient of friction is 0.01. What tangential force applied to the rim will cause rotation? The diameter of the shaft on which the flywheel is mounted is 150 mm $= 0.15$ m.

Solution:

$$W = m \cdot g = (920 \text{ kg})(9.81 \text{ m/s}^2) = 9020 \text{ N}$$
$$F_m = 0.01(9020) = 90.2 \text{ N}$$

Taking moments about the axis of the shaft gives us

$$P(0.9) = 90.2(0.075)$$
$$P = \frac{90.2(0.075)}{0.9} = 7.5 \text{ N}$$

This is the force that may be assumed to be sufficient to turn the wheel. There is a slight omission in that the actual normal force is equal to $9020 + P$, but in this case it can be seen that there is no appreciable error.

Sample Problem 10   A pulley 27 in in diameter weighs 80 lb. It is mounted on a 2-in axle. A weight of 1000 lb is to be lifted. If $f = 0.1$, what vertical pull will be needed? The value of f implies poor lubrication.

Solution:

$$N = 80 + 1000 + P$$
$$F_m = 0.1(80 + 1000 + P)$$

Taking moments about the center of the axle gives us

$$0.1(80 + 1000 + P)1 = P(13.5) - 1000(13.5)$$
$$80 + 1000 + P = (P - 1000)(135) = 135P - 135\,000$$
$$135P - P = 135\,000 + 1080 = 136\,080$$
$$134P = 136\,080$$
$$P = \frac{136\,080}{134} = 1015 \text{ lb}$$

6-8   JACKSCREW

A jackscrew has a threaded screw which turns in a stationary frame or base (Fig. 6-25). As the force is applied to the lever, the screw turns, moves upward, and lifts the load. The thread in the frame is an inclined plane. The load supported by the screw is transmitted to the inclined plane. As shown in Fig. 6-26, a single thread is set out as a free body. (Note that the x and y axes are respectively horizontal and vertical in this analysis.) From $\Sigma F_x = 0$

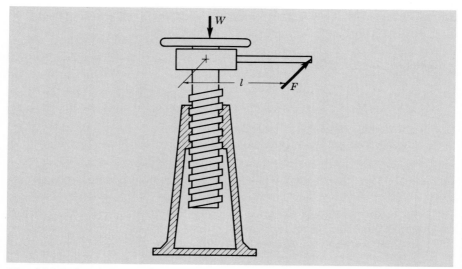

Fig. 6-25 Jackscrew.

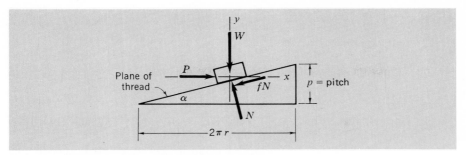

Fig. 6-26 Free-body diagram of a jackscrew thread.

$$P = N \sin \alpha + fN \cos \alpha$$

From $\Sigma F_y = 0$ $\qquad N \cos \alpha - fN \sin \alpha = W$

$$\frac{P}{W} = \frac{\sin \alpha + f \cos \alpha}{\cos \alpha - f \sin \alpha} = \tan (\alpha + \theta_m)$$

To produce motion, the screw must be rotated by the force F. Then

$$Fl = Pr = Wr \tan (\alpha + \theta_m)$$

where $r =$ mean radius of screw thread
$\qquad \alpha =$ helix angle of the thread
$\qquad \theta_m = \tan^{-1} f.$

*Sample Problem 11 A jackscrew with a mean radius of 18 mm has a pitch of 6 mm. With $f = 0.1$, what force applied to the end of a 500-mm lever will lift 1.84 metric tons (*Note:* 1 metric ton $= 1000$ kg)?

Solution:

$$W = m \cdot g = (1840 \text{ kg})(9.81 \text{ m/s}^2) = 18.05 \text{ kN}$$

$$\tan \alpha = \frac{6}{2\pi(18)} = 0.053$$

$$\alpha = 3°2' = 3.03°$$

$$\tan \theta_m = 0.1$$

$$\theta_m = 5°42' = 5.7°$$

$$F(500) = 18(18.05)(\tan 8.73°)$$

$$F = \frac{18(18.05)(0.1536)}{500} = 0.1 \text{ kN} = 100 \text{ N}$$

PROBLEMS

6-1. Find the force required, parallel to an inclined plane, to just start a body weighing 100 lb down the plane. The slope of the plane is 12° with the horizontal, and $f = 0.3$.

*6-2. Find the force required, parallel to an inclined plane, to start a body having a mass of 46 kg down the plane. The slope of the plane is 20° with the horizontal, and $f = 0.3$.

6-3. If, in Fig. 6-9, P is a horizontal force, prove that, for motion up the plane, $P = W \tan (\alpha + \theta)$. See Sec. 6-3.

6-4. If, in Fig. 6-9, P is a horizontal force, prove that, for motion down the plane, $P = W \tan (\alpha - \theta)$. See Sec. 6-3.

6-5. Show that the least force P necessary to start a body up a plane whose slope is α is given by the equation $P = W \sin (\theta + \alpha)$. This result shows that the least force is required when it acts at an angle θ with the direction of impending motion.

6-6. Parcels are to slide down a sheet-metal chute. If the coefficient of friction is 0.15, find the minimum angle of inclination of the chute.

6-7. A body weighing 100 lb rests on a plane making an angle of 10° with the horizontal. A horizontal force of 30 lb is just sufficient to cause motion up the plane. What is the coefficient of friction?

*6-8. A body having a mass of 46 kg rests on an inclined plane. A horizontal force of 135 N is just sufficient to cause motion up the plane. If $f = 0.1$, what angle does the plane make with the horizontal?

6-9. Will the bodies shown in Fig. Prob. 6-9 slide? What is the tension in the cord connecting the two bodies? What force is needed to start the 200-lb body to the right? The bodies are brass, and the planes are steel.

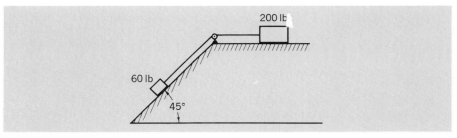

FIGURE PROBLEM 6-9

*6-10. In the friction drive shown in Fig. Prob. 6-10, the wheel A transmits a turning moment of 22 N·m to the wheel B. If the coefficient of friction between the surfaces of the wheels in contact is 0.35, what must be the force P between the wheels and what are the vertical reactions at the bearings C and D?

6-11. A load of 1000 lb is to be raised by means of a wedge which is moved by a horizontal force P (Fig. Prob. 6-11). The load is compelled to move up the slope because of the horizontal force F. What are the amounts of the horizontal forces P and F if all surfaces are wood?

6-12. What load Q can be supported by the force of 600 lb if all surfaces are dry steel (Fig. Prob. 6-12)?

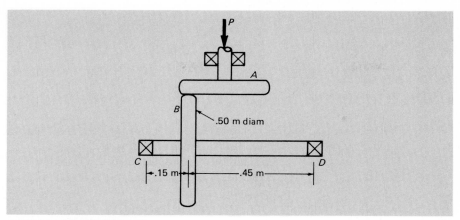

FIGURE PROBLEM 6-10

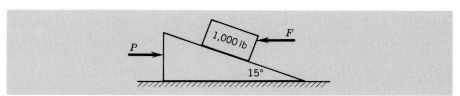

FIGURE PROBLEM 6-11

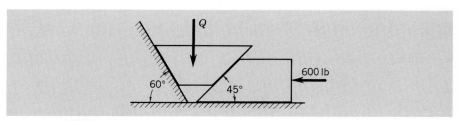

FIGURE PROBLEM 6-12

*6-13. A ladder 3 m long and having a mass of 18 kg is placed with the lower end resting on a rough floor and the upper end against a smooth wall at a point 2.4 m above the floor. If $f = 0.2$, will the ladder remain in position? The weight of the ladder is assumed to act at the midpoint of the length of the ladder.

6-14. A turbine-generator rotor weighing 5 tons on a 14-in-diameter shaft can be turned with a force of 80 lb applied at a radius of 4 ft. Find the coefficient of friction. There are two bearings.

6-15. Block A, weighing 250 lb, rests on a surface whose coefficient of friction is 0.25. A force $F = 68$ lb is applied to a cable which passes over a 4-in-diameter sheave (with no slip) as shown in Fig. Prob. 6-15. The

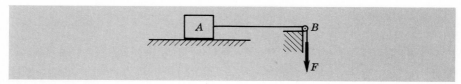

FIGURE PROBLEM 6-15

sheave weighs 2 lb and is mounted on a 1-in-diameter shaft with a coefficient of friction of 0.03. Will block *A* be moved?

*6-16. The mean radius of the screw of a square-threaded jackscrew is 25 mm. The pitch of the threads is 7.5 mm. If the coefficient of friction is 0.12, what force applied to the end of a lever 600 mm long is needed to raise a load having a mass of 1.84 metric tons?

7

Simple Stresses

7-1 STRESS

When a restrained body is subjected to external forces, there is a tendency for the shape of the body to be deformed or changed. Since materials are not perfectly rigid, the applied forces will cause the body to deform. The *internal resistance* to deformation of the fibers of a body is called *stress.*

Stress can be considered either as total stress or unit stress. *Total stress* represents the total resistance to an external effect, and it is expressed in pounds, kips, newtons, or other force dimension. *Unit stress* represents the resistance developed by a unit area of cross section, and it is usually expressed in one of the following: pounds per square inch (psi), kips per square inch (ksi), newtons per square millimeter (N/mm^2), or megapascals (MPa). Note that, since the prefix mega (M) represents 10^6, 1 MPa = 10^6 Pa. Note also, that 1 MPa = 1 N/mm^2. In the remainder of this book, the word *stress* will be used to signify *unit stress.*

The various types of stress can be classified as:

1. Simple or direct stress
   (a) Tension[1]
   (b) Compression[1]
   (c) Shear
   (d) Bearing

---

[1] See Sec. 2-8.

2. Indirect stress
    (a) Bending (see Chap. 12)
    (b) Torsion (see Chap. 13)
3. Combined stress (see Chap. 14)
    (a) Any possible combination of types 1 and 2

This chapter deals with simple stresses only.

7-2 SIMPLE STRESS

Simple stress is often called *direct stress* because it develops under direct loading conditions. That is, simple tension and simple compression occur when the applied force (called *load*) is in line with the axis of the member (*axial* loading; see Figs. 7-1 and 7-3), and simple shear occurs when equal, parallel, and opposite forces tend to cause a surface to slide relative to the adjacent surface (see Fig. 7-5).

There are many loading situations in which the stresses that develop are not simple stresses. For example, referring to Fig. 11-7, the member is subjected to a load which is perpendicular to the axis of the member (*transverse* loading). This will cause the member to bend, resulting in deformation of the material and stresses being developed internally to resist the deformation. All three types of stresses — tension, compression, *and* shear — will develop, but they will not be simple stresses because they were not caused by direct loading.

When any type of *simple* stress develops, we can calculate the magnitude of the stress by

$$s = \frac{F}{A} \qquad (7\text{-}1)$$

where s = average unit stress, psi; N/mm² = MPa

F = external force causing stress to develop, lb; N

A = area over which stress develops, in²; mm²

If a member is in tension because of the action of force F, as in Fig. 7-1, the stress on each cross section is the same because all the cross-sectional areas, A_1, A_2, A_3, etc., are equal. However, the stresses on the various cross sections in Fig.

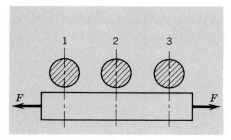

Fig. 7-1 Tension member — uniform cross section.

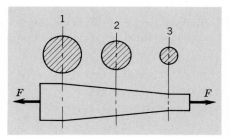

Fig. 7-2 Tension member — varying cross section.

7-2 are different because the areas are different. The largest stress occurs where the force must be resisted by the smallest area. In this case, section 3 carries the largest stress. That is, $s_1 = F/A_1 < s_2 = F/A_2 < s_3 = F/A_3$.

A similar analysis applies to members in compression. Figure 7-3 shows a compression member in which the same stress exists on all cross sections.

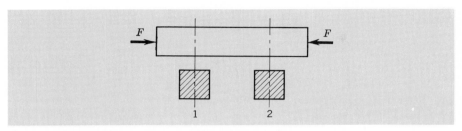

Fig. 7-3 Compression member—uniform cross section.

The compression member shown in Fig. 7-4 carries different stresses at different levels because of the changing areas. Of course, the stress is maximum where the area is minimum, since the same load must be transmitted by all sections.

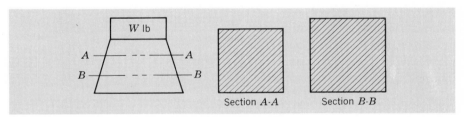

Fig. 7-4 Compression member—varying cross section.

It should be noted that, in cases of either simple tension or simple compression, the areas which resist the load are *perpendicular* to the direction of the forces.

When a member is subjected to simple shear, the resisting area is parallel to the direction of the force. Common situations causing shear stresses are shown in Figs. 7-5 and 7-6.

A bolt connecting two flat plates, as in Fig. 7-5*a*, is subjected to shear stress on the circular area *aa*. Thus, the magnitude of this stress is $s_s = F/A$. If the bolt material is not strong enough to provide the necessary shear stress, the failure might occur as shown in Fig. 7-5*b*. A similar situation is encountered in riveted joints.

Stamping and blanking operations depend upon the material to fail cleanly in shear. Figure 7-6*a* shows a circular blanking punch approaching the work. The extent of the slug to be sheared out is indicated by broken lines. Figure 7-6*b* shows the punch penetrating the plate and the slug falling out below. Figure 7-6*c* indicates the shear area as the cylindrical surface of the slug.

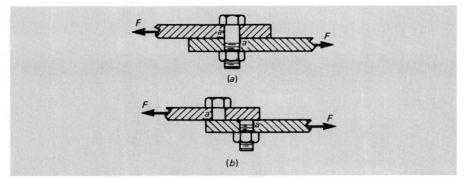

Fig. 7-5 (*a*) Bolt resisting shear. (*b*) Bolt failure due to shear.

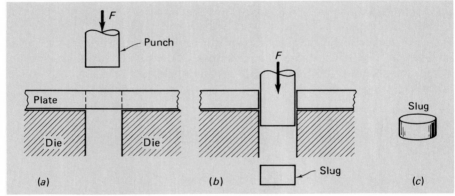

Fig. 7-6 (*a*) Punch approaching plate. (*b*) Punch shearing plate. *(c)* Slug showing sheared area.

Sample Problem 1 A steel rod 1 in in diameter is subjected to a pull of 6000 lb. What is the average stress?

Solution:

$$F = 6000 \text{ lb}$$

$$A = \frac{\pi(1^2)}{4} = 0.785 \text{ in}^2$$

$$s = \frac{F}{A} = \frac{6000}{0.785} = 7640 \text{ psi (tension)}$$

*Sample Problem 2 A 100- by 100-mm bar that is 150 mm long is under a compression force of 45 kN. Find the average stress.

Solution:

$$F = 45 \text{ kN} = 45(10^3) \text{ N}$$
$$A = (100 \text{ mm})(100 \text{ mm}) = 1 \times 10^4 \text{ mm}^2$$
$$s = \frac{F}{A} = \frac{45(10^3)}{1 \times 10^4} = 4.5 \text{ N/mm}^2 = 4.5 \text{ MPa (compression)}$$

Sample Problem 3   In Fig. 7-7, what will be the average shearing stress on the area $ABCD$? $F = 2000$ lb, $AB = 8$ in, and $BC = 4$ in.

Solution:

$$F = 2000 \text{ lb}$$
$$A = 32 \text{ in}^2$$
$$s = \frac{F}{A} = \frac{2000}{32} = 62.5 \text{ psi (shear)}$$

Sample Problem 4   Figure 7-8 shows a common type of bolt. If a tensile load is applied to the bolt, the greatest tensile stress will be developed across the root area (section AA) rather than the shank area (section BB), since the root area is smaller.

A 1″-8 UNC hex-head bolt carries a tensile load of 6000 lb. Find the stresses on the root and shank areas.

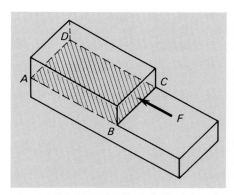

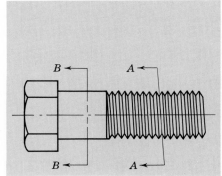

Fig. 7-7 Diagram for Sample Problem 3.   Fig. 7-8 Diagram for Sample Problem 4.

Solution: From Table 3A, App. B, the root diameter is found to be 0.8647 in.[1]

---

[1] The complete table (ANSI B1.1-1982) from which Table 3A, App. B, is abstracted gives a tensile stress area for each of the thread sizes. The root area is slightly smaller than the stress area. Thus, calculations based on the root area result in a more conservative design.

$$F = 6000 \text{ lb}$$

$$\text{Root area} = \frac{\pi}{4} (0.8647^2) = 0.587 \text{ in}^2$$

$$s = \frac{6000}{0.587} = 10\ 220 \text{ psi (tension at root)}$$

$$\text{Shank area} = \frac{\pi(1^2)}{4} = 0.785 \text{ in}^2$$

$$s = \frac{6000}{0.785} = 7640 \text{ psi (tension on shank)}$$

*Sample Problem 5[2]   The rod and yoke shown in Fig. 7-9 are subjected to a tensile load of 29 kN. The pin diameter is 12 mm. The rod diameter at section AA is 22 mm. Find the average shearing stress in the pin and the average tensile stress in the rod at section AA.

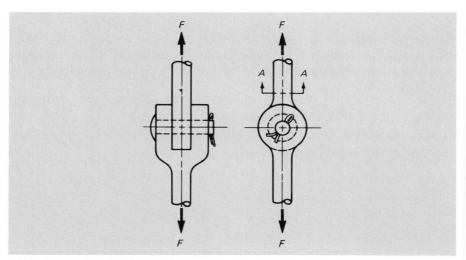

Fig. 7-9 Diagram for Sample Problem 5.

Solution: The pin is in double shear; that is, it tends to fail simultaneously on two separate areas. The total area resisting shear is

$$A_s = 2\left(\frac{\pi d^2}{4}\right) = 2\left[\frac{\pi(12^2)}{4}\right] = 2(113) = 226 \text{ mm}^2$$

The average shearing stress in the pin is

$$s_s = \frac{F}{A_s} = \frac{29(10^3)}{226} = 128 \text{ N/mm}^2 = 128 \text{ MPa}$$

[2] See also Sample Computer Program 9, Chap. 17.

At section AA, the area resisting tension is

$$A_t = \frac{\pi d^2}{4} = \frac{\pi(22^2)}{4} = 380 \text{ mm}^2$$

The average tensile stress in the rod is

$$s_t = \frac{F}{A_t} = \frac{29(10^3)}{380} = 76 \text{ N/mm}^2 = 76 \text{ MPa}$$

*Sample Problem 6 A circular brass rod is to carry a tensile load of 45 kN, and its average stress is not to exceed 52 MPa. What diameter rod is required?

Solution:

$$s = \frac{F}{A} \quad \text{or} \quad A = \frac{F}{s}$$

$$A = \frac{45(10^3)}{52} = 0.86 \times 10^3 \text{ mm}^2$$

$$= 860 \text{ mm}^2 \text{ (required tensile area)}$$

$$A = \frac{\pi}{4} d^2 \quad \text{or} \quad d = \sqrt{\frac{4A}{\pi}}$$

$$d = \sqrt{\frac{4(860)}{\pi}} = \sqrt{1090} = 33 \text{ mm}$$

*Sample Problem 7 If the rod diameter in the preceding problem is doubled, what tensile load can be carried?

Solution:

$$s = \frac{F}{A} \quad \text{or} \quad F = As$$

$$A = \frac{\pi d^2}{4} = \frac{\pi(66^2)}{4} = 3420 \text{ mm}^2$$

$$F = 3420(52) = 178 \text{ kN}$$

Note: If the calculations had not been rounded off, F would equal 180 kN because doubling the diameter results in an increase of 4 times the load.

PROBLEMS

7-1. A bar of steel 1 by $1\frac{1}{2}$ in in cross section and 6 in long is subjected to a pull of 6000 lb. What is the average tensile stress in the bar?

*7-2. A short rod of steel 50 mm in diameter supports a load of 9 kN. What is the average compressive stress?

7-3. What load F will develop an average shearing stress of 8800 psi in the shank of the $\frac{3}{4}''$-10 UNC bolt shown in Fig. 7-5a?

*7-4. A structure having a mass of 180 metric tons is supported equally by 24 short posts that are 150 mm in diameter. What average compressive stress will be developed?

*7-5. In Prob. 7-4, what diameter posts may be used if an average compressive stress of 7.5 MPa is permitted?

7-6. A $1''$-8 UNC hex-head bolt is subjected to a tensile load of 10 000 lb. The height of the head is $\frac{3}{4}$ in. Find the average tensile stress in the bolt and the average shearing stress in the bolt head (Fig. Prob. 7-6).

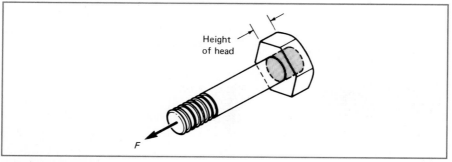

FIGURE PROBLEM 7-6

7-7. In Prob. 7-6, what height of head is necessary so that the bolt will be stressed twice as much in tension as the head is stressed in shear?

7-8. Select an American Standard fine-thread bolt to carry a tensile load of 1000 lb. An average tensile stress of 10 000 psi must not be exceeded.

*7-9. What force is required to punch a hole 50 mm in diameter through a 16-mm-thick aluminum-alloy plate? An average stress of 265 MPa will cause this material to shear.

*7-10. A steel rod 30 mm in diameter and 18 m long is suspended vertically. Calculate the average stress in the rod that is due to its own weight at a section 9 m from the top. What is the maximum average stress in the rod that is due to its weight, and on what section does it act? ($\rho = 7830$ kg/m³ for steel.)

7-11. A platform 16 by 24 ft is to support a uniformly distributed load of 450 psf of surface. How many square timber posts whose nominal size is 4 by 4 in (see Table 11, App. B, for actual sizes) must be used to support the platform if an average stress of 1200 psi in compression is permitted? Show how you would arrange these posts.

7-12. The lower one-third of a wall consists of concrete, and the upper two-thirds consists of brick. What total height of wall would cause an average stress of 100 psi at the base? ($\rho = 150$ lb/ft³ for concrete; $\rho = 120$ lb/ft³ for brick.)

7-13. The maximum gas pressure in an 8-in-diameter diesel-engine cylinder is 750 psi. Six $\frac{7}{8}''$-9 UNC cap screws fasten the cylinder head to the engine block. In order to provide a gas-tight joint, the cap screws have been tightened so that each screw has an initial tensile load of 14 000 lb. Find the maximum average tensile stress in the cap screws when the engine is operating. Ignore the effect of the gasket.

*7-14. What maximum load F can the link shown in Fig. Prob. 7-14 withstand if the average tensile stress must not exceed 75 MPa? All dimensions are in millimeters.

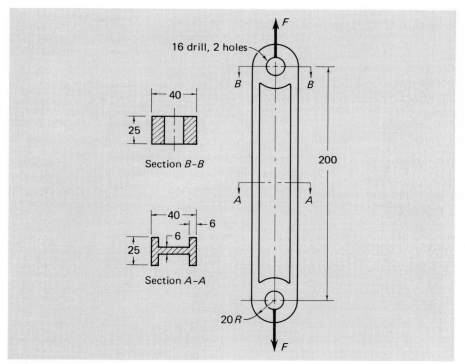

FIGURE PROBLEM 7-14

*7-15. A hollow, cylindrical, short steel compression member is required to support an axial load of 180 kN. Specifications limit the average compressive stress to 95 MPa. The cylinder is 200 mm long.
   a. Find the inside and outside diameters if the outside diameter is $1\frac{1}{2}$ times the inside diameter.
   b. Find the inside and outside diameters if the wall thickness is 6 mm.

7-16. For the machine-part connection shown in Fig. Prob. 7-16, describe (in words) the possible ways in which failure might occur, indicate what kind of simple stress causes the failures described, and calculate (by using the symbols in the figure) the area that resists each failure.

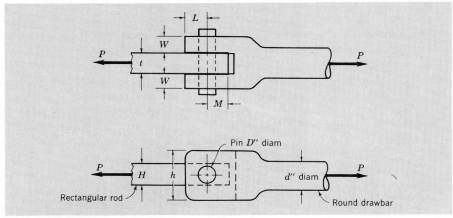

FIGURE PROBLEM 7-16

8

Properties of Materials

8-1 USE OF PROPERTIES OF MATERIALS IN DESIGN

Designers of machinery and structures must know both the action of the forces that are applied and the strength of the various members working together to resist the forces successfully. It is essential for the members to be strong enough, but any part that is unnecessarily strong means a waste of material. It is therefore highly important for the designer to know the mechanical and physical properties of the ferrous (e.g., steel) and nonferrous (e.g., aluminum) metals, timber, concrete, plastics, brick, stone, etc., with which he or she works. In the solution of problems, a knowledge of the laws of mechanics alone is insufficient and must be accompanied by a substantial knowledge of the various properties of the materials.

Many experiments are conducted in the laboratory to determine required properties. One of the most important tests which gives the designer many properties of a material is the tension test. A typical tension test is described below.

8-2 TENSION TEST

In order to determine the tensile properties of metals, standard tests have been devised by the American Society for Testing and Materials (ASTM) and other groups. Figure 8-1 shows the dimensions of a steel test specimen.

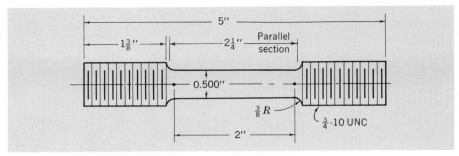

Fig. 8-1 Tension test specimen.

The specimen is mounted in a testing machine, and an axial tensile load is applied. The length increases as the load is increased. This change in length is measured by an extensometer which is attached to the specimen. The data in Fig. 8-2a were obtained in a typical tension test.

With reference to the data, some explanation is in order. Under Initial data, the material is specified as AISI 1020 cold-rolled steel. The American Iron and Steel Institute (AISI) uses a code, similar to the Society of Automotive Engineers (SAE) code, to identify various steels. A steel designated as AISI 1020 is a low-carbon steel containing approximately 0.20 percent carbon. If a steel is designated as AISI 1045, it is a medium-carbon steel containing approximately 0.45 percent carbon. Also under Initial data, the hardness of the steel specimen is given as Rockwell B75 (commonly written as RB75). The Rockwell test establishes the hardness of a material by forcing a specific-shaped indentor below the surface of the specimen. The depth of penetration is then measured and specified. The hardness test can be conducted with a combination of different magnitudes of forces and indentor shapes, which result in different Rockwell scales such as A, B, C, etc. Rockwell B75 indicates that the B scale was used. Note that the hardness under Final data is Rockwell B76, which indicates that the specimen became harder as the test progressed.[1]

8-3   STRESS AND STRAIN[2]

In order to permit comparison with standard values, the data must be converted to a unit basis. The load is expressed in terms of load per unit area, which is unit stress (henceforth called *stress*). The deformation (in this case elongation) is expressed in terms of deformation per unit length, which is unit strain (henceforth called *strain*).

---

[1] For a complete discussion of hardness testing and the classification systems used for steels and other metals, refer to P. A. Thornton and V. J. Colangelo, *Fundamentals of Engineering Materials,* Prentice-Hall, Inc., Englewood Cliffs, N.J., 1985.

[2] See also Sample Computer Program 10, Chap. 17.

TENSION TEST

Laboratory Data (a)			Calculated Data (b)	
Initial data:	Load kip	Elongation in 2 in, in $\times 10^{-4}$	Stress, ksi	Strain, in/in $\times 10^{-4}$
Materials: AISI 1020 cold-rolled steel				
Diameter: 0.503 in				
Gage length: 2.000 in	F	δ	s	ϵ
Hardness: Rockwell B75				
Speed of testing:	0.5	0	2.5	0
0.025 in/min	1.5	3	7.5	1.5
	2.5	6	12.6	3.0
	3.5	10	17.6	5.0
	4.5	13	22.6	6.5
	5.5	16	27.7	8.0
	6.5	20	32.7	10.0
	7.5	23	37.7	11.5
	8.5	26	42.8	13.0
	9.5	30	47.8	15.0
	*10.0	32	50.3	16.0
	10.5	34	52.8	17.0
	11.0	36	55.4	18.0
	11.5	39	57.9	19.5
	12.0	43	60.4	21.5
	12.5	46	62.9	23.0
	13.0	51	65.4	25.5
	13.5	60	67.9	30.0
	14.0	73	70.5	36.5
	14.5	88	73.0	44.0
	15.0	118	75.5	59.0
Ultimate	15.83		79.15	
Breaking	12.50		62.50	
Final data:				
Diameter: 0.312 in			Elastic-limit stress: 52 800 psi	
Gage length: 2.328 in			Yield stress: 73 000 psi	
Hardness: Rockwell B76			Ultimate stress: 79 150 psi	
Character of fracture: $\frac{3}{4}$ cup-cone			Percent elongation in 2 in: 16.4%	
			Percent reduction in area: 61.8%	
			Modulus of elasticity $E = 29.9 \times 10^6$ psi	

* Data used for sample calculations on next page.

Fig. 8-2 Data for typical tension test.

Strain can be calculated by

$$\epsilon = \frac{\delta}{l} \qquad (8\text{-}1)$$

where ϵ = unit strain, in/in; mm/mm

δ = deformation, in; mm

l = original length, in; mm

In Fig. 8-3, three applications of the use of Eq. (8-1) for calculating unit strain ϵ are shown. Strains related to the three simple stresses — tension, compression, and shear — are shown. Note that, in all cases, the deformation δ is measured in the same direction as the applied force.

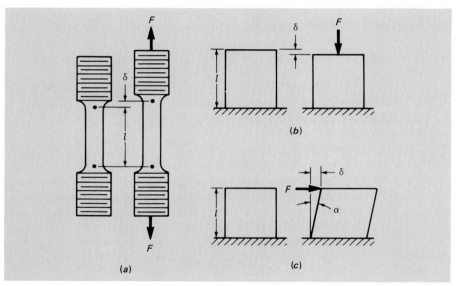

Fig. 8-3 (*a*) Tensile specimen before loading and after a load F is applied. (*b*) Short compression member before loading and after a load F is applied. (*c*) Member before loading and after a shearing load F is applied. (*Note:* $\epsilon = \delta/l = \tan \alpha$)

The following sample calculations will demonstrate conversions from load and deformation to stress and strain. The values used correspond to the starred data in Fig. 8-2.

$$\text{Stress} = \frac{\text{load}}{\text{area}} \qquad s = \frac{F}{A} \qquad\qquad \text{Strain} = \frac{\text{elongation}}{\text{original length}} \qquad \epsilon = \frac{\delta}{l}$$

$$A = \frac{\pi d^2}{4} = \frac{\pi (0.503^2)}{4} = 0.199 \text{ in}^2 \qquad\qquad \epsilon = \frac{0.0032}{2.000} = 0.0016 \text{ in/in}$$

$$s = \frac{10\,000}{0.199} = 50\,300 \text{ psi} = 50.3 \text{ ksi}$$

By following this procedure, the test data are converted to stress and strain as tabulated in Fig. 8-2b.

These results are represented graphically by plotting stress vs. strain (Figs. 8-4 and 8-5). The American Society for Testing and Materials (ASTM) test specifications require that an initial load be applied to the specimen, after which the extensometer dial is set to zero. This ensures that the specimen is securely gripped by the machine before data are recorded. For this reason, point A is plotted at the initial stress (corresponding to the initial load) and zero strain.

The straight-line portion of the curve AB' indicates a direct proportion between stress and strain in this range. Point B' represents the limit of proportionality, and thus it is called the *proportional limit*.

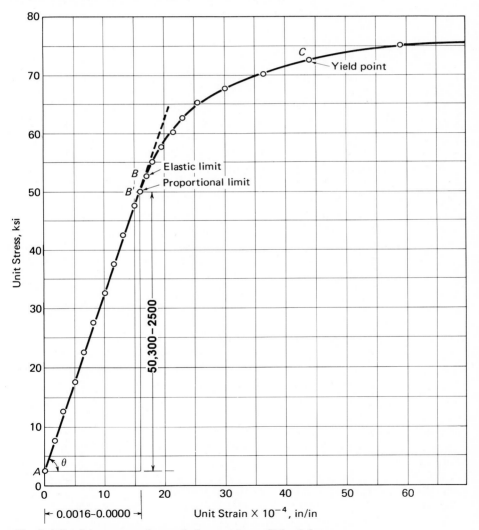

Fig. 8-4 Partial stress-strain graph for test data of Fig. 8-2.

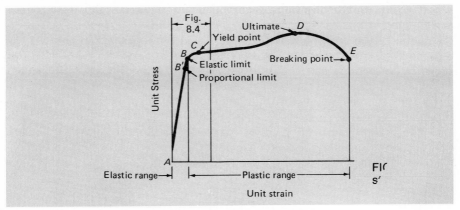

Fig. 8-5 Complete stress-strain diagram for data of Fig. 8-2.

At any stress up to point *B*, the specimen will return to its original dimensions upon removal of the load. At stresses higher than *B*, the specimen will not return to its original dimensions. Point *B* represents the limit of elasticity, since a permanent deformation, or *set*, occurs when the material is stressed beyond that point. Thus, point *B* is called the *elastic limit.*

For ductile materials, the elastic limit *B* and the proportional limit *B'* are sufficiently close that they can be assumed to be located at the same point without serious error. Henceforth, this point will be referred to as the elastic limit. However, if it is necessary to establish the accurate location of the elastic limit, the specimen will have to be loaded and unloaded until the condition described in the preceding paragraph is reached.

In the course of testing the material beyond the elastic limit, a point is reached where, without a significant increase in the load, the specimen continues to elongate. This is point *C* on the stress-strain curve (Figs. 8-4 and 8-5), and it is called the *yield point.* For materials which do not show an apparent yielding, this point can also be found by the 0.2 percent offset method which is discussed in Sec. 8-7.

The stress at point *D* (Fig. 8-5) is the *ultimate strength* of the material and the maximum stress which the specimen can withstand. In the neighborhood of this point the specimen begins to *neck down;* that is, its cross-sectional area decreases rapidly (Fig. 8-6), and failure occurs at point *E*.

8-4 DEFINITIONS

The following terms are among those which are associated with properties of materials.

Hardness.   The ability of a material to resist wear or penetration. The strength of a metal varies directly as its hardness.

Malleability.   The ability of a material to deform appreciably in a hammering or rolling operation before rupture.

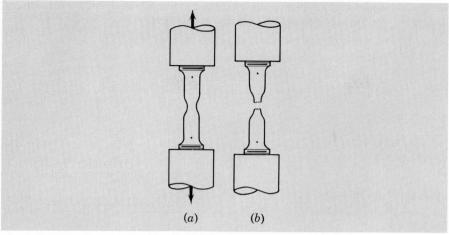

Fig. 8-6 (*a*) Tension specimen necking down before failure. (*b*) Tension specimen after failure.

Toughness. The ability of a material to withstand high-impact (shock) loads. The area under the stress-strain curve is a measure of toughness.

Ductility. The ability of a material to deform appreciably under a tensile load before rupture; high percent elongation and percent reduction of area indicate ductility.

Brittleness. The tendency of a material to fail with very little deformation; low percent elongation and percent reduction of area indicate brittleness.

Elasticity. The ability of a material to return to its original size and shape upon removal of the load; high elastic limits indicate materials with good elasticity.

Plasticity. The ability of a material to deform nonelastically without rupturing; large plastic ranges indicate materials with good plasticity.

Stiffness. The ability of a material to resist deformation; stiffness of a material varies directly as the modulus of elasticity of the material.

Alloy. A mixture of two or more elements, at least one of which is a metal, exhibiting metallic properties.

8-5 MODULUS OF ELASTICITY

The slope of the *straight-line* portion of the stress-strain curve is the ratio of stress to strain.

$$\text{Slope of } AB = \frac{s}{\epsilon} = \tan \theta$$

This ratio is often considered a constant for a material and is called the *modulus of elasticity,* represented by E. Actually, E varies with temperature, but the variation will not be considered in this book.

$$\tan \theta = E = \frac{s}{\epsilon} \quad \text{or} \quad s = E\epsilon \tag{8-2}$$

In U.S. customary units, the unit of modulus of elasticity is pounds per square inch (psi) but is usually expressed in terms of 10^6 psi. In SI metric units, E is usually expressed in terms of 10^3 MPa, or GPa(GPa = MPa $\times$ 10^3).

Another form of the equation can be obtained by substituting F/A for s and δ/l for ϵ.

$$E = \frac{s}{\epsilon} = \frac{F/A}{\delta/l} = \frac{Fl}{A\delta} \tag{8-3}$$

The reader should recall that the ASTM specifications require an initial load on the specimen before elongation is measured. Thus, the initial stress is s_a at zero strain. Therefore, to determine the modulus of elasticity from the stress-strain curve (Fig. 8-4),

$$\tan \theta = E = \frac{s_b' - s_a}{\epsilon_b' - \epsilon_a} \tag{8-4}$$

In the application of Eq. (8-4), points A and B' may be any two widely separated points on the *straight-line* portion of the stress-strain curve.

From the data in Fig. 8-2b, ϵ_a is zero, and

$$E = \frac{50\ 300 - 2500}{0.0016 - 0} = \frac{47\ 800}{0.0016} = 29\ 900\ 000 \text{ psi}$$

This value of E is within the expected range for steel, as indicated in Table 1A, App. B.

The modulus of elasticity E for a material is the same in tension and compression. However, the modulus of elasticity in shear, called the *modulus of rigidity* and designated as G (not to be confused with G, which is used in the SI metric system to represent the prefix giga), is smaller in value. The values of G are found experimentally and are indicated in Table 1A, App. B.

8-6   DUCTILITY

In manufacturing processes, such as rolling, drawing, and extrusion, it is important to know the relative ductility of the material. Percent reduction in area and percent elongation are measures of ductility. These terms are defined by the following equations.

$$\text{Percent reduction in area}^1 = \frac{\text{original area} - \text{final area}}{\text{original area}} (100) \quad (8\text{-}5)$$

$$\text{Percent elongation} = \frac{\text{change in gage length}}{\text{original gage length}} (100) \quad (8\text{-}6)$$

By using Eqs. (8-5) and (8-6) with preceding test data, we obtain

$$\text{Original area} = 0.199 \text{ in}^2$$

$$\text{Final area} = \frac{\pi}{4}(0.312^2) = 0.076 \text{ in}^2$$

$$\text{Percent reduction in area} = \frac{0.199 - 0.076}{0.199}(100) = 61.8 \text{ percent}$$

$$\text{Percent elongation} = \frac{2.328 - 2.000}{2.000}(100) = 16.4 \text{ percent}$$

This elongation is an average value for the 2-in gage length. Actually, the elongation is more severe near the point of failure.

8-7 BRITTLE METALS

Members made of brittle materials, such as cast iron, do not readily deform before breaking. Therefore, they do not show an apparent yield point in physical tests. Figure 8-7 is a stress-strain diagram for a brittle specimen in tension. Since the yield strength is often necessary in design, it can be arbitrarily determined by the 0.20 percent offset method.[2] The offset method establishes the yield strength by drawing a line parallel to the straight-line portion of the stress-strain curve as indicated in Fig. 8-7. Note that 0.20 percent of 1 in of length results in a strain of 0.0020 in/in. (If metric units are used, the offset is 0.0020 mm/mm.)

8-8 ALLOWABLE STRESSES

The strength properties of a material can be determined from physical tests such as the tension test described above. In most design situations it is important not only to prevent failure of the material but also to avoid permanent deformation. This means that the maximum stress that is considered safe for a material to carry must fall within the elastic range. The values used depend upon the application to which the building materials or machine parts are to be put.

---

[1] An alternative form of this equation may be written:

$$\text{Percent reduction in area} = \frac{(\text{original diameter})^2 - (\text{final diameter})^2}{(\text{original diameter})^2}(100)$$

[2] Theodore Baumeister, Eugene A. Avalone, and Theodore Baumeister III: *Marks' Standard Handbook for Mechanical Engineers,* 8th ed., p. 5-2, McGraw-Hill Book Company, New York, 1978.

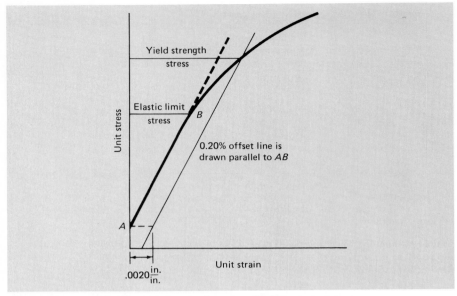

Fig. 8-7 Stress-strain diagram for tension test of brittle material.

These stresses, called *allowable, safe working,* or *design stresses,* are decided upon by designers and engineers through years of experience and experiment.

8-9 FACTOR OF SAFETY

Allowable stresses depend primarily upon the type of loading and the material used. Values of the physical properties of materials given in various tables are, at best, average values. In addition, the actual loading may vary somewhat from predicted conditions. Therefore, to ensure a safe design, the allowable stress must be made low enough to cover the uncertainties. Allowable stresses are usually based on either the ultimate strength or the yield stress. The *factor of safety* (N_u, when based on the ultimate stress) is defined as the ratio of the ultimate stress to the allowable stress. (Note that ultimate stress is frequently referred to as ultimate strength.)

$$N_u = \frac{\text{ultimate stress}}{\text{allowable stress}}$$

$$\text{(8-7)}$$

or $$\text{Allowable stress} = \frac{\text{ultimate stress}}{N_u}$$

Similarly, the factor of safety N_y based on the yield stress is given by

$$N_y = \frac{\text{yield stress}}{\text{allowable stress}}$$

or $\qquad$ Allowable stress $= \dfrac{\text{yield stress}}{N_y}$ $\qquad$ (8-8)

(Note that yield stress is frequently referred to as yield strength.)

Values of N_u and N_y for various conditions are given in Table 2, App. B. In certain fields, allowable stresses are directly specified by codes which have been established by supervising organizations or by legislation.

8-10 POISSON'S RATIO

One of the interesting properties of materials is the deformation that takes place in a transverse direction when the piece is subjected to a force in a longitudinal direction. If a tensile force is applied, the piece will elongate and the transverse dimensions will decrease (Fig. 8-8). Conversely, a compressive force will

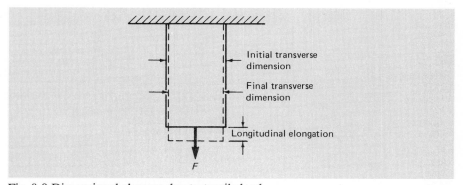

Fig. 8-8 Dimensional changes due to tensile load.

shorten the piece and increase the transverse dimensions. The ratio of the transverse strain to the longitudinal strain is known as *Poisson's ratio* (μ). Table 8-1 gives average values of Poisson's ratio for some materials.

The modulus of rigidity G (see Sec. 8-5) is related to the modulus of elasticity E through Poisson's ratio as given by Eq. (8-9).

$$G = \frac{E}{2(1 + \mu)} \qquad (8\text{-}9)$$

Sample Problem 1 A steel rod 2 in in diameter is subjected to a pull of 60 000 lb. What are the longitudinal and transverse strains?

TABLE 8-1 POISSON'S RATIO (μ)

Material	μ
Aluminum alloy	0.330
Brass	0.340
Bronze	0.350
Cast iron	0.270
Concrete	0.200
Copper	0.355
Glass	0.244
Monel	0.315
Steel	0.288
Stainless steel	0.305
Wrought iron	0.278

Solution:

$$F = 60\ 000 \text{ lb}$$

$$A = \frac{\pi(2^2)}{4} = 3.14 \text{ in}^2$$

$$E = 30 \times 10^6 \text{ psi}$$

Longitudinal:

$$s = \frac{F}{A} = \frac{60\ 000}{3.14} = 19\ 100 \text{ psi (tension)}$$

$$E = \frac{s}{\epsilon}$$

$$\epsilon = \frac{s}{E} = \frac{19\ 100}{30\ 000\ 000} = 0.000637 \text{ in/in}$$

Transverse:

$$\mu = \frac{\epsilon_{transverse}}{\epsilon_{longitudinal}}$$

$$\epsilon_{transverse} = \mu\epsilon_{longitudinal} = 0.288(0.000637) = 0.000183 \text{ in/in}$$

8-11 THERMAL EXPANSION

All materials tend to change their dimensions when subjected to changes in temperature. Table 8-2 lists values of *coefficients. of thermal linear expansion* (α) for several common materials in units of inches per inch per degree Fahrenheit and millimeters per millimeter per degree Celsius. That is, α is the change in length for each unit of original length for each degree of temperature change.

TABLE 8-2   COEFFICIENTS OF LINEAR EXPANSION (α)

Material	10^{-6} in/in/°F	10^{-6} mm/mm/°C
Aluminum alloy	12.8	23.0
Brass	10.4	18.7
Bronze	10.0	18.0
Cast iron	6.3	11.3
Concrete	5.5	9.9
Copper	9.2	16.6
Glass	5.0	9.0
Monel	7.8	14.0
Polymers (plastics)		
Thermosets		
Phenolic	40.0	72.0
Polyester (reinforced)	50.0	90.0
Thermoplastics		
Nylon 6/6	40.0	72.0
Polyethylene (high-density)	120.0	216.0
Styrene-acrylonitrile (SAN)	36.0	65.0
Steel	6.5	11.7
Wrought iron	6.4	11.5

For example, a structural steel bar 8 ft 4 in long (100 in) subjected to a *change in temperature* of 200°F will change its length by 6.5×10^{-6} in/in/°F, or the total change in length for the bar will be $6.5 \times 10^{-6} \times 100 \times 200 = 0.13$ in. If the change in temperature is an *increase,* the length of the bar will *increase* by 0.13 in. Similarly, if the temperature change is a *decrease* of 200°F, the length of the bar will be *decreased* (shortened) by 0.13 in. Equation (8-10) gives the expression for the total deformation or change in length due to a temperature change.

$$\delta = \alpha l \, \Delta t \qquad (8\text{-}10)$$

where δ = total deformation, in; mm

    α = coefficient of thermal linear expansion, in/in/°F; mm/mm/°C (Table 8-2)

    l = original length, in; mm

    Δt = change in temperature, °F; °C ($\Delta t = t_2 - t_1$)

Equation (8-10) gives the actual total deformation for members that are *free* to change their dimensions. When a member deforms freely because of a temperature change, no stresses due to Δt occur in the material. If, however, a member is *prevented* from deforming freely when a temperature change occurs, the restraining forces which prevent the deformation will produce stresses in the material. Stresses due to restrained deformation will be discussed below, after the following applications of *free* thermal expansion.

Sample Problem 2 Steel rails 60 ft long are to be laid with a small gap between the end of one section and the beginning of the next. How large should the gap be at 20°F so that the rail ends will just touch when the temperature is 110°F?

Solution: Assume that the rails are free to expand longitudinally. Since each rail length will undergo equal expansion and will expand at both ends, one rail may be assumed to close half the gap at each end. However, an adjacent rail will close the other half of the gap, so that we can calculate the elongation of one rail as if it closed one full gap rather than two halves of a gap. From Eq. (8-10),

$$\delta = \alpha l \, \Delta t$$

$$\alpha = 6.5 \times 10^{-6} \text{ in/in/°F (Table 8-2)}$$

$$l = 60 \, \cancel{ft} \times 12 \frac{\text{in}}{\cancel{ft}} = 720 \text{ in}$$

$$\Delta t = t_2 - t_1 = 110 - 20 = 90°F$$

$$\delta = (6.5 \times 10^{-6})(720)(90) = 0.42 \text{ in}$$

(total elongation of one 60-ft rail section)

Therefore, the gap should be 0.42 in, or about 27/64 in.

*Sample Problem 3 A continuous Monel shaft is to be designed for connecting two power units in a ship. The shaft will be supported by suitable bearings and will be 25 m long. The system is to be designed for a minimum temperature of −32°C and a maximum temperature of 77°C. Determine the amounts of contraction and expansion that must be provided for if the system is to be installed at 27°C.

Solution:

Contraction:

$$\Delta t = t_2 - t_1 = (-32) - 27 = -59°C \text{ (temperature drop)}$$

$$\alpha = 14.0 \times 10^{-6} \text{ mm/mm/°C (Table 8-2)}$$

$$l = 25 \text{ m}$$

$$\delta = \alpha l \, \Delta t = (14.0 \times 10^{-6})(25 \times 10^3)(-59) = -21 \text{ mm}$$

(total contraction from installed position)

The minus (−) sign for δ indicates negative expansion (i.e., contraction).

Expansion:

$$\Delta t = 77 - 27 = 50°C \text{ (temperature rise)}$$

$$\delta = (14.0 \times 10^{-6})(25 \times 10^3)(50) = 18 \text{ mm}$$

(total expansion from installed position)

Note that the total change in shaft length that must be designed for is 21 + 18 = 39 mm.

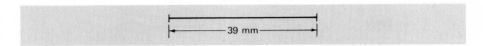

8-12 THERMAL STRESSES

In Sec. 8-11, the total deformation of a member due to a temperature change was defined by Eq. (8-10). It was further stated that the value of δ given by Eq. (8-10) would represent the actual change in length of a member that is *free* to change its dimensions. There are many situations, however, in which a member undergoing a temperature change is *prevented* from expanding or contracting. When thermal expansion or contraction is restrained, the restraining forces cause new stresses in the member. These stresses are just as real as if equivalent externally applied loads instead of the restraining forces were acting. Let us develop an expression for these new stresses (or changes in stresses) that are due to a temperature change of a restrained member.

From Eq. (8-2),

$$s = E\epsilon \qquad \text{where } \epsilon = \frac{\delta}{l}$$

From Eq. (8-10),

$$\frac{\delta}{l} = \alpha \, \Delta t$$

Therefore, $$\epsilon = \alpha \, \Delta t$$

By substituting this expression into Eq. (8-2), we obtain

$$s = E\alpha \, \Delta t \qquad\qquad (8\text{-}11)$$

where s = thermal stress due to prevented deformation, psi; MPa

E = modulus of elasticity, psi; MPa

α = coefficient of thermal linear expansion, in/in/°F; mm/mm/°C

Δt = temperature change °F; °C

In the cases considered here, the thermal stress will be assumed to be axial in either tension or compression. *Tension* occurs when the member should contract but is prevented from doing so; *compression* occurs when the member should expand but is prevented from doing so.

Sample Problem 4 A Class 20 cast-iron pipe 12 ft long is rigidly fastened between two walls. Find the stress induced in the pipe that is due to a temperature rise of 25°F.

Solution:

$$E = 11 \times 10^6 \text{ psi (Table 1A, App. B)}$$
$$\alpha = 6.3 \times 10^{-6} \text{ in/in/}°\text{F (Table 8-2)}$$
$$\Delta t = 25°\text{F}$$

By Eq. (8-11),

$$s = E\alpha \, \Delta t = (11 \times 10^6)(6.3 \times 10^{-6})(25) = 1730 \text{ psi (compression)}$$

*Sample Problem 5[1]   A steel tie rod is subjected to an axial pull of 18 kN. The allowable tensile stress is specified as 160 MPa.

(a)   Find the minimum diameter required.
(b)   A temperature drop of 38°C is anticipated. If the ends of the tie rod are fixed, what is the minimum required diameter?

Solution a:

$$A = \frac{F}{s} = \frac{18(10^3)}{160} = 112 \text{ mm}^2$$

$$A = \frac{\pi d^2}{4}$$

$$d = \sqrt{\frac{4A}{\pi}} = \sqrt{\frac{4(112)}{\pi}} = \sqrt{143}$$

$$= 11.96 \text{ mm} \qquad \text{use 12-mm rod}$$

Solution b:

$$E = 200 \times 10^9 \text{ Pa} = 200 \times 10^3 \text{ MPa (Table 1A, App. B)}$$
$$\alpha = 11.7 \times 10^{-6} \text{ mm/mm/}°\text{C (Table 8-2)}$$
$$s = E\alpha \, \Delta t = (200 \times 10^3)(11.7 \times 10^{-6})(38)$$
$$= 89 \text{ MPa (tension)}$$

Since the allowable stress = 160 MPa, $160 - 89 = 71$ MPa is the maximum stress which can be developed by the 18-kN load.

$$A = \frac{F}{s} = \frac{18(10^3)}{71} = 254 \text{ mm}^2$$

$$d = \sqrt{\frac{4A}{\pi}} = 18 \text{ mm} \qquad \text{use 18-mm rod}$$

8-13   MEMBERS COMPOSED OF TWO MATERIALS IN PARALLEL

The term *parallel* as used in this section and the term *series* as used in Sec. 8-14 refer to the similarity between the paths of force transmission and the corresponding current flow in simple electric circuits.

---

[1] See also Sample Computer Program 11, Chap. 17.

Consider a tension member composed of two different materials (Fig. 8-9). Let F be the tensile load carried by this member. Since materials A and B act as a single member, their elongation must be the same. Therefore, $\delta_a = \delta_b$ and, since the length is the same for both materials, $\epsilon_a = \epsilon_b$.

Fig. 8-9 Member composed of two different materials in parallel.

But

$$\epsilon_a = \frac{S_a}{E_a} \qquad \epsilon_b = \frac{S_b}{E_b}$$

Therefore,

$$\frac{S_a}{E_a} = \frac{S_b}{E_b} \tag{8-12}$$

Also, the total load $F = F_a + F_b$, the sum of the force carried by A and the force carried by B. Therefore,

$$F = A_a S_a + A_b S_b. \tag{8-13}$$

The use of Eqs. (8-12) and (8-13) is demonstrated in the following Sample Problem.

Sample Problem 6

A short 8- by 8-in timber post is reinforced with two 7- by $\frac{1}{2}$-in steel plates bolted to the post so that the parts act as a single member (Fig. 8-10). Note that the dressed size of an 8- by 8-in post is actually $7\frac{1}{2}$ by $7\frac{1}{2}$ in (see Table 11, App. B).

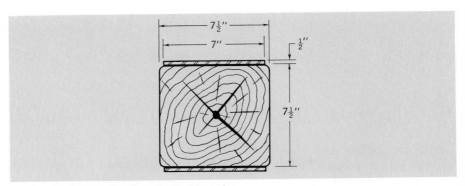

Fig. 8-10 Diagram for Sample Problem 6.

Find the compressive load that this post can carry if the stress in the timber is not to exceed 1000 psi. Assume that $E = 1\,600\,000$ psi for timber.

Solution: Use $E = 30 \times 10^6$ psi for steel. By Eq. (8-12),

$$S_{steel} = S_{timber}\frac{E_{steel}}{E_{timber}} = 1000\left(\frac{30 \times 10^6}{1.6 \times 10^6}\right)$$

$$= 18\ 750\ \text{psi} \qquad \text{(stress in steel when timber is stressed to 1000 psi)}$$

By applying Eq. (8-13), we have

$$F = (As)_{timber} + (As)_{steel}$$
$$= 7.5(7.5)(1000) + 2(7)(0.5)(18\ 750)$$
$$= 56\ 200 + 131\ 000 = 187\ 200\ \text{lb}$$

Say, 187 000 lb is the safe compressive load.

8-14   MEMBERS COMPOSED OF TWO MATERIALS IN SERIES

Consider a member in compression composed of two different materials, as in Fig. 8-11. Since the load F must be transmitted from one end of the member to the other, $F = F_a = F_b$. When the cross-sectional areas of the two materials are equal, $s_a = s_b$.

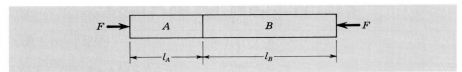

Fig. 8-11 Member composed of two different materials in series.

The total strain (shortening) is shared by the materials, so that $\delta = \delta_a + \delta_b$. But

$$\delta_a = \frac{Fl_a}{A_a E_a} \qquad \text{and} \qquad \delta_b = \frac{Fl_b}{A_b E_b}$$

Therefore,

$$\delta = F\left(\frac{l_a}{A_a E_a} + \frac{l_b}{A_b E_b}\right) \tag{8-14}$$

Also, the total load

$$F = A_a s_a = A_b s_b \tag{8-15}$$

The use of Eqs. (8-14) and (8-15) is demonstrated in the following example.

Sample Problem 7   A steel bar that is 1 in square and 8 in long is set end to end with a cast-iron (CI) Class 40 bar that is 2 in square and 4 in long between two immovable supports. What stress will develop in each material because of a temperature rise of 50°F?

Solution: From Table 1A, App. B,

$$E_s = 30 \times 10^6 \text{ psi (for steel)}$$
$$E_c = 16 \times 10^6 \text{ psi (for CI Class 40)}$$

From Table 8-2,

$$\alpha_s = 6.5 \times 10^{-6} \text{ in/in/}°\text{F (for steel)}$$
$$\alpha_c = 6.3 \times 10^{-6} \text{ in/in/}°\text{F (for CI)}$$

If the member were not restrained, each material would elongate by amounts of

$$\delta_s = \alpha_s l_s \, \Delta t = (6.5 \times 10^{-6})(8)(50) = 0.00260 \text{ in}$$
$$\delta_c = \alpha_c l_c \, \Delta t = (6.3 \times 10^{-6})(4)(50) = 0.00126 \text{ in}$$

The total elongation δ that would occur would equal $\delta = \delta_s + \delta_c = 0.00260 + 0.00126 = 0.00386$ in.

Since the member is not free to elongate, the materials are placed in compression by a force F, which can be found by using Eq. (8-14):

$$F = \cfrac{\delta}{\cfrac{l_s}{A_s E_s} + \cfrac{l_c}{A_c E_c}} = \cfrac{0.00386}{\cfrac{8}{1(30 \times 10^6)} + \cfrac{4}{4(16 \times 10^6)}}$$

$$F = \frac{0.00386}{0.27 \times 10^{-6} + 0.06 \times 10^{-6}} = \frac{0.00386}{0.33 \times 10^{-6}} = 11\,700 \text{ lb}$$

By applying Eq. (8-15), we have

$$S_s = \frac{F}{A_s} = \frac{11\,700}{1} = 11\,700 \text{ psi (compression in steel)}$$

$$S_c = \frac{F}{A_c} = \frac{11\,700}{4} = 2\,900 \text{ psi (compression in cast iron)}$$

PROBLEMS

8-1. Find the total elongation and the stress in a 1-in-diameter AISI 1045 steel rod 12 ft long under a pull of 12 000 lb.

*8-2. The total elongation of a square wrought-iron bar 50 mm on a side and 3 m long is 1.5 mm. What is the load?

*8-3. What is the strain of Monel when stressed in tension to 196 MPa? If the original length was 992 mm, what is the final length?

8-4. A wrought-iron rod 0.80 in^2 in cross section is elongated 0.06 in by a load of 16 000 lb. What was the original length?

*8-5. A bar of AISI 1020 structural steel 30 mm in diameter is under successive tensile loads of 198, 210, and 270 kN. Calculate the stress that develops at each load. What are the strains for each load? Are they all true strains? If not, why not?

8-6. The following data were obtained from a tension test of 6061-T6 aluminum alloy.

    a. Plot a stress-vs.-strain diagram for these data.

    b. Estimate the elastic-limit stress from the stress-vs.-strain diagram.

    c. From the stress-vs.-strain diagram, determine the modulus of elasticity.

Load, lb	Elongation, in	Data	
500	0	Original gage length	2.000 in
1000	0.0004	Final gage length	2.400 in
1500	0.0009	Original diameter	0.5038 in
2000	0.0013	Final diameter	0.3850 in
2500	0.0018	Yield point	9140 lb
3000	0.0022	Ultimate load	11 940 lb
3500	0.0027	Rockwell	B53.8 before
4000	0.0033	Rockwell	B57.2 after
4500	0.0038		
5000	0.0043		
5500	0.0048		
6000	0.0053		
6500	0.0058		
7000	0.0065		
7500	0.0073		
8000	0.0084		
8500	0.0103		

    d. Calculate the yield stress and the ultimate stress.

    e. Calculate the percent elongation and the percent reduction in area.

8-7. The following data were obtained from a tension test of yellow brass.

Load, lb	Elongation, in	Data	
500	0	Original gage length	2.000 in
1000	0.0004	Final gage length	2.438 in
1500	0.0007	Original diameter	0.5048 in
2000	0.0011	Final diameter	0.3660 in
2500	0.0015	Yield point	11 600 lb
3000	0.0018	Ultimate load	12 300 lb
3500	0.0022	Rockwell	B59.2 before
4000	0.0026	Rockwell	B59.6 after
4500	0.0030		
5000	0.0034		
5500	0.0039		
6000	0.0044		
6500	0.0049		
7000	0.0055		
7500	0.0063		
8000	0.0072		
8500	0.0081		
9000	0.0092		
9500	0.0140		

    a. Plot a stress-vs.-strain diagram for these data.

    b. Estimate the elastic limit stress from the stress-vs.-strain diagram.

    c. From the stress-vs.-strain diagram, determine the modulus of elasticity.

    d. Calculate the yield stress and the ultimate stress.

    e. Calculate the percent elongation and the percent reduction in area.

*8-8. In an experiment in the testing laboratory, a specimen 20 mm in diameter and 250 mm long reached the elastic limit under a tensile load of 120 kN and failed under a load of 175 kN. The length of the bar at its elastic limit was 251.2 mm, and at failure it was 333.8 mm. Compute the stress at the elastic limit, the ultimate tensile stress, and the strain for the elastic limit.

8-9. A round Class 20 cast-iron post 2 ft long, 8 in OD, and 6 in ID supports a load of 25 tons. What is the total decrease in length?

*8-10. How much will an alloy steel punch 25 mm in diameter and 75 mm long be compressed while punching a hole through a 12-mm AISI 1045 steel plate?

*8-11. A wire made of copper alloy is 4.5 m long and 1 mm in diameter. It is stretched 14.5 mm by a load of 340 N. What is the modulus of elasticity of this alloy?

8-12. An AISI 1020 steel rod 32 ft long is used as a hanger for a theater balcony. The rod has to support 52 000 lb. What should be its diameter if its total elongation does not exceed 0.15 in and the stress does not exceed 12 000 psi?

8-13. A short round member has a cross-sectional area of 4.20 in² and carries a compressive load of 10 tons. Find the factor of safety N_u if the member is *(a)* nylon 6/6 and *(b)* reinforced polyester.

*8-14. The shearing force on a hard-steel (AISI 1095) pin 30 mm in diameter is 25 kN. What is the factor of safety N_u? The pin is in single shear.

8-15. Find the diameter of a Monel rod in tension required to carry a shock load of 7500 lb:

    a. Based on ultimate strength

    b. Based on yield strength

*8-16. A Monel bar is 60 by 35 mm in cross section. By using a safety factor N_u of 5, find the tensile load the bar will support.

8-17. A square timber member carries a compressive load of 73 000 lb. What should be its nominal cross-sectional dimensions. Use a safety factor of 5 based on the ultimate strength of 3500 psi, and use Table 11, App. B, to select a proper member.

8-18. How many 2- by 3-in SAN (plastic) bars are required to carry a steady tensile load of 255 000 lb? Determine the allowable stress based on $N_y = 2$.

*8-19. What space must be left between ends of AISI 1095 steel rails, each 9.9 m long and at a temperature of 10°C, so that the ends can be in contact at a temperature of 38°C? If these rails had been in contact at 10°C, what stress would result from the rise in temperature?

8-20. An AISI 1045 steel tie rod 24 ft long and 1 in in diameter is used to tie two walls of a building together. The rod is screwed up to a tension of 9000 lb. What will be the stress in the rod if the temperature rises 30°F? If it falls 30°F?

8-21. How wide a gap should be allowed between the 50-ft-long paving slabs of concrete in a street so that the slabs will touch each other at a temperature of 90°F? The pavement is laid at a temperature of 65°F. What would be the compressive stress in the slabs if the temperature rose to 110°F? (E for concrete is 3 000 000 psi.)

8-22. A surveyor's steel measuring tape is exactly 100 ft long at 60°F when subjected to a pull of 10 lb. Determine the correction to be made in reading the tape if the tape is used at 100°F under a pull of 20 lb. The tape is $\frac{1}{32}$ in thick and $\frac{3}{8}$ in wide. Express the correction in inches per foot of length and indicate whether it is added to or subtracted from the reading.

*8-23. A 40-mm-diameter bronze rod is 1.8 m long and is subjected to a temperature rise of 60°C. $E = 100$ GPa.
  a. Find the total elongation and the stress if the ends are not fixed.
  b. Find the total elongation and the stress if the ends are rigidly fixed.
  c. For the conditions in part b, determine the force in the rod.

8-24. A refrigerator freezer part made of polyethylene (high-density) with a cross-sectional area of 0.75 in² and a length of 14 in is installed at 75°F. When the refrigerator is in operation, the temperature in the freezer will drop to 0°F.
  a. Find the change in length and the stress if the ends are not fixed.
  b. Find the change in length and the stress if the ends are rigidly fixed.

8-25. A short CI Class 40 square column (Fig. Prob. 8-25) is filled with plain concrete. A load F is applied so as to cause the parts to act as a single member. If the modulus of elasticity of concrete is 2 000 000 psi, what is the compressive load that the short column can carry if the stress in the concrete is not to exceed 900 psi?

8-26. A concrete cylinder 4 in in diameter and 2 ft long supports a structural steel rod 1 in in diameter and 6 in long. What is the maximum compressive load that can be applied to the top of the steel rod such that the total shortening of the two cylinders does not exceed 0.005 in? Use $E = 3.0 \times 10^6$ psi for concrete.

8-27. A 6061-T6 aluminum-alloy rod 4 in long is supported at one end so that its other end provides a gap of 0.002 in with the end of a yellow brass rod 6 in long supported in a like manner and in line with it. The cross-sectional area of each rod is 2 in². It is anticipated that the rods will be subjected to a maximum temperature increase of 100°F.
  a. What maximum stress will develop in each of the rods if the supports are immovable?

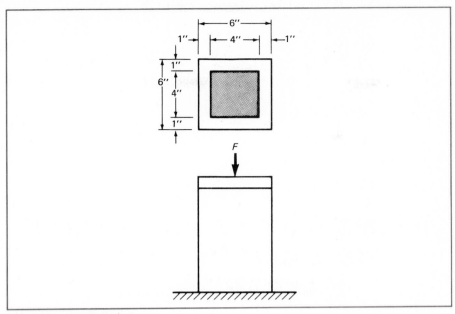

FIGURE PROBLEM 8-25

    b. If both rods are made of aluminum alloy, what stress will develop in each of them?

    c. If both rods are made of yellow brass, what stress will develop in each of them?

*8-28. A surveyor's steel measuring tape is exactly 30 m long at 15°C when subjected to a pull of 45 N. Determine the correction to be made in reading the tape if the tape is used at 35°C under a pull of 70 N. The tape is 0.8 mm thick and 10 mm wide. Express the correction in millimeters per meter of length and indicate whether it is added to or subtracted from the reading.

9

Bolted, Riveted, and Welded Joints and Thin-Walled Pressure Vessels

9-1 INTRODUCTION

In the construction of a steam generator or the erection of a building or bridge, the individual members or parts must be securely connected in order to safely carry and transmit forces and moments due to the loadings. Such connections are achieved by means of bolts, rivets, pins, and welds.

This chapter will be concerned with the analysis and design of several types of bolted, riveted, and welded joints which occur in the construction and mechanical fields. However, the emphasis will be on bolted and welded joints, since rivets are seldom used in those fields today.

9-2 BOLTED JOINTS

Bolted joints are designated either lap or butt joints. When two plates are placed with the ends overlapping, they form a *lap joint*. Figure 9-1 represents the top and sectional views of a lap joint with a single row of bolts, and Fig. 9-2 represents one with a double row of bolts. Rows of bolts are perpendicular to the applied load.

When two plates are placed edge to edge and are connected by bolts, as in Fig. 9-3, they form a *butt joint*. Since there is one row of bolts on either side of the centerline of the joint, the joint shown in Fig. 9-3 is called a single-bolted butt joint. Similarly, Fig. 9-4 shows a double-bolted butt joint (two rows of bolts on each side of the joint). It should be noted that bolted butt joints require the use of cover plates.

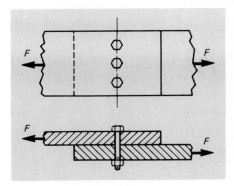

Fig. 9-1 Single-bolted lap joint.

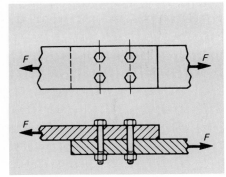

Fig. 9-2 Double-bolted lap joint.

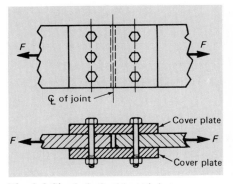

Fig. 9-3 Single-bolted butt joint.

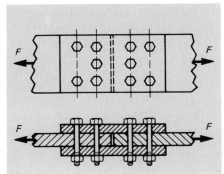

Fig. 9-4 Double-bolted butt joint.

9-3 TYPES OF FAILURE IN BOLTED JOINTS

The actual stresses developed in bolted joints are quite complex, but in practice they are treated as *simple* tension, compression, and shearing stresses.

The analysis of bolted joints involves the continued use of

$$s = \frac{F}{A} \tag{9-1}$$

where s represents the particular stress to be determined.

It is usually assumed that each bolt will carry its proportional share of the total load on the joint. In the lap joint shown in Fig. 9-2, each bolt transmits one-fourth of the load from one plate to the other. The transmission of load by the bolts can be represented as in Fig. 9-5.

A butt joint is composed of two identical halves. Each half is designed to carry the full load. Therefore, in analyzing butt joints, only one half of the joint is considered. The joint shown in Fig. 9-4 is analyzed as shown in Fig. 9-6. In this butt joint each bolt carries one-fifth of the load and transmits one-half of it to each cover plate (Fig. 9-6).

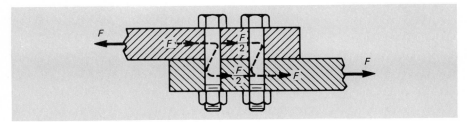

Fig. 9-5 Force transmission in lap joint in Fig. 9-2.

To understand the types of failure, consider the lap joint shown in Fig. 9-1. The joint may fail by shearing the bolts, as shown in Fig. 9-7. Another possibility is that the plate may pull apart at its weakest section. This section occurs on the centerline of the row of bolt holes in Fig. 9-1, and the failure will appear as shown in Fig. 9-8.

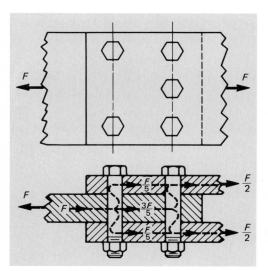

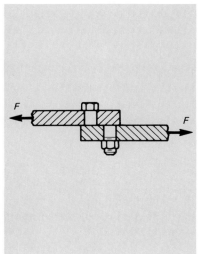

Fig. 9-6 Force transmission in butt joint in Fig. 9-4.

Fig. 9-7 Bolt shear failure in lap joint in single shear.

A third mode of failure may occur by crushing of the bolts or the plate material in contact with the bolts. This type of failure is called a *bearing,* or compression, failure. Such a failure is shown in Fig. 9-9. (*Note:* A bolted joint can be so designed that the possibility of failure in bearing need not be considered. This type of joint is referred to as a *friction-type connection,* and it will be discussed later in the chapter.)

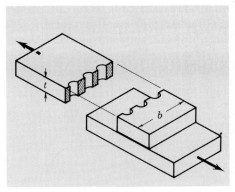

Fig. 9-8 Plate tension failure in lap joint.

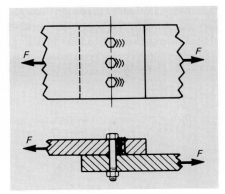

Fig. 9-9 Bearing failure in lap joint.

9-4 STRESSES IN A BOLTED JOINT

The joints discussed in this chapter involve only direct (axial) loading. Eccentric loading, which tends to cause rotation of the joint, will be dealt with in Chap. 14. Shear, tension, and bearing stresses in axially loaded bolted joints are each represented by

$$s = \frac{F}{A} \tag{9-1}$$

or $$F = As \tag{9-1a}$$

Shear: When the shear stress in a bolted joint is to be computed, Eq. (9-1) gives the relation between the external force F, which produces a shearing stress s_s, and the total area resisting shear A_s. Since the total shear area A_s is composed of a number of circular bolt cross sections, each having a diameter d, the special forms of Eqs. (9-1) and (9-1a) for shear are

$$s_s = \frac{F}{A_s} = \frac{F}{n\left(\dfrac{\pi d^2}{4}\right)} \tag{9-2}$$

or $$F = \left[n\left(\frac{\pi d^2}{4}\right)\right] s_s \tag{9-2a}$$

where F = external force on joint, lb

s_s = shear stress in bolt material, psi

d = bolt diameter at shear plane, in

n = number of shear areas resisting force F

In a lap joint, such as that shown in Fig. 9-2, each bolt is in *single shear.* Thus, there are four shear areas in this joint, and $n = 4$ in Eqs. (9-2) and (9-2a). Each bolt in one half of a butt joint is in *double shear.* That is, each bolt tends to fail on two cross sections simultaneously, as shown in Fig. 9-10. Therefore, for the joint shown in Fig. 9-4, the value of n in Eqs. (9-2) and (9-2a) is 10.

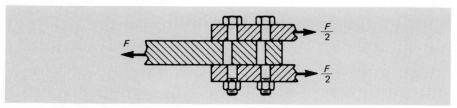

Fig. 9-10 Bolt shear failure in butt joint in double shear.

Tension: This analysis deals with the possibility of the main plate failing in tension, as shown in Fig. 9-8. Equation (9-1) applies, where s_t is the tensile stress in the plate on cross-sectional area A_t. This area resists a total force F. For the lap joint shown in Fig. 9-8, the effective area of the plate cross section through the center of the bolts is reduced because of the bolt holes.

$$s_t = \frac{F}{A_t} \qquad F = A_t s_t$$
$$A_t = bt - nDt = (b - nD)t$$
$$F = [(b - nD)t]s_t \tag{9-3}$$

where F = external force on joint, lb

s_t = tensile stress in plate material at cross section in question, psi

b = width of plate, in

t = thickness of plate, in

D = diameter of bolt hole, in

n = number of holes in cross section in question

Note that bt is the cross-sectional area of the plate without holes and Dt is the area of material removed from the plate cross section by each hole. Also, the term $b - nD$ represents the effective width of plate remaining after n holes are punched or drilled.

Bearing: With reference to Fig. 9-11, it is evident that, when the force F is applied to the plates, each plate is forced against the curved surface of the bolt and tends to crush the bolt and the plate. Thus, the compressive stress that is produced varies in intensity at various points on the curved surface. However, it is customary in figuring bearing stress to take a section through the axis of the bolt perpendicular to the direction of the applied forces. The section is a rectangle of height t, the thickness of the plate; and width d, the diameter of the bolt. Then the bearing area for one bolt is td. If s_c is the compressive, or bearing,

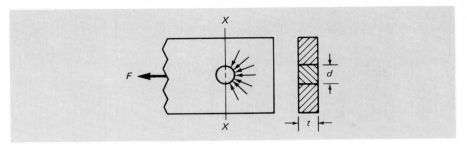

Fig. 9-11 Bearing forces and bearing area.

stress,

$$s_c = \frac{F}{A_c} \qquad F = A_c s_c$$
$$A_c = ntd$$
$$F = [ntd]s_c \qquad\qquad (9\text{-}4)$$

where F = external force on joint, lb
s_c = bearing (compressive) stress in plate and bolt material, psi
t = thickness of plate, in
d = bolt diameter, in
n = number of bearing areas (usually one per bolt in a given plate)

9-5 TERMINOLOGY AND CODES FOR BOLTED JOINTS

The distance between centers of adjacent bolts, within a row, is called *pitch*.

In the case of a long or continuous bolted joint subjected to uniform tension for its entire length, the strength of the joint can be computed from a repeating section. A repeating section is the shortest length of joint which contains a complete bolt pattern. In Fig. 9-12 the distance *l* is the length of the repeating section. Usually, the length of the repeating section is equal to the pitch of the bolts in row 1 (outer row).

The distance *g* between rows of bolts is called the *gage*.

The distance from the center of a bolt to any plate edge is called the *edge distance*. Such a distance is indicated in Fig. 9-12 by *e*.

The design of bolted joints for specific applications is usually governed by codes which have been established by local legislation or technical groups. One of the most commonly used codes is the American Institute of Steel Construction (AISC) code for connections. Some of the provisions of this code are given in Table 9-1. For other codes or more complete information on the above code, the reader is referred to the technical agency concerned.

It should be noted that welding has replaced bolting in continuous joints for boilers and pressure vessels. Consequently, the American Society of Mechanical Engineers (ASME) no longer specifies a code for these joints.

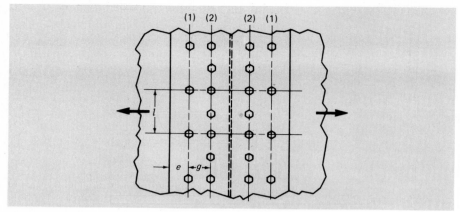

Fig. 9-12 Continuous bolted joint.

TABLE 9-1 AISC CODE FOR CONNECTIONS (RIVETS, BOLTS, AND THREADED PARTS)

Type of Stress and Material Specification	Allowable Stress, psi	
	Friction-Type Connections	Bearing-Type Connections
Shear on fasteners:		
A 502, grade 1, hot-driven rivets		17 500
A 502, grades 2 and 3, hot-driven rivets		22 000
A 307, low-carbon steel bolts		10 000
A 325, high-strength bolts; threads occur at the shear plane	17 500	21 000
A 325, high-strength bolts; threads do not occur at the shear plane	17 500	30 000
A 490, high-strength bolts; threads occur at the shear plane	22 000	28 000
A 490, high-strength bolts; threads do not occur at the shear plane	22 000	40 000
Threaded parts of other steels: threads occur at the shear plane		$0.17\,s_u$
Threaded parts of other steels: threads do not occur at the shear plane		$0.22\,s_u$
Bearing on projected area (bearing-type connections only)*:		$1.5\,s_u$
A 36, carbon structural steel; ($s_y = 36\,000$ psi; $s_u = 58\,000$ psi)		87 000

TABLE 9-1 *(continued)*

Type of Stress and Material Specification	Allowable Stress, psi	
	Friction-Type Connections	Bearing-Type Connections
A 242, corrosion-resistant high-strength low-alloy structural steel; ($s_y = 50\,000$ psi; $s_u = 70\,000$ psi)		105 000
A 441, high-strength low-alloy structural steel; ($s_y = 46\,000$ psi; $s_u = 67\,000$ psi)		101 000
A 572, grade 50, high-strength low-alloy steel; ($s_y = 50\,000$ psi; $s_u = 65\,000$ psi)		97 500
A 588, corrosion-resistant high-strength low-alloy structural steel; thickness 4 in and less ($s_y = 42\,000$ psi; $s_u = 63\,000$ psi)		94,500
Tension on net section (bearing- and friction-type connections)*:	$0.5\,s_u$	$0.5\,s_u$
A 36, carbon structural steel; ($s_y = 36\,000$ psi)	29 000	29 000
A 242, corrosion-resistant high-strength low-alloy structural steel; ($s_y = 50\,000$ psi)	35 000	35 000
A 441, high-strength low-alloy structural steel; ($s_y = 46\,000$ psi)	33 500	33 500
A 572, grade 50, high-strength low-alloy steel; ($s_y = 50\,000$ psi)	32 500	32 500
A 588, corrosion-resistant high-strength low-alloy structural steel; thickness 4 in and less ($s_y = 50\,000$ psi)	31 500	31 500

* For steels other than those listed, refer to the *Manual of Steel Construction of AISC.*
Additional provisions:
a. Minimum preferable pitch (center to center between adjacent fasteners) = 3d to provide clearance for tools in assembling connections.
b. Minimum edge distance varies approximately from $1\frac{1}{4}\,d$ to $1\frac{3}{4}\,d$.
c. For purpose of design, hole diameters are taken as $\frac{1}{16}$ in larger than fastener diameter; $D = d + \frac{1}{16}''$.
Note: The AISC code *(Manual of Steel Construction)* also establishes allowable stresses for rivets, bolts, and threaded parts subjected to direct tension.

Referring to Table 9-1, the AISC code for connections, the allowable shear stress given for most of the listed fasteners distinguishes between friction-type and bearing-type connections.

The designer of a connection (joint) may use either of the two types. The *friction type* assumes no movement between adjacent surfaces of the plate material and, as a result, bearing stress is not considered. However, *shear* on the fastener and *tension* on the plates are taken into account. The *bearing type* does assume the possibility of movement between adjacent surfaces. Thus, the possibility of failure in shear, bearing, and tension are all considered.

Sample Problem 1[1] A single-bolted lap joint with 12- by $\frac{1}{2}$-in A36 plates contains four $\frac{7}{8}$-in-diameter A325 bolts (Fig. 9-13). Based on the AISC code, friction-type connection, determine the largest load the joint can safely carry.

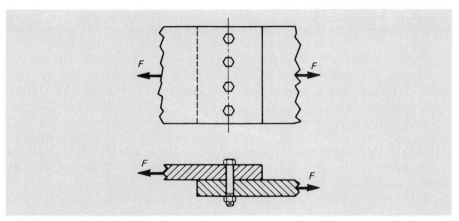

Fig. 9-13 Diagram for Sample Problem 1.

Solution: To find the safe load on the joint, calculate the safe loads in shear and tension. Bearing is not considered because the joint is a friction-type connection. The largest safe load on the joint will be the *least* of the two calculated loads, since it must satisfy both conditions.

Shear (Eq. 9-2a): Note that each bolt is in single shear; $n = 4$.

$$F_s = \left[n \left(\frac{\pi d^2}{4} \right) \right] s_s = 4 \left[\frac{\pi (0.875^2)}{4} \right] (17\ 500)$$
$$= 4(0.601)(17\ 500) = 42\ 100 \text{ lb}$$

Tension (Eq. 9-3): There are four holes in the plate along the weak section; $n = 4$, $D = d + \frac{1}{16}$ (AISC code).

$$F_t = [(b - nD)t]s_t = [12 - 4(\tfrac{15}{16})]0.5(29\ 000)$$
$$= 8.25(0.5)(29\ 000) = 120\ 000 \text{ lb}$$

The largest safe load on the joint is 42 100 lb. This is called the *strength of the joint.*

Sample Problem 2[2] A double-bolted butt joint, with four 1-in-diameter A490 bolts on each side and two bolts per row, joins 10- by $\frac{3}{4}$-in A441 plates (Fig. 9-14). The cover plates are $\frac{1}{2}$ in thick. Based on the AISC code, bearing-

---

[1] See also Sample Computer Program 12, Chap. 17.

[2] See also Sample Computer Program 13, Chap. 17.

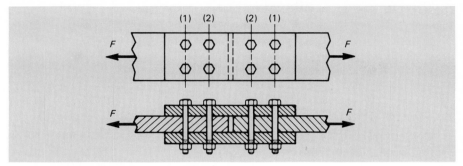

Fig. 9-14 Diagram for Sample Problem 2.

type connection, find the strength of the joint (maximum safe load). Assume that the bolt threads do not occur at the shear plane.

Solution:

Shear: Note that each bolt is in double shear and that only one half of the joint is considered; $n = 8$.

$$F_s = \left[n\left(\frac{\pi d^2}{4}\right) \right] s_s = 8\left[\frac{\pi(1)^2}{4} \right](40\ 000)$$
$$= 8(0.785)(40\ 000) = 251\ 000\ \text{lb}$$

Tension: The weak section in the main plate occurs at row 1 (Fig. 9-14). This section must transmit the full load, and it has two bolt holes. The section at row 2 also has two holes, but it is required to transmit only half of the external force (see Sec. 9-3). For the cover plates in tension, the critical section is at row 2. It is usually assumed that each cover plate transmits one-half of the external force to the other side of the joint. In this problem, it is not necessary to check the safe load on the cover plates because the total cover plate thickness exceeds that of the main plate and the number of bolt holes in row 2 is the same as for row 1. For the main plate at row 1, $n = 2$ and $D = d + \frac{1}{16}$ (AISC code)

$$F_t = [(b - nD)t]s_t = [10 - 2(1\tfrac{1}{16})](0.75)(33\ 500)$$
$$= 7.875(0.75)(33\ 500) = 198\ 000\ \text{lb}$$

Bearing: Each bolt contributes one bearing area in the main plate; $n = 4$.

$$F_c = [ntd]s_c = 4(0.75)(1)(101\ 000)$$
$$= 3(101\ 000) = 303\ 000\ \text{lb}$$

The strength of the joint is 198 000 lb.

9-6   EFFICIENCY OF A BOLTED JOINT

The area of the cross section of a plate, *bt*, is called the *gross area.*

The quotient obtained by dividing the tensile strength of the gross area into the strength of the bolted joint is called the *efficiency* of the joint and is represented by η (the greek letter *eta*). That is, the efficiency is

$$\eta = \frac{\text{strength of the joint}}{\text{tensile strength of the gross area}} \times 100 \qquad (9\text{-}5)$$

In determining the efficiency, the repeating section width of plate (*l*) may be used instead of the entire width.

Sample Problem 3[1]   Find the efficiency of a single-bolted lap joint with $\frac{1}{2}$-in-thick plates (A36), $\frac{7}{8}$-in-diameter bolts (A325), and a pitch width (repeating section) of 3 in. Use the AISC code, bearing-type connection; bolt threads do not occur at the shear plate.

Solution:

Shear:

$$n = 1$$

$$F_s = 1\left[\frac{\pi(0.875^2)}{4}\right](30\ 000) = 18\ 000\ \text{lb}$$

Tension:

$$n = 1$$

$$F_t = [3 - 1(\tfrac{15}{16})](\tfrac{1}{2})(29\ 000) = 29\ 900\ \text{lb}$$

Bearing:

$$n = 1$$

$$F_c = 1(0.875)(\tfrac{1}{2})(87\ 000) = 38\ 100\ \text{lb}$$

From the above calculations, the strength of the joint is 18 000 lb. The tensile strength of the gross area is

$$F_g = (lt)s_t = 3(\tfrac{1}{2})(29\ 000) = 43\ 500\ \text{lb}$$

Therefore,

$$\eta = \frac{18\ 000}{43\ 500} = 0.414 = 41.4\%$$

Sample Problem 4   Find the efficiency of a triple-bolted butt joint with $\frac{3}{4}$-in main plates, $\frac{1}{2}$-in cover plates, and 1-in-diameter A490 steel bolts arranged as shown in Fig. 9-15. The plates are A441. Use the AISC code, bearing-type connection; bolt threads do not occur at the shear plane.

---

[1] See also Sample Computer Program 12, Chap. 17.

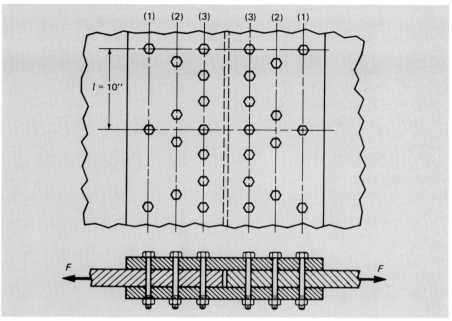

Fig. 9-15 Diagram for Sample Problem 4.

Solution: The bolt-hole size of $1\frac{1}{16}$ in (1.0625 in) is used for tension calculations (see Table 9-1).

All calculations refer to a 10-in repeating section, as indicated in Fig. 9-15.

Shear:

$$n = 12 \ (6 \text{ bolts in double shear})$$

$$F_s = \left[n \left(\frac{\pi d^2}{4} \right) \right] s_s = 12(0.785)(1^2)(40\ 000) = 377\ 000 \text{ lb}$$

Bearing:

$$n = 6$$

$$F_c = [ntd]s_c = 6(0.75)(1)(101\ 000) = 455\ 000 \text{ lb}$$

Tension (Row 1):

$$n = 1$$

$$F_t = [(l - nD)t]s_t = [10 - 1(1.0625)](0.75)(33\ 500)$$
$$= 8.9375(0.75)(33\ 500) = 225\ 000 \text{ lb}$$

Tension (Row 2):

$$n = 2$$

The purpose of these calculations is to evaluate the external force F which a 10-in repeating section can carry. This force is transmitted by the bolts to the cover plates, assuming that each bolt transmits its share of the force (see Sec. 9-3 and Fig. 9-6). The tensile force on the plate at row 2 is smaller than the external force F by the amount which the bolt in row 1 has transmitted to the cover plates. Thus, the force on row 2 in this problem is $\frac{5}{6}$ of the external force.

$$\tfrac{5}{6}F_t = [(l - nD)t]s_t = [10 - 2(1.0625)](0.75)(33\ 500)$$
$$= (7.8750)(0.75)(33\ 500) = 197\ 900$$
$$F_t = \tfrac{6}{5}(197\ 900) = 237\ 000\ \text{lb}$$

Tension   (Row 3):

$$n = 3$$
$$\tfrac{3}{6}F_t = [(l - nD)t]s_t = [10 - 3(1.0625)](0.75)(33\ 500)$$
$$= (6.8125)(0.75)(33\ 500) = 171\ 200$$
$$F_t = \tfrac{6}{3}(171\ 200) = 342\ 000\ \text{lb}$$

From the above calculations, the strength of the joint is 225 000 lb. The tensile strength of the gross area is

$$F_g = (lt)s_t = 10(0.75)(33\ 500) = 251\ 000\ \text{lb}$$

and
$$\eta = \frac{225\ 000}{251\ 000} = 0.896 = 89.6\% \qquad \text{say, } 90\%$$

Check the cover plate size. Assume one-half of the external force is transmitted to each cover plate. This load must be transmitted by the cover plate from the inner row centerline (row 3) to the inner row in the other half of the joint. Since the safe load is 225 000 lb, each cover plate must transmit 112 500 lb across its weakest section (row 3).

$$\text{Allowable } F_{cp} = [(l - nD)t]s_t = [10 - 3(1.0625)](0.5)(33\ 500)$$
$$= 6.8125(0.5)(33\ 500) = 114\ 000\ \text{lb}$$
$$\text{Actual } F_{cp} = 112\ 500\ \text{lb} \qquad \text{therefore OK}$$

9-7   BOLTED JOINTS OF MAXIMUM EFFICIENCY

The procedure for design of bolted joints of maximum efficiency is similar for either lap or butt joints. A practical maximum-efficiency joint is so designed that plate tension in the outer row is the limiting factor. The procedure can be outlined as follows.

1.   Determine the allowable load on the main plate at row 1 by subtracting one bolt hole from the gross plate section.
2.   Determine the allowable shear force on one bolt. Use one area if single shear or two areas if double shear.
3.   Determine the allowable bearing force on one bolt.

4. The smaller numerical result from step 2 or 3 is the limiting force on one bolt. Determine the minimum number of bolts required by dividing the allowable plate load (from step 1) by the limiting bolt force.

$$\text{Minimum number of bolts} = \frac{\text{allowable plate load}}{\text{limiting bolt force}}$$

5. Select an appropriate bolt pattern.
6. Check the plate stress at each row in the pattern.
7. Revise the pattern if the stress on any row exceeds that of row 1.
8. Calculate the joint efficiency.

Suggested trial patterns for lap and butt joints of maximum efficiency are given in Figs. 9-16 and 9-17, respectively.[1] The number shown below each pattern is the number of bolts required to develop the allowable plate stress in row 1.

[1] David Singer, *Basic Structural Design,* Pelex Publishers Inc., New York, 1952.

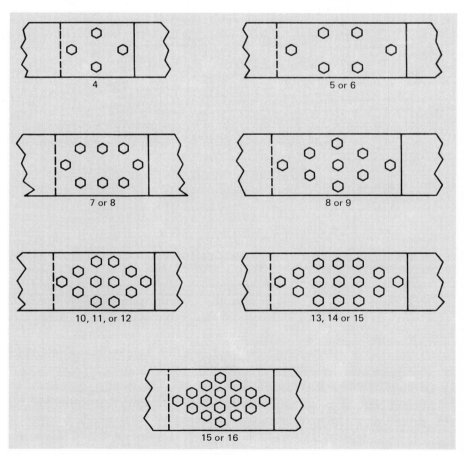

Fig. 9-16 Suggested trial patterns—lap joints.

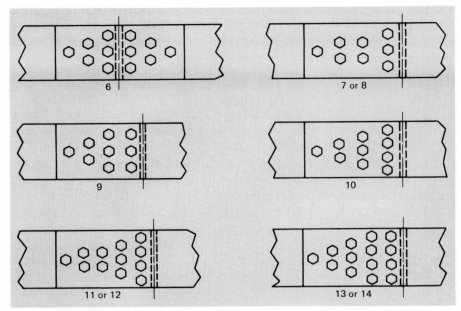

Fig. 9-17 Suggested trial patterns—butt joints.

Sample Problem 5   Two plates (A36) 10 in wide by $\frac{1}{4}$ in thick are to be connected by a lap joint with $\frac{5}{8}$-in-diameter bolts (A325). Design a joint of maximum efficiency by using the AISC code, bearing-type connection. The bolt threads do not occur at the shear plane.

Solution: Allowable load on main plate at row 1 using $\frac{11}{16}$-in-diameter hole:

$$F_t = [(b - nD)t]s_t = [10 - 1(\tfrac{11}{16})](0.25)(29\ 000)$$
$$= 9.31(0.25)(29\ 000) = 67\ 500\ \text{lb}$$

Allowable shear force on one bolt (single shear):

$$F_s = \left[n\left(\frac{\pi d^2}{4}\right) \right] s_s = 1(0.785)(0.625^2)(30\ 000)$$
$$= 0.307(30\ 000) = 9200\ \text{lb}$$

Allowable bearing force on one bolt:

$$F_c = [ntd]s_c = 1(0.25)(0.625)(87\ 000)$$
$$= 13\ 600\ \text{lb}$$

Limiting force on one bolt = 9200 lb.
Minimum number of bolts required:

$$\frac{67\ 500}{9200} = 7.3\ \text{bolts} \qquad \text{Use 8 bolts}$$

Referring to Fig. 9-16, select the trial pattern for seven or eight bolts, as in Fig. 9-18. Check the trial pattern.

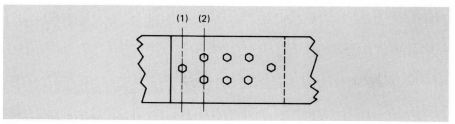

Fig. 9-18 Diagram for Sample Problem 5.

Force transmitted by each bolt:

$$\frac{\text{Allowable plate load}}{\text{Number of bolts}} = \frac{67\ 500}{8} = 8440 \text{ lb}$$

Tension at row 2:

$$\tfrac{7}{8}F_t = [(b - nD)t]s_t = [10 - 2(\tfrac{11}{16})](0.25)(29\ 000)$$
$$= 8.625(0.25)(29\ 000) = 62\ 530$$
$$F_t = \tfrac{8}{7}(62\ 530) = 71\ 500 \text{ lb}$$

The pattern is satisfactory, since the allowable load on the main plate at row 1 is less than on row 2.

The joint efficiency is

$$\eta = \frac{67\ 500}{10(\tfrac{1}{4})(29\ 000)} = \frac{67\ 500}{72\ 500} = 0.931 = 93.1\%$$

9-8 RIVETED JOINTS

The procedures discussed earlier in this chapter for investigating and designing bolted joints may be used for riveted joints as well. Check the appropriate code specifications for design procedures and allowable stresses.

With improvements in welding technology and the availability of economical high-strength bolts, riveting for structures has practically disappeared. Although rivets are still used in lightweight and light-load applications, adhesives and other fastening materials are successfully competing with them.

9-9 WELDED JOINTS

For many purposes, welded connections are found to be more suitable than bolted or riveted connections. Welding is a process of joining two pieces of metal by fusion. In the types of welded joints discussed here, the pieces to be joined are heated, and molten material is deposited between them. Although a great variety of welding processes are in use today, the heating is most frequently accomplished by means of an electric arc. This section will deal with lap and butt joints (called groove joints), although there are several other types of welded joints of some importance. Figure 9-19 shows what is known as a

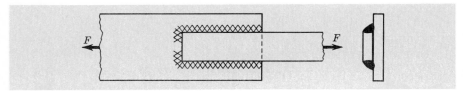

Fig. 9-19 Side fillet welds (with corner returns).

side-fillet lap weld with a corner return. The fused metal is deposited in the angle between the two pieces. Shearing stresses are developed in this weld when either tensile or compressive forces are applied to the long axis of the plates. Figure 9-20 shows a joint using *end* lap welds with corner returns. This type of weld is usually used in conjunction with side welds.

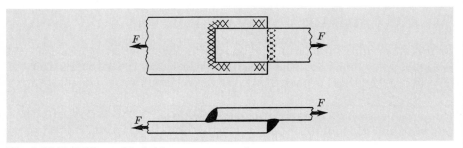

Fig. 9-20 End fillet welds (with corner returns).

Practically all metals can be welded successfully, although special techniques are required for some. Commonly encountered materials, such as cast iron, cast steel, structural steel, tool steel, bronze, aluminum, and copper, are capable of being welded. Figure 9-21 shows various types of butt (groove) welds, most of which require advance preparation of the plate edges at the joint.

The size of fillet welds is designated by the length of the leg, as shown in Fig. 9-22a. Fillet welds resist load on the throat area, which is the minimum cross section of the weld. The throat area is the product of the throat thickness and the effective length of the weld. The throat thickness is determined from Fig. 9-22b as Leg · sin 45° or 0.707 Leg. For 1 in of effective weld length, we can determine the allowable load for welds of given sizes at certain allowable shearing stresses. Thus, for $s_s = 21\ 000$ psi, a $\frac{3}{16}$-in weld with an effective length of 1 in can carry an allowable load of $0.707(\frac{3}{16})(1)(21\ 000) = 2800$ lb. In the same way, allowable loads per inch of effective length can be calculated for various weld sizes.

The AISC code specifies allowable stresses for welds on several types of structural steel by different methods. These specifications conform to Ameri-

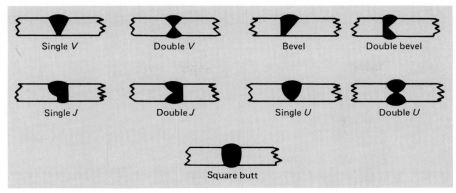

Fig. 9-21 Types of butt (groove) welds.

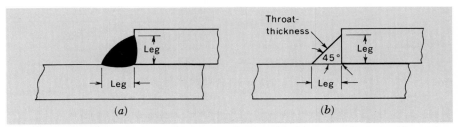

Fig. 9-22 (*a*) Cross section of a fillet weld. (*b*) Simplified cross section of a fillet weld.

can Welding Society (AWS) standards. Both AISC and AWS specifications should be consulted for complete information. Table 9-2 gives appropriate parts of the AISC specifications for the purposes of this book.

Table 9-3 is a list of allowable loads on welds of various sizes for two different levels of allowable shearing stress.

TABLE 9-2 AISC CODE FOR WELDS (MANUAL ARC METHOD)

Fillet Welds	
Type of Steel and Electrode	Allowable Shearing Stress on Throat Area of Weld, psi
A36 steel with Class E 60 series electrodes	18 000
A36, A242, A441, A572, and A588 steels with Class E 70 series electrodes	21 000

Additional provisions: [see footnote at end of table on next page]:*

The effective area of a fillet weld shall be considered as the effective length of the weld times the effective throat thickness.

Maximum size of a fillet weld applied to a square edge of plate shall be $\frac{1}{16}$ in less than the plate thickness, for plate thickness $\frac{1}{4}$ in or more, and equal to the plate thickness for plate thickness less than $\frac{1}{4}$ in.

TABLE 9-2   *(continued)*

Butt (Groove) Welds—Complete Penetration

The allowable stresses permitted for the connected plate material shall apply to complete-penetration butt-weld stresses in tension, compression, bending, and bearing. For shear, use the same allowable shear stress as is used for fillet welds.

The effective area of a butt weld shall be considered as the effective length of the weld times the effective throat thickness.

The effective length of a butt weld shall be the width of the part joined.

The effective throat thickness of a complete-penetration butt weld shall be the thickness of the thinner part joined.

---

* The following provisions may be of interest, but they have not been uniformly applied to the problems in this book. The reader is referred to the *Steel Construction Manual* of the AISC for full specifications.

Minimum effective length of a fillet weld shall not be less than 4 times the nominal size of the weld.

Minimum width of laps on lap joints shall be 5 times the thickness of the thinner part joined and not less than 1 in.

Side or end fillet welds should be returned continuously around the corners for a distance not less than twice the nominal size of the weld.

Lap joints joining plates or bars shall be fillet-welded along the edge of both lapped parts . . . to prevent opening of the joint under maximum loading.

TABLE 9-3   ALLOWABLE LOADS FOR FILLET WELDS

Weld Size (Leg), in	Allowable Load (lb) per Inch of Effective Weld Length	
	For $s_s = 18\,000$ psi	For $s_s = 21\,000$ psi
3/16 (minimum size, AISC)	2 400	2 800
1/4	3 200	3 700
5/16	4 000	4 600
3/8	4 800	5 600
7/16	5 600	6 500
1/2	6 400	7 400
9/16	7 200	8 400
5/8	8 000	9 300
11/16	8 800	10 200
3/4	9 600	11 100

The total allowable force which a given fillet weld can withstand may be expressed by

$$F = F'(L) \tag{9-6}$$

where F = total allowable force on entire weld, lb

F' = allowable force per inch of effective weld length, lb, from Table 9-3

L = total effective length of fillet welding, in

Sample Problem 6   A lap-welded joint of A36 steel is to be designed for a total load of 50 000 lb. The plates are $\frac{3}{8}$ in thick, and $\frac{5}{16}$-in fillet welds will be used. Find the total length of welding required to carry this load if Class E 60 series electrodes are used.

Solution: From Table 9-2, $s_s = 18\ 000$ psi for A36 steel with Class E 60 electrodes. The allowable load per inch of weld $F' = 4000$ lb for $\frac{5}{16}$-in welding at $s_s = 18\ 000$ psi from Table 9-3. By using Eq. (9-6), we have

$$F = F'(L)$$

$$L = \frac{F}{F'} = \frac{50\ 000}{4000} = 12.5 \text{ in of weld required (minimum)}$$

Note: This welding would be placed symmetrically on the joint.

Sample Problem 7   If the plates in the preceding problem are overlapped by 3 in and the narrower plate is 5 in wide, design two possible arrangements for the fillet welding.

Solution: *(a)* Using full side fillets, turned corners, and partial end fillets, the weld pattern would include 3 in along each side (6 in total) + end fillets both top and bottom extending for 6.5/4 = 1.625 = $1\frac{5}{8}$ in from each corner toward the centerline (6.5 in total end fillet welding). This would leave a gap of $5 - 3\frac{1}{4} = 1\frac{3}{4}$ in on each end which would not be welded. See Fig. 9-23a.

(b) Using full end fillets, turned corners, and partial side fillets, the weld pattern would include 5 in along each end (10 in total) + side fillets extending for $\frac{3}{8}$ in from each corner along the sides ($2\frac{1}{2}$ in total side fillet welding). This

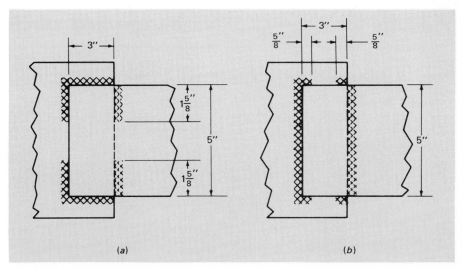

(a)   (b)

Fig. 9-23 (*a*) Diagram for Sample Problem 7, solution *a*. (*b*) Diagram for Sample Problem 7, solution *b*.

would leave an unwelded gap of $3 - 1\frac{1}{4} = 1\frac{3}{4}$ in on each side (see Fig. 9-23b). Note that minimum corner returns of twice the weld size are provided with this arrangement.

9-10   THIN-WALLED PRESSURE VESSELS

A thin-walled pressure vessel has a diameter at least 10 times its wall thickness. When a thin shell, such as a boiler drum, is subjected to steam pressure or water pipes are subjected to the pressure of the water, the forces tending to rupture the vessel act normally (perpendicular) to the surface of the shell. Internal stresses are developed in the metal of the walls, and they must not exceed the safe working stress of the metal. These internal pressures tend to rupture the vessel in two ways: first, along seams parallel to the elements of the shell and, second, along a seam corresponding to a circumference of the shell. Figure 9-24 shows a longitudinal section of a cylindrical shell. The internal pressure p acts normal to the curved surface of the cylinder.

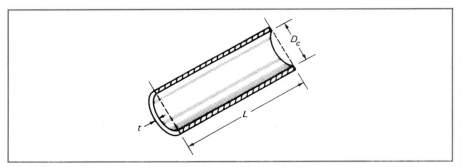

Fig. 9-24 Longitudinal section of a thin-walled cylinder.

Longitudinal Section: Figure 9-25a shows a free-body diagram with the pressure acting normal to the curved surface and held in equilibrium by the resistance of the material in the longitudinal section. The pressure distributes

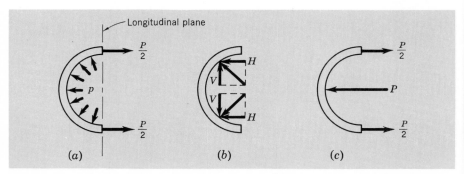

Fig. 9-25 (a) Free-body diagram of half of cylinder. (b) Pressure vectors and components. (c) Resultant pressure force.

uniformly on the curved surface, but only components which are perpendicular to the longitudinal plane tend to cause failure. Figure 9-25b shows two pressure vectors and their components. Note that the two vertical components V act in opposite directions, thus canceling each other, and both horizontal components act in the same direction. Therefore, the total pressure force on the curved surface may be represented by a single force P, as shown in Fig. 9-25c, which is the sum of all H components. The force P is calculated by multiplying the pressure p and the area of the curved surface *projected on the longitudinal plane*. The projected area in Fig. 9-24 is $D_c \times L$. Therefore,

$$P = pD_cL$$

The resistance of the material in the longitudinal section holds the pressure force in equilibrium. Assuming the resistance is shared equally by both sections in the longitudinal plane (see Fig. 9-25c), from force equals area times stress, the total resistance is

$$2\left(\frac{P}{2}\right) = (2Lt)s_t \qquad \text{or} \qquad P = (2Lt)s_t$$

where $2Lt$ = total resistive area, in²; mm²
$\qquad s_t$ = tensile stress in material, psi; N/mm² = MPa

Equating the resistance to the pressure force (for equilibrium) gives

$$2Lts_t = pD_cL \qquad \text{or} \qquad s_t = \frac{pD_c}{2t} \tag{9-7}$$

where s_t = average tensile stress in material on longitudinal section, psi; N/mm² = MPa
$\quad p$ = internal pressure in cylinder, psi; N/mm² = MPa
$\quad D_c$ = inside cylinder diameter, in; mm
$\quad t$ = thickness of plate material, in; mm

Note that the pressure p is actually the difference between the absolute internal pressure and the absolute external pressure. Usually, the external pressure is atmospheric. In such cases, the correct value for p is the internal *gage* pressure.

It is also of interest to determine the force which a given length of longitudinal joint must resist. This relation has application in designing bolted joints and lap welds for longitudinal seams on cylindrical pressure vessels. The total force which the material in the longitudinal section must withstand is

$$P = pD_cl$$

Since each of the two areas cross-hatched in Fig. 9-26 is assumed to carry one-half the total force P, the force F on *each* seam is

$$F = \frac{pD_cl}{2} \tag{9-8}$$

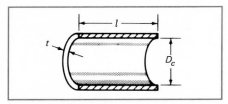

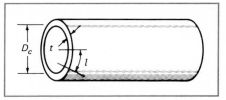

Fig. 9-26 Dimensions of longitudinal section of cylinder.

Fig. 9-27 Transverse section of a thin-walled cylinder.

where F = force on one longitudinal seam of length l, lb; N
p = internal pressure in cylinder, psi; N/mm² = MPa
D_c = inside cylinder diameter, in; mm
l = length of seam (not necessarily entire length), in; mm

Transverse Section: The stresses which exist in the material on a transverse section (the transverse plane is perpendicular to the axis of the cylinder) may be analyzed from Figs. 9-27 and 9-28. In Fig. 9-28 the internal pressure p is shown acting against the closed circular end of the cylinder. The total pressure force P, which acts to the right, is

$$P = pA \qquad \text{or} \qquad P = p\left(\frac{\pi D_c^2}{4}\right)$$

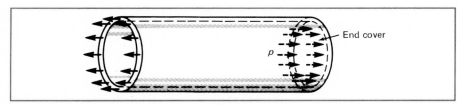

Fig. 9-28 Free-body diagram of transverse section.

This force must be balanced by resistive forces in the thin ring of metal on the transverse section. The area of metal in this ring is approximately equal to its circumference πD_c times its thickness t. Therefore, the total resistive force will equal P, and this will also equal the tensile stress in the transverse section s_t times the metal area, or

$$P = (\pi D_c t)s_t$$

Equating the resistive force to the pressure force (for equilibrium) gives

$$\pi D_c t s_t = p\left(\frac{\pi D_c^2}{4}\right)$$

$$s_t = \frac{pD_c}{4t} \tag{9-9}$$

where s_t = average tensile stress in material on transverse section, psi; $N/mm^2 = MPa$

p = internal pressure in cylinder, psi; $N/mm^2 = MPa$

D_c = inside cylinder diameter, in; mm

t = thickness of plate, in; mm

Again, it may be of interest to find the force acting on a given length l of transverse joint (Fig. 9-27). The force on this length of joint is given by

$$F = \frac{pD_c l}{4} \qquad (9\text{-}10)$$

where F = force on a transverse seam of length l, lb; N

l = length of seam (not necessarily entire circumference), in; mm

D_c = inside cylinder diameter, in; mm

The derivation of Eq. (9-10) is left as an exercise for the reader. It should be noted that Eqs. (9-9) and (9-10) apply to any cross section through the center of a spherical pressure vessel.

*Sample Problem 8   What water pressure will burst a 400-mm-diameter ASTM Class 60 cast-iron water pipe if the wall thickness is 16 mm?

Solution: From Table 1A, App. B, the ultimate tensile strength of Class 60 cast iron is 410 MPa. In a pipe, failure due to internal pressure tends to occur on longitudinal rather than on transverse sections. This is evident from a comparison of Eqs. (9-7) and (9-9). Therefore, by Eq. (9-7) for a longitudinal section,

$$s_t = \frac{pD_c}{2t} \qquad \text{or} \qquad p = \frac{2ts_t}{D_c}$$

$$p = \frac{2(16)(410)}{400} = 32.8 \text{ N/mm}^2 = 32.8 \text{ MPa}$$

*Sample Problem 9   A 3.5-m-diameter sphere of 6061-T6 aluminum alloy is to contain nitrogen gas under a pressure of 6.0 MPa. The wall thickness is 20 mm. A minimum safety factor of 5 (based on ultimate) is specified. Will this spherical tank meet specifications?

Solution: From the discussion in Sec. 9-10, Eq. (9-9) applies to the sphere. The stress in the sphere on any transverse section is given by

$$s_t = \frac{pD_c}{4t}$$

$$= \frac{6.0(3500)}{4(20)} = 263 \text{ MPa}$$

The factor of safety is

$$N_u = \frac{\text{ultimate stress}}{\text{actual stress}}$$

The ultimate stress is 310 MPa from Table 1A, App. B.

$$N_u = \frac{310}{263} = 1.18$$

This tank does not meet specifications. There are several alternatives which could produce a satisfactory solution. They include *(a)* increasing the wall thickness *(b)* reducing the tank diameter, and *(c)* changing to a stronger material.

Sample Problem 10   A cylindrical tank 3 ft in diameter and 10 ft long is to contain compressed air. The tank is made of $\frac{3}{4}$-in steel plate with an allowable tensile strength of 12 600 psi. What is the maximum safe pressure in the tank if the plate is to be welded as follows:

(a)   A single bead of lap (fillet) welding (Fig. 9-29); allowable shear strength = 18 000 psi
(b)   A complete-penetration butt weld
(c)   A double bead of fillet welding (Fig. 9-30); allowable shear strength = 18 000 psi

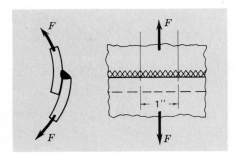

Fig. 9-29 Diagram for Sample Problem 10*a*.

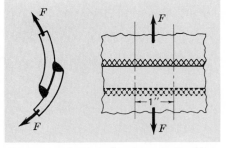

Fig. 9-30 Diagram for Sample Problem 10*c*.

Solution a: The tank plates are $\frac{3}{4}$ in thick; hence, a fillet weld with an $\frac{11}{16}$-in leg may be placed (see Table 9-2). Assuming a 1-in length of joint, Equation (9-6) can be adapted to give

$$F = F'(L) = 8800(1) = 8800 \text{ lb (maximum safe force on a 1-in length of weld)}$$

The stress in the plate material due to this tensile force is

$$s_t = \frac{F}{A} = \frac{F}{tl} = \frac{8800}{0.75(1)} = 11\ 730 \text{ psi}$$

The maximum safe pressure in the tank is determined from Eq. (9-7).

$$S_t = \frac{pD_c}{2t}$$

$$p = \frac{2ts_t}{D_c}$$

$$= \frac{2(0.75)(11\ 730)}{36} = 489 \text{ psi (maximum safe pressure in tank)}$$

or, from Eq. (9-8),

$$p = \frac{2F}{D_cl} = \frac{2(8800)}{36(1)} = 489 \text{ psi}$$

Solution b: The butt-welded joint is allowed to be stressed in tension equal to the allowable plate stress. Thus, $s_t = 12\ 600$ psi. From Eq. (9-7) the pressure is

$$p = \frac{2ts_t}{D_c} = \frac{2(0.75)(12\ 600)}{36} = 525 \text{ psi (maximum safe pressure in tank)}$$

Solution c: The double-bead weld will withstand twice the force which the single-bead weld of part a resists. Thus, $F = 17\ 600$ lb is the maximum safe force on two 1-in lengths of weld. However, the stress in the plate material is

$$s_t = \frac{F}{tl} = \frac{17\ 600}{0.75(1)} = 23\ 500 \text{ psi}$$

which exceeds the allowable tensile stress of 12 600 psi. Consequently, the value of 12 600 psi must be used as the plate stress. This gives a safe tank pressure of 525 psi, as in part b. In this case, the weld is not being used to its full load-carrying capacity.

Sample Problem 11 An open cylindrical water tank is 25 ft high and 40 ft in diameter. What plate thickness must be used if the longitudinal joints are to be fillet-welded with a single bead? The AISC code applies. Plate material is A242 steel welded with Class E 70 series electrodes.

Solution: Assume the tank to be filled with water to the full 25-ft height. The pressure at a given depth in a fluid is equal to the depth times the fluid density, or

$$p = \frac{h\rho}{144} \tag{9-11}$$

where p = fluid pressure, psi

h = depth from top surface, ft

ρ = fluid density, lb/ft³

The density of water may be taken as 62.4 lb/ft³ and considered to remain constant. From Eq. (9-11) it is evident that the greatest pressure occurs at the greatest depth of water. Thus, the lower end of the longitudinal (vertical) weld is subjected to the largest force. Consider the last inch of longitudinal weld at the bottom of the tank. The water pressure at the bottom is

$$p = \frac{hp}{144} = \frac{25(62.4)}{144} = 10.83 \text{ psi}$$

The tensile force on 1 in of longitudinal section near the bottom is

$$F = \frac{pD_c l}{2} = \frac{10.83(40)(12)(1)}{2} = 2600 \text{ lb}$$

The required height of leg of a 1-in-long fillet weld can be found from Table 9-3. To resist 2600 lb, a $\frac{3}{16}$-in weld leg is required; thus, $\frac{3}{16}$-in-thick plates are needed. (The AISC code specifies that weld size be equal to plate thickness for plate thickness less than $\frac{1}{4}$ in.) Check the stress in the plate material against the AISC allowable tensile stress of 35 000 psi (Table 9-1).

$$s_t = \frac{F}{A} = \frac{F}{tl} = \frac{2600}{\frac{3}{16}(1)} = 13\ 900 \text{ psi}$$

Use $\frac{3}{16}$-in plates.

PROBLEMS

9-1. A double-bolted lap joint is made by connecting two 4- by $\frac{1}{2}$-in A36 steel plates with two $\frac{3}{4}$-in A325 high-strength bolts (threads do not occur at the shear plane). What safe load will the joint carry? Use the AISC code, bearing-type connection.

9-2. Two A441 steel plates 5 in wide and $\frac{1}{2}$ in thick are connected as a lap joint by three $\frac{5}{8}$-in A490 steel bolts (threads do not occur at the shear plane) arranged in one row. Find the strength of the joint by using the AISC code, bearing-type connection.

9-3. Two A36 plates, each 10 in wide and $\frac{1}{2}$ in thick, are connected by $\frac{3}{4}$-in A325 steel bolts (threads occur at the shear plane) to form a lap joint. Based on shear, how many bolts will be needed if the pull on the joint is 15 000 lb? Use the AISC code, friction-type connection.

9-4. Two plates forming a lap joint carrying 5000 lb in tension are connected by two A307 steel bolts. What should be the diameter of the bolts based on the AISC code? Consider shear only.

9-5. Two 12-in-wide by $\frac{3}{4}$-in-thick A36 steel plates are joined by a single-bolted butt joint with $\frac{1}{2}$-in cover plates. The plates are connected by four $\frac{7}{8}$-in A490 steel bolts on each side. Find the strength of the joint by using the AISC code. The bolt threads do not occur at the shear plane. Assume friction-type connection.

9-6. Two A36 steel plates 2 in wide and $\frac{3}{8}$ in thick are formed into a double-bolted butt joint with two $\frac{5}{16}$-in cover plates. There are two $\frac{1}{2}$-in A307 steel bolts on each side. If the load is 15 600 lb, find the shear and bearing stresses and the maximum tensile stress in the plate. Use the AISC code to determine the hole size, and compare the calculated stresses with the stresses allowed by the code.

9-7. Figure Problem 9-7 shows the lower joint of a roof truss. Eleven $\frac{5}{8}$-in A307 steel bolts are used to fasten member A to the $\frac{1}{2}$-in gusset plate. The tensile force in member A is 56 000 lb. Is the bolting safe? What may be the compressive force on member B, on the assumption that the bolts are used to their full capacity? Use the AISC code. The angles and plate are of A36 steel.

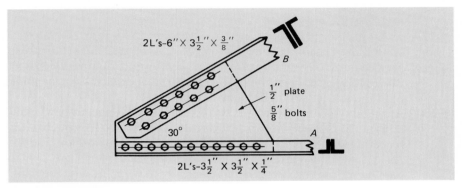

FIGURE PROBLEM 9-7

9-8. In a butt joint connecting $\frac{3}{8}$-in A242 steel plates, how many $\frac{3}{4}$-in A325 steel bolts (threads do not occur at the shear plane) are needed to carry 80 000 lb safely? What would be the number of bolts for a lap joint? Use the AISC code, bearing-type connection.

9-9. The joint at a point in the lower chord of a roof truss is shown in Fig. Prob. 9-9. If the compression in member A is 18 000 lb and the tension in member B is 14 000 lb, how many $\frac{5}{8}$-in bolts (bearing-type connection) are needed to fasten these members to the $\frac{3}{8}$-in gusset plate? Use

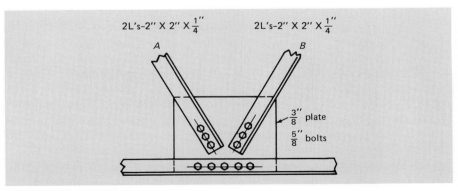

FIGURE PROBLEM 9-9

the AISC code. The angles and gusset plate are of A36 steel. The bolts are of A325 steel with the stress developing at the threaded portion of the shank.

9-10.   By using the AISC code, find the strength and efficiency for a repeating section of a single-bolted lap joint (friction-type connection) if the A242 steel plates are $\frac{1}{2}$ in thick and the $\frac{3}{8}$-in A490 steel bolts are placed four bolt diameters apart.

9-11.   Determine the strength and efficiency of a double-bolted two-cover-plate butt joint. The A36 plates are 10 in wide and $\frac{3}{4}$ in thick. The A490 steel bolts (threads do not occur at the shear plane) are 1 in in diameter. There is one bolt in the first row, and there are three bolts in the second row. Use the AISC code, bearing-type connection.

9-12.   Two $\frac{3}{4}$-in-thick plates are joined by a lap joint. The A588 steel plates are 12 in wide. The joint contains a total of 12 A490 steel bolts (threads do not occur at the shear plane) arranged in four rows of three bolts each. The bolts are $\frac{3}{8}$ in in diameter. Use the AISC code, friction-type joint. Find the strength and efficiency of the joint.

9-13.   A triple-bolted butt joint with two cover plates has one, two, and four bolts in the first, second, and third rows, respectively. The pitch of the repeating section is 10 in. The A36 steel plates are $\frac{3}{8}$ in thick, and the A325 steel bolts (threads do not occur at shear plane) are $\frac{3}{4}$ in in diameter. Find the strength and efficiency of the joint. Use the AISC code, bearing-type connection.

9-14.   Two A36 steel plates $\frac{1}{2}$ in thick are connected by a triple-bolted two-cover-plate butt joint (bearing-type connection). The bolts are $\frac{3}{8}$ in in diameter, and the length of the repeating section is 12 in. There is one bolt in the first row; there are three bolts in the second row; and there are six bolts in the third row. Find the strength and efficiency of the joint. Use the AISC code. The bolts are A325 steel, and the stress develops at the unthreaded portion of the shank.

9-15.   In Fig. 9-19 the narrow plate is 6 by $\frac{1}{2}$ in and resists a pull of 16 000 lb. The plate material is A36 steel; the electrodes are Class E 60 series; and the weld size is $\frac{7}{16}$ in (see Sec. 9-9).

   a.   What must be the length of each side weld?

   b.   Could a single end weld resist the pull?

9-16.   In a lap weld, the narrow plate is 6 by $\frac{1}{2}$ in and is subjected to a pull of 60 000 lb. What must be the length of each $\frac{7}{16}$-in side weld if a total of 4 in of end weld is also to be used? The plates are A242 steel, and Class E 70 series electrodes are used.

9-17.   a.   Two 5- by $\frac{1}{4}$-in plates are joined by top and bottom end fillet welding. What allowable load will the $\frac{3}{16}$-in weld carry? Check the plate strength in tension. The plate material is A36 steel, and Class E 60 series electrodes will be used.

   b.   What safe load would the joint carry if the plates were butt-welded?

9-18.   a.   The narrower plate in Fig. 9-20 is 6 by $\frac{1}{2}$ in, and the plates are to be top and bottom end fillet-welded. Find the safe tensile load this joint can carry if the plates are A441 steel and Class E 70 series electrodes develop $\frac{7}{16}$-in welds.

b. What length of plate overlap is required to carry the same load if only side fillet welds are used?

9-19. The plates in Fig. Prob. 9-19 are to be lap-welded. The thickness of each plate is $\frac{5}{16}$ in and the weld size is $\frac{1}{4}$ in. The plates are A36 steel and Class E 70 series electrodes will be used.

    a. What overlap is required to carry a load of $P = 32\,000$ lb if fillet welding is used on sides A and C only?

    b. What maximum safe load can be carried if an overlap of 12 in is used and sides A, B, C, and D are fillet-welded for the full length?

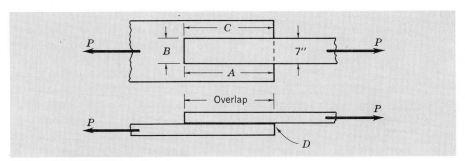

FIGURE PROBLEM 9-19

9-20. **a.** Two A441 steel plates $3\frac{1}{2}$ in wide are to carry a load of 25 000 lb. They are to be top and bottom end fillet-welded with Class E 70 series electrodes. What is the required thickness of the plate?

    b. Assuming a complete-penetration butt weld, what would be the required thickness?

    c. Assuming 1-in A307 steel bolts are to be used, find how many would be required in shear.

9-21. A cylindrical pressure vessel is subjected to an internal pressure of 70 psi. It is 7 ft in diameter, and the plates are $\frac{3}{8}$ in thick.

    a. Find the maximum stress.

    b. If it is made of Class 40 cast iron, is it safe? Assume steady-stress conditions.

*9-22. What should be the thickness of the plate for a 254-mm steam pipe to resist a pressure of 550 kPa? The stress in the pipe should not exceed 62 MPa.

9-23. What safe internal pressure should be used for a steel water pipe that is 6 in in diameter and has a $\frac{1}{2}$-in shell if the allowable stress is 11 000 psi?

*9-24. Nitrogen gas is to be placed in a sphere 4.5 m in diameter under 800-kPa pressure. If the material is aluminum alloy 6061-T6, what thickness of plate will be required? Assume varying load conditions.

9-25. If an 18-in Class 20 cast-iron vertical pipe is subjected to a head of

140 ft of water, what thickness of plate will be required? Assume varying stress conditions.

*9-26.   A tank 2.4 m in diameter and 1.8 m deep is hooped with 12-mm round wrought-iron rods. What must be the spacing of the rods if the tank contains water? (*Note:* Mass density of water = 1000 kg/m^3.)

9-27.   Find the necessary spacing of wrought-iron hoops $\frac{3}{4}$ in in diameter for a wood-stave water tank 20 ft in diameter and 16 ft deep. Assume s_t = 10 000 psi for wrought iron.

9-28.   A butt-welded A36 steel water pipe 2 ft in diameter is made of $\frac{3}{8}$-in thick plate. What is the allowable pressure?

9-29.   A butt-welded spherical gas container made of $\frac{3}{8}$-in A242 steel plates must maintain gas at a pressure of 100 psi. Determine the maximum permissible diameter of the sphere.

9-30.   A cylindrical oil tank is 100 ft in diameter and 28 ft high. If butt-welded $\frac{7}{8}$-in A36 steel plates are used, will the tank be safe? (Density of the oil is 56 lb/ft^3.)

*9-31.   A cylindrical water tank (axis vertical) has a longitudinal double-V butt weld. The tank diameter is 6.5 m. The plates are 10 mm thick. The allowable tensile stress on the weld is 110 MPa. What maximum height of water can be safely maintained in this tank? (*Note:* Mass density of water = 1000 kg/m^3.)

9-32.   A pipe 2 ft in diameter is constructed from $\frac{1}{4}$-in A242 steel plates. What is the allowable working pressure on the assumption that longitudinal complete-penetration butt-welded joints are used?

9-33.   **a.**   A gas tank 3 ft in diameter is to be operated under a pressure of 250 psi. What will be the thickness of A36 steel plate if the longitudinal seam is lap-welded with Class E 70 series electrodes?

   b.   If the seams are butt-welded, what will be the thickness?

9-34.   A boiler 3 ft in diameter is made of A36 steel plates $\frac{1}{4}$ in thick.

   a.   If a single-bead lap weld is used, find the safe internal pressure if the $\frac{3}{16}$-in weld has an allowable shear stress of 18 000 psi.

   b.   If a butt weld is used, find the safe internal pressure.

9-35.   A water tank 18 ft in diameter is made of $\frac{3}{8}$-in A36 steel plates. What will be the safe height of water in the tank if:

   a.   A single-bead lap weld is made by using Class E 60 series electrodes?

   b.   A butt-welded joint is used?

*9-36.   A seamless pipe 600 mm in diameter is to be used to manufacture a pressure vessel designed to carry steam under a pressure of 2 MPa. The vessel is made up of a continuous length of pipe with end covers that are butt-welded to the pipe on transverse sections. The pipe material has an ultimate tensile strength of 900 MPa, and a safety factor (based on ultimate) of 5 is specified. The allowable stress for the butt weld is 110 MPa. What should the wall thickness of the pipe be?

10

Center of Gravity, Centroids, and Moment of Inertia

10-1 CENTER OF GRAVITY OF A BODY

Every part of a body possesses weight. Weight is the force of attraction between a body and the earth, and it is proportional to mass of the body. The weights of all parts of a body can be considered as parallel forces directed toward the center of the earth. Therefore, they may be combined into a resultant force whose magnitude is equal to their algebraic sum. If a supporting force, equal and opposite to the resultant, is applied to the body along the line of action of the resultant, the body will be in equilibrium. This line of action will pass through the center of gravity of the body.

The center of gravity of some objects can be found by balancing the object on a point. Take a thin plate of the thickness t, shown in Fig. 10-1. Draw the diagonals of the upper and lower faces to intersect at J and K, respectively. If the plate is placed on a pivot at K, the plate will not fall. That is, it is balanced. If suspended from J, the plate will hang horizontally. The center of gravity of the plate is at the center of the line JK.

If we suspend a uniform rod by a string (Fig. 10-2) and move the position of the string until the rod hangs horizontally, we can determine that the center of gravity of the rod lies at the center. Through the use of similar procedures it can be established that a body which has an axis, or line, of symmetry has its center of gravity located on that line, or axis. Of course, if a body has more than one axis of symmetry, the center of gravity must lie at the intersection of the axes.

201

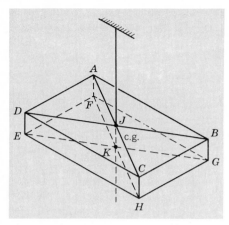

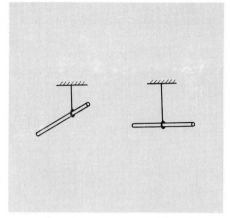

Fig. 10-1 Locating the center of gravity of a plate.

Fig. 10-2 Locating the center of gravity of a rod.

10-2 CENTER OF GRAVITY OF AREA — CENTROID

Strictly speaking, there is no center of gravity of an area, for an area does not have weight. The point J at the intersection of the diagonals of the face $ABCD$ of Fig. 10-1 is the center of area of a rectangle. Because of its relation to the center of gravity of the plate, it is spoken of as the center of gravity, or *centroid,* of the rectangle.

The center of gravity of the area of the cross section of an irregular body can be found by a simple experiment. Cut a piece of cardboard to the shape of the cross section and find the balancing point. This will give the position of the center of gravity for all practical purposes.

Another method for determining the center of gravity is by suspension. Take an object, the section of which is shown in Fig. 10-3. Suspend it from some point A. The body will not come to rest until its resultant weight is vertically downward from A. Through A, draw a vertical line AC. Then suspend the body

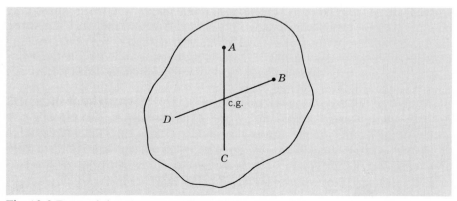

Fig. 10-3 Determining the center of gravity by suspension.

from a point B, and let it come to rest. Through B, draw a vertical line BD. The point of intersection of AC and BD is the position of the center of gravity, or centroid.

The centroids, or centers of gravity, for some simple areas are given in Table 10-1. The horizontal and vertical distances to the centroid are indicated by $\bar{x}$ (read as x bar) and $\bar{y}$ (y bar).

TABLE 10-1   CENTROIDS OF SIMPLE AREAS

Shape	Area	$\bar{x}$	$\bar{y}$	
Rectangle	bh	$\dfrac{b}{2}$	$\dfrac{h}{2}$	
Triangle	$\dfrac{bh}{2}$	$\dfrac{b}{3}$	$\dfrac{h}{3}$	
Circle	$\dfrac{\pi d^2}{4}$	$\dfrac{d}{2}$	$\dfrac{d}{2}$	
Semicircle	$\dfrac{\pi d^2}{8}$	$\dfrac{d}{2}$	$\dfrac{4r}{3\pi}$ $(0.425r)$	
Quadrant	$\dfrac{\pi d^2}{16}$	$0.425r$	$0.425r$	

10-3 MOMENT OF AN AREA

In order to determine the location of the centroid of a plane figure, the term *moment of an area* must be understood. Imagine a force (perpendicular to the page) of as many pounds as there are square inches in the area applied at the centroid of the rectangle in Fig. 10-4. Then the moment of the rectangular area about the y axis will be the product of the area and the distance from the centroid to the y axis. That is,

$$M_y = A(\bar{x})$$

In a like manner, the moment of the area about the x axis will be the product of the area and the distance from the centroid to the x axis. That is,

$$M_x = A(\bar{y})$$

Sample Problem 1 Determine the moment of the triangular area shown about the x axis and the y axis, Fig. 10-5.

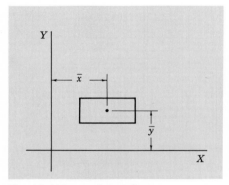

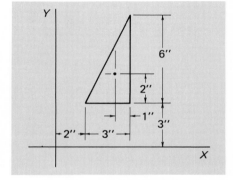

Fig. 10-4 Determining the moment of an area.

Fig. 10-5 Diagram for Sample Problem 1.

Solution:

$$\text{Area} = \frac{3(6)}{2} = 9 \text{ in}^2$$

The distance from the centroid to the horizontal base of the triangle is $h/3 = \frac{6}{3} = 2$ in. Therefore, the distance from the centroid to the x axis is $2 + 3 = 5$ in. Then

$$M_x = A(\bar{y}) = 9(5) = 45 \text{ in}^3$$

The distance from the centroid to the vertical edge of the triangle is $b/3 = \frac{3}{3} = 1$ in. Therefore, the distance from the centroid to the y axis is $2 + 2 = 4$. Then

$$M_y = A(\bar{x}) = 9(4) = 36 \text{ in}^3$$

10-4   CENTROIDS OF COMPOSITE AREAS

The location of the centroid of a plane figure can be thought of as the average distance of the area to an axis. Usually, the axes involved will be the x and y axes.

In determining the location of the centroid, the reader will find it advantageous to place the x axis through the lowest point and the y axis through the left edge of the figure. This places the plane area entirely within the first quadrant, where x and y distances are positive (Fig. 10-6). Then divide the area into simple areas such as rectangles and triangles (Fig. 10-7). Take the moment of each

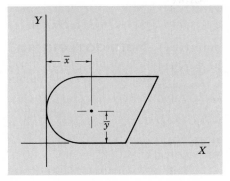

Fig. 10-6 Centroid of a composite area.

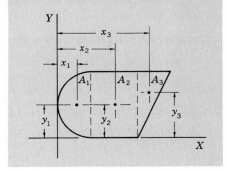

Fig. 10-7 Composite area divided into simple areas.

simple area about the x axis. Sum up the moments about the x axis. Since the centroid of the composite figure is the point at which the entire area is assumed to be concentrated, the moment of the entire area about the x axis must be equal to the sum of the moments of its component parts about the x axis. That is,

$$(A_1 + A_2 + \cdots + A_n)\bar{y} = A_1 y_1 + A_2 y_2 + \cdots + A_n y_n$$

Therefore,
$$\bar{y} = \frac{A_1 y_1 + A_2 y_2 + \cdots + A_n y_n}{A_1 + A_2 + \cdots + A_n}$$

or
$$\bar{y} = \frac{\Sigma A y}{\Sigma A} \qquad (10\text{-}1)$$

Following the same procedure for moments about the y axis gives us

$$\bar{x} = \frac{A_1 x_1 + A_2 x_2 + \cdots A_n x_n}{A_1 + A_2 + \cdots + A_n}$$

or
$$\bar{x} = \frac{\Sigma A x}{\Sigma A} \qquad (10\text{-}2)$$

If a hole in the plane figure exists, treat it as a negative area. The moment of a negative area will be negative, provided that the entire figure lies in the first quadrant.

*Sample Problem 2   Locate the centroid of the cross section of an angle 150 by 100 by 12 mm (Fig. 10-8).

Solution: Divide the figure into two rectangles, 1 and 2.

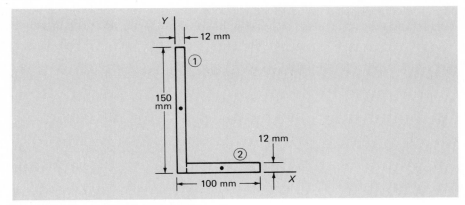

Fig. 10-8 Diagram for Sample Problem 2.

$$A_1 = 150(12) = 1800 \text{ mm}^2$$
$$A_2 = 88(12) = \underline{1056} \text{ mm}^2$$
$$\Sigma A = 2856 \text{ mm}^2$$

By using Eqs. (10-1) and (10-2), we have

$$\bar{x} = \frac{1800(6) + 1056(56)}{2856} = \frac{10\ 800 + 59\ 140}{2856} = \frac{69\ 940}{2856} = 24.5 \text{ mm}$$

$$\bar{y} = \frac{1800(75) + 1056(6)}{2856} = \frac{135\ 000 + 6340}{2856} = \frac{141\ 340}{2856} = 49.5 \text{ mm}$$

Some readers may prefer to systematize the procedure as follows:

Area	Dimen.	A	x	Ax	y	Ay
1	150 × 12	1 800	6	10 800	75	135 000
2	88 × 12	1 056	56	59 140	6	6 340
		$\Sigma A = 2\ 856$		$\Sigma Ax = 69\ 940$		$\Sigma Ay = 141\ 340$

$$\bar{x} = \frac{\Sigma Ax}{\Sigma A} = \frac{69\ 940}{2856} = 24.5 \text{ mm}$$

$$\bar{y} = \frac{\Sigma Ay}{\Sigma A} = \frac{141\ 340}{2856} = 49.5 \text{ mm}$$

Sample Problem 3[1]   Locate the centroid of the piece of sheet metal shown in Fig. 10-9.

Solution: Divide the figure into three simple areas.

[1] See also Sample Computer Program 14, Chap. 17.

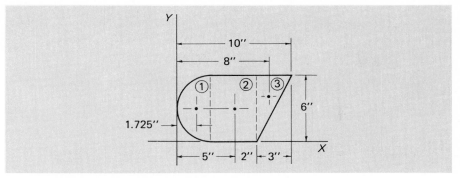

Fig. 10-9 Diagram for Sample Problem 3, $\bar{x}$ calculation.

1. Semicircle:

$$A_1 = \frac{\pi d^2}{8} = \frac{\pi(6^2)}{8} = 14.14 \text{ in}^2$$

2. Rectangle:

$$A_2 = bh = 4(6) = 24.00 \text{ in}^2$$

3. Triangle:

$$A_3 = \frac{bh}{2} = \frac{3(6)}{2} = 9.0 \text{ in}^2$$

$$\Sigma A = 47.14 \text{ in}^2$$

Taking moments of areas about the y axis (Fig. 10-9) gives us

$$\bar{x} = \frac{14.14(1.725) + 24(5) + 9(8)}{47.14}$$

$$= \frac{24.4 + 120 + 72}{47.14}$$

$$= \frac{216.4}{47.14} = 4.59 \text{ in} \quad \text{say, } 4.6 \text{ in}$$

Taking moments about the x axis (Fig. 10-10) gives us

$$\bar{y} = \frac{14.14(3) + 24(3) + 9(4)}{47.14}$$

$$= \frac{42.42 + 72 + 36}{47.14}$$

$$= \frac{150.42}{47.14} = 3.19 \text{ in} \quad \text{say, } 3.2 \text{ in}$$

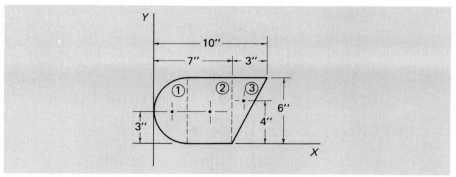

Fig. 10-10 Diagram for Sample Problem 3, $\bar{y}$ calculation.

Area	Dimen.	A	x	Ax	y	Ay
1	$d = 6$	14.14	1.73	24.4	3.0	42.42
2	6×4	24.0	5.0	120.0	3.0	72.0
3	$b = 3$	9.0	8.0	72.0	4.0	36.0
	$h = 6$					
		$\Sigma A = \overline{47.14}$		$\Sigma Ax = \overline{216.4}$		$\Sigma Ay = \overline{150.42}$

$$\bar{x} = \frac{\Sigma Ax}{\Sigma A} = \frac{216.4}{47.14} = 4.59 \text{ in} \quad \text{say, } 4.6 \text{ in}$$

$$\bar{y} = \frac{\Sigma Ay}{\Sigma A} = \frac{150.42}{47.14} = 3.19 \text{ in} \quad \text{say, } 3.2 \text{ in}$$

*Sample Problem 4   Determine the location of the centroid of the plane figure shown in Fig. 10-11.

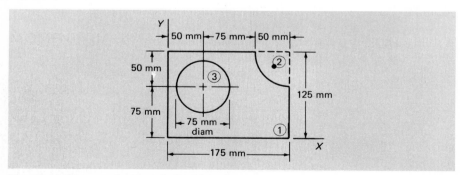

Fig. 10-11 Diagram for Sample Problem 4.

Solution: Divide the composite figure into three simple areas: a rectangle 175 by 125 mm, a quadrant with a 50-mm radius, and a circle with a 75-mm diameter. The rectangle is a positive area. The quadrant and hole are treated as negative areas.

$$A_1 = bh = 175(125) = 21\ 880\ \text{mm}^2$$

$$A_2 = -\frac{\pi d^2}{16} = \frac{-\pi(100^2)}{16} = -1963\ \text{mm}^2$$

$$A_3 = \frac{-\pi d^2}{4} = \frac{-\pi(75^2)}{4} = -4418\ \text{mm}^2$$

$$\Sigma A = 15\ 500\ \text{mm}^2$$

$$\bar{x} = \frac{\Sigma Ax}{\Sigma A} = \frac{21\ 880(87.5) - 1963(153.75) - 4418(50)}{15\ 500}$$

$$\bar{x} = \frac{1915(10^3) - 302(10^3) - 221(10^3)}{15\ 500} = \frac{1392(10^3)}{15.5(10^3)} = 89.8\ \text{mm}$$

$$\bar{y} = \frac{\Sigma Ay}{\Sigma A} = \frac{21\ 880(62.5) - 1963(103.75) - 4418(75)}{15\ 500}$$

$$\bar{y} = \frac{1368(10^3) - 204(10^3) - 331(10^3)}{15\ 500} = \frac{833(10^3)}{15.5(10^3)} = 53.7\ \text{mm}$$

or:

Area	Dimen.	A	x	Ax	y	Ay
1	175×125	21 880	87.5	1 915(10³)	62.5	1 368(10³)
2	$r = 50$	$-1\ 963$	153.75	$-302(10^3)$	103.75	$-204(10^3)$
3	$d = 75$	$-4\ 418$	50	$-221(10^3)$	75	$-331(10^3)$
		$\Sigma A = 15\ 500$		$\Sigma Ax = 1\ 392(10^3)$		$\Sigma Ay = 833(10^3)$

$$\bar{x} = \frac{\Sigma Ax}{\Sigma A} = \frac{1392(10^3)}{15.5(10^3)} = 89.8\ \text{mm}$$

$$\bar{y} = \frac{\Sigma Ay}{\Sigma A} = \frac{833(10^3)}{15.5(10^3)} = 53.7\ \text{mm}$$

10-5 CENTER OF GRAVITY OF SIMPLE SOLIDS

The weight of a body is a force acting at its own center of gravity and directed toward the center of the earth. The position of the center of bodies weighing W_1, W_2, W_3, etc., is found in the same manner as the resultant of parallel forces.

$$\bar{x} = \frac{\Sigma Wx}{\Sigma W}$$

$$\bar{y} = \frac{\Sigma Wy}{\Sigma W} \tag{10-3}$$

$$\bar{z} = \frac{\Sigma Wz}{\Sigma W}$$

If all the bodies are of the same material weighing ρ lb/ft^3, then

$$W_1 = \rho V_1 \qquad W_2 = \rho V_2 \qquad W_3 = \rho V_3 \qquad \text{etc.}$$

By substituting in Eqs. (10-3), we have

$$\bar{x} = \frac{\Sigma \rho V x}{\Sigma \rho V} = \frac{\Sigma V x}{\Sigma V}$$

$$\bar{y} = \frac{\Sigma \rho V y}{\Sigma \rho V} = \frac{\Sigma V y}{\Sigma V} \qquad\qquad (10\text{-}4)$$

$$\bar{z} = \frac{\Sigma \rho V z}{\Sigma \rho V} = \frac{\Sigma V z}{\Sigma V}$$

That is, if the bodies are made of the same material and are of the same density throughout, the centers of gravity of the bodies are their centers of volume. If the bodies are of the same cross section but perhaps of different lengths,

$$V_1 = al_1 \qquad V_2 = al_2 \qquad V_3 = al_3 \qquad \text{etc.}$$

Substituting in Eqs. (10-4) produces

$$\bar{x} = \frac{\Sigma alx}{\Sigma al} = \frac{\Sigma lx}{\Sigma l}$$

$$\bar{y} = \frac{\Sigma aly}{\Sigma al} = \frac{\Sigma ly}{\Sigma l} \qquad\qquad (10\text{-}5)$$

$$\bar{z} = \frac{\Sigma alz}{\Sigma al} = \frac{\Sigma lz}{\Sigma l}$$

If the bodies are parts of a wire, pipe, or rod of constant cross section, their centers of gravity can be found from the centers of their lengths.

The center of gravity of a right circular cylinder is at the center of the geometric axis of the cylinder, Fig. 10-12.

The center of gravity of a right circular cone is on the geometric axis $\frac{1}{4} h$ from the base or $\frac{3}{4} h$ from the vertex, where h is the altitude of the cone, Fig. 10-13.

The center of gravity of a hemisphere of radius r is $\frac{3}{8} r$ from the center of the sphere, Fig. 10-14.

*Sample Problem 5   A thin wire is bent into a shape shown in Fig. 10-15. Find the coordinates of its center of gravity.

Solution: Since lengths are involved, Eqs. (10-5) will be used. It will be seen that for A, the center of the 200-mm length,

$$x = 0 \qquad y = 0 \qquad z = 100$$

For B, $\qquad\qquad\qquad x = 125 \qquad y = 0 \qquad z = 0$

For C, $x = 250 + 150 \cos 45° = 356 \qquad y = 150 \sin 45° = 106 \qquad z = 0$

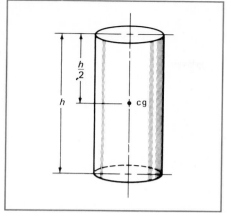

Fig. 10-12 Center of gravity of a cylinder.

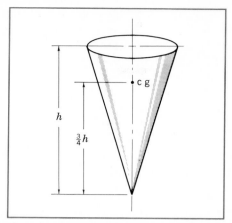

Fig. 10-13 Center of gravity of a cone.

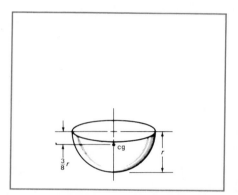

Fig. 10-14 Center of gravity of a hemisphere.

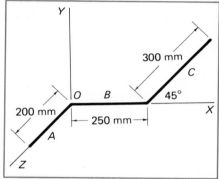

Fig. 10-15 Diagram for Sample Problem 5.

l	*x*	*y*	*z*	*lx*	*ly*	*lz*
200	0	0	100	0	0	20 000
250	125	0	0	31 250	0	0
300	356	106	0	106 800	31 800	0
750				138 050	31 800	20 000

Then

$$\bar{x} = \frac{138\ 050}{750} = 184.1 \text{ mm}$$

$$\bar{y} = \frac{31\ 800}{750} = 42.4 \text{ mm}$$

$$\bar{z} = \frac{20\ 000}{750} = 26.7 \text{ mm}$$

Note that the center of gravity does not lie on the wire.

Sample Problem 6   A cylinder for which $r = 2$ in and $h = 6$ in has an inverted cone cut from it (Fig. 10-16). The cone, for which $r = 2$ in and $h = 3$ in, has its base in the upper base of the cylinder. Find the center of gravity of the part remaining.

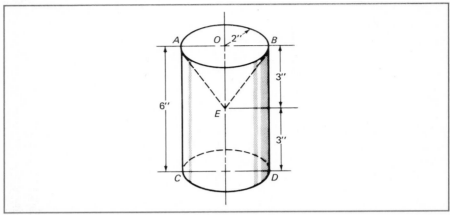

Fig. 10-16 Diagram for Sample Problems 6 and 7.

Solution: Choose O as the origin. Since the figure is symmetrical with respect to the axis OE, the center is on the axis and we need only find $\bar{y}$. The volume of the cylinder is

$$V_1 = \frac{\pi d^2 h}{4} = \frac{\pi(4^2)}{4}(6) = 24\pi \text{ in}^3$$

and for the cone,

$$V_2 = -\frac{1}{3}\left(\frac{\pi d^2}{4}\right)h = \frac{-\pi(4^2)}{12}(3) = -4\pi \text{ in}^3$$

V_2 is negative because it is volume removed. Then, from Eqs. (10-4),

$$\bar{y} = \frac{\Sigma V y}{\Sigma V} = \frac{24\pi(3) - 4\pi(\frac{1}{4})(3)}{24\pi - 4\pi}$$

$$= \frac{72 - 3}{20} = 3.45 \text{ in}$$

Sample Problem 7   Suppose the cylinder in Fig. 10-16 is made of steel ($\rho = 490$ lb/ft^3) and, after the conical hole is bored out, it is filled with concrete ($\rho = 150$ lb/ft^3). Find the center of gravity of the solid.

Solution: The weight of the steel cylinder after the conical hole is bored, but before concrete is added, is

$$W_s = V_s \rho_s = (24\pi - 4\pi)\left(\frac{490}{1728}\right) = 17.82 \text{ lb}$$

The center of gravity for the steel is 3.45 in from point O (see Sample Problem 6).

The weight of the concrete cone is

$$W_c = V_c \rho_c = 4\pi \left(\frac{150}{1728} \right) = 1.09 \text{ lb}$$

The center of gravity for the concrete is 0.75 in from point O. From Eqs. (10-3),

$$\bar{y} = \frac{\Sigma W y}{\Sigma W} = \frac{17.82(3.45) + 1.09(0.75)}{17.82 + 1.09} = 3.3 \text{ in}$$

10-6 AREAS AND VOLUMES — CENTROID METHOD

Since the center of gravity of an area or a body is the point at which the area or mass of the body may be assumed to be concentrated, it can be said that the distance through which an area or a body moves is the same as the distance described by its center of gravity. This relation is used in finding areas and volumes.

Thus, a line moving parallel to its original position is said to generate an area that is equal to the length of the line multiplied by the distance through which its centroid moves. That is, area is equal to length times width.

Also, an area moving parallel to its original position is said to develop the volume of a prism that is equal to the area multiplied by the distance through which the centroid moves. That is, a volume is equal to the area of the base times the altitude.

Similarly, a line rotating about one end will develop the area of a circle. A right triangle rotating about either leg will develop the volume of a cone. In each case, a line or an area moves through a distance equal to the length of a path described by the centroid of either the line or the area. Many determinations of areas or volumes are simplified by the use of this method.

Sample Problem 8 Find the volume of a vertical cylinder by the method of revolving an area (Fig. 10-17).

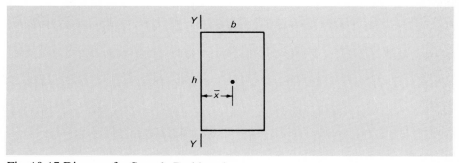

Fig. 10-17 Diagram for Sample Problem 8.

Solution: The vertical section of the cylinder is a rectangle. It can be seen that the volume of this cylinder can be developed by revolving a rectangle about a vertical edge YY.

$$A_{\text{rectangle}} = bh$$

where b = radius of cylinder
h = height of cylinder

The distance from edge YY to the centroid is $\bar{x} = b/2$. In one revolution, the centroid moves a distance

$$l = 2\pi \frac{b}{2} = \pi b$$

Then, $$V_{\text{cylinder}} = A_{\text{rectangle}}(l) = bh(\pi b) = \pi b^2 h$$

This result is the same as that for volume determined by the ordinary methods, where πb^2 is the area of the base and h is the altitude.

10-7 MOMENT OF INERTIA

In determining the strength of members which are loaded in bending, such as beams, a term called the *moment of inertia* appears in the strength equations. It is necessary to have an understanding of the moment of inertia of an area before these members can be analyzed.

Figure 10-18 indicates a rectangle $ABCD$ with an axis parallel to the base passing through the centroid. The figure is divided into a number of small rectangles as shown. Each area is represented by a with a subscript. The distance from the axis XX to the centroid of each rectangle is represented by y with a subscript corresponding to the area. Thus, for area a_1, there is a moment arm y_1, etc. If each area is multiplied by the square of its moment arm and the sum of all such terms is taken, because the number of small areas into which the

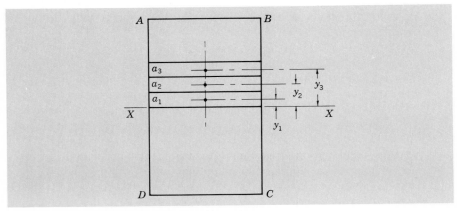

Fig. 10-18 Determining the moment of inertia.

rectangle $ABCD$ is divided becomes very great, the result is called the moment of inertia I_x of the rectangle about the centroidal axis XX. Thus,

$$I_x = \sum_{i=1}^{n} a_i y_i^2 \qquad (10\text{-}6)$$

The derivation of formulas for I_x is a problem in calculus and is beyond the scope of this text. Fortunately, the areas involved in engineering problems are often rectangles, triangles, circles, or combinations of one or more of these figures. Thus, computation of moment of inertia is not always a difficult problem.

The formulas for centroidal moments of inertia for some common areas (Table 10-2) have been developed from the calculus.

TABLE 10-2   CENTROIDAL MOMENTS OF INERTIA FOR SIMPLE AREAS

Shape	Moment of Inertia	
Rectangle	$I_x = \dfrac{bh^3}{12}$	
Triangle	$I_x = \dfrac{bh^3}{36}$	
Circle	$I_x = \dfrac{\pi d^4}{64}$	
Semicircle	$I_x = 0.11 r^4$ $I_y = \dfrac{\pi d^4}{128}$	
Quadrant	$I_x = 0.055 r^4$	

Sample Problem 9 Find the moment of inertia about the centroidal axes of the rectangle shown in Fig. 10-19.

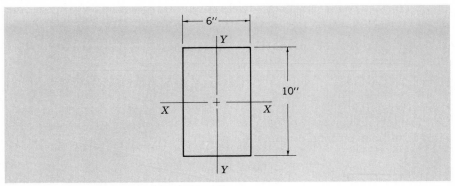

Fig. 10-19 Diagram for Sample Problem 9.

Solution: The dimension parallel to the axis is b; and h is the dimension perpendicular to the axis

$$I_x = \frac{bh^3}{12} \qquad b = 6 \text{ in}, h = 10 \text{ in}$$

$$I_x = \frac{6(10^3)}{12} = 500 \text{ in}^4$$

$$I_y = \frac{bh^3}{12} \qquad b = 10 \text{ in}, h = 6 \text{ in}$$

$$I_y = \frac{10(6^3)}{12} = 180 \text{ in}^4$$

Moments of inertia are expressed in inches to the fourth power or millimeters to the fourth power, units which have no apparent physical significance.

The results of Sample Problem 9 show that the moment of inertia of this rectangle is greater about the axis XX than about axis YY. The moment of inertia is then seen to depend on the arrangement of the area with reference to the axis.

*Sample Problem 10 Find the moment of inertia of the triangle (shown in Fig. 10-20) with reference to centroidal axis XX.

Solution:

$$I_x = \frac{bh^3}{36} \qquad b = 200 \text{ mm}, h = 150 \text{ mm}$$

$$I_x = \frac{200(150^3)}{36} = 18.8(10^6) \text{ mm}^4$$

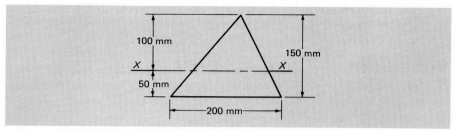

Fig. 10-20 Diagram for Sample Problem 10.

Sample Problem 11 Find the moment of inertia for an 8-in diameter circle about any centroidal axis.

Solution:

$$I_x = \frac{\pi d^4}{64} \qquad d = 8 \text{ in}$$

$$I_x = \frac{\pi(8^4)}{64} = 201 \text{ in}^4$$

Note that the circle is symmetrical about any centroidal axis.

*Sample Problem 12 Find the moment of inertia of the semicircle shown in Fig. 10-21 about the horizontal and vertical centroidal axes.

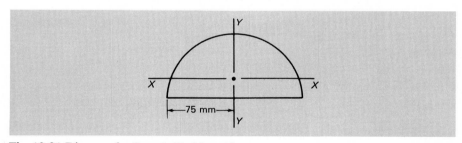

Fig. 10-21 Diagram for Sample Problem 12.

Solution:

$$I_x = 0.11r^4 = 0.11(75^4) = 3.48(10^6) \text{ mm}^4$$

$$I_y = \frac{\pi d^4}{128} = \frac{\pi(150^4)}{128} = 12.4(10^6) \text{ mm}^4$$

0-8 TRANSFER FORMULA

It is often desirable to obtain the moment of inertia of an area about some axis parallel to a centroidal axis but not passing through the centroid of the area. This can be accomplished by calculating the moment of inertia about the gravity axis of the figure and adding to this value the product of the area of the figure and the square of the distance between the gravity axis and the other axis. Referring to Fig. 10-22,

$$I_{a-a} = I_x + Ad^2 \qquad (10\text{-}7)$$

Equation (10-7) is referred to as the *transfer formula*. Transfer can be made only between parallel axes.

Sample Problem 13   Find the moment of inertia of a 6- by 10-in rectangle about its 6-in base (Fig. 10-23).

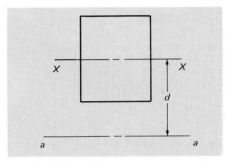

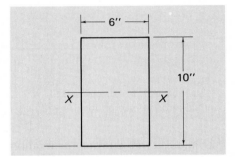

Fig. 10-22 Transfer of moment of inertia.

Fig. 10-23 Diagram for Sample Problem 13.

Solution: Let XX be a gravity axis parallel to the base.

$$I_x = \frac{bh^3}{12} = \frac{6(10^3)}{12} = 500 \text{ in}^4$$

$$d = 5 \text{ in} \quad \text{(the perpendicular distance between the axes)}$$

Then, by Eq. (10-7),

$$I_{a-a} = I_x + Ad^2 = 500 + 60(5^2) = 2000 \text{ in}^4$$

10-9   MOMENTS OF INERTIA OF COMPOSITE AREAS[1]

The transfer principle discussed above can now be used to determine the moment of inertia of a composite area; that is, an area which can be broken into simple areas. The T section shown in Fig. 10-24 is such an area.

To calculate the moment of inertia of this area about its horizontal and vertical gravity axes, the centroid must be located. The composite section can be divided into two rectangles which will be identified by 1 and 2. Because of the symmetry of the figure, it can be seen that

$$\bar{x} = 3 \text{ in (from left edge)}$$

[1] See also Sample Computer Program 15, Chap. 17.

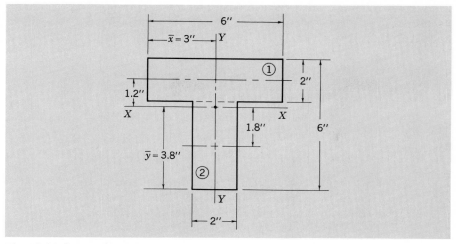

Fig. 10-24 Composite area.

Calculating $\bar{y}$,

$$\bar{y} = \frac{\Sigma A y}{\Sigma A} = \frac{12(5) + 8(2)}{12 + 8} = \frac{60 + 16}{20} = \frac{76}{20} = 3.80 \text{ in} \qquad \text{(from base)}$$

The moment of inertia of this area about a gravity axis will be found by computing the moment of inertia of each rectangle about its own centroidal axis, transferring the moment of inertia of each rectangle to the parallel gravity axis of the entire figure, and then summing the transformed moments of inertia.

Moment of inertia about horizontal gravity axis XX:

For Area 1:

$$I_{x,1} = I_1 + A_1 d^2$$
$$I_1 = \frac{bh^3}{12} = \frac{6(2^3)}{12} = 4 \text{ in}^4$$
$$I_{x,1} = 4 + 12(1.2^2) = 4 + 12(1.44)$$
$$= 4 + 17.28 = 21.28 \text{ in}^4$$

For Area 2:

$$I_{x,2} = I_2 + A_2 d^2$$
$$I_2 = \frac{bh^3}{12} = \frac{2(4^3)}{12} = 10.67 \text{ in}^4$$
$$I_{x,2} = 10.67 + 8(1.8^2) = 10.67 + 8(3.24)$$
$$= 10.67 + 25.92 = 36.59 \text{ in}^4$$
$$I_x = I_{x,1} + I_{x,2} = 21.28 + 36.59 = 57.87 \text{ in}^4$$

Moment of inertia about vertical gravity axis YY:

For Area 1:

$$I_{y,1} = I_1 + A_1 d^2$$

$$I_1 = \frac{bh^3}{12} = \frac{2(6^3)}{12} = 36 \text{ in}^4$$

$$I_{y,1} = 36 + 12(0^2) = 36 + 0 = 36 \text{ in}^4$$

For Area 2:

$$I_{y,2} = I_2 + A_2 d^2$$

$$I_2 = \frac{bh^3}{12} = \frac{4(2^3)}{12} = 2.67 \text{ in}^4$$

$$I_{y,2} = 2.67 + 8(0^2) = 2.67 + 0 = 2.67 \text{ in}^4$$

$$I_y = I_{y,1} + I_{y,2} = 36.00 + 2.67 = 38.67 \text{ in}^4$$

The calculations performed above can be systematized through use of the following tabular form.

I_x:

Area	I	A	d	d^2	Ad^2	$I + Ad^2$
1	4	12	1.2	1.44	17.28	21.28
2	10.67	8	1.8	3.24	25.92	36.59
						$I_x = 57.87$ in^4

I_y:

Area	I	A	d	d^2	Ad^2	$I + Ad^2$
1	36	12	0	0	0	36
2	2.67	8	0	0	0	2.67
						$I_y = 38.67$ in^4

In the preceding problem, it can be seen that the vertical gravity axis of the entire figure coincides with the vertical centroidal axis of each rectangle. The transfer distance is therefore zero.

°Sample Problem 14 Determine the moment of inertia of the area shown in Fig. 10-25.

(a) About the vertical gravity axis YY
(b) About a horizontal axis a-a 50 mm below the base

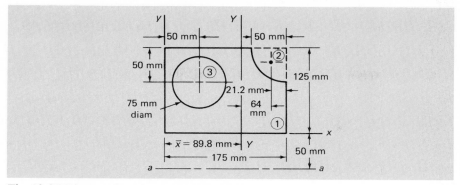

Fig. 10-25 Diagram for Sample Problem 14a.

In calculating the moment of inertia of an area that has holes or cutouts in it, treat the areas and moments of inertia of the holes and cutouts as negative values.

Solution a: From Sample Problem 4, $\bar{x} = 89.8$ mm from the left edge. About vertical axes,

$$I_1 = \frac{bh^3}{12} = \frac{125(175^3)}{12} = 55.83(10^6) \text{ mm}^4$$

$$I_2 = -0.055r^4 = -0.055(50^4) = -0.34(10^6) \text{ mm}^4$$

$$I_3 = \frac{-\pi(d^4)}{64} = \frac{-\pi(75^4)}{64} = 1.55(10^6) \text{ mm}^4$$

Area	I	A	d	d^2	Ad^2	$I + Ad^2$
1	$55.83(10^6)$	21 880	2.3	5.29	$0.12(10^6)$	$55.95(10^6)$
2	$-0.34(10^6)$	$-1\,963$	64	4 096	$-8.04(10^6)$	$-8.38(10^6)$
3	$-1.55(10^6)$	$-4\,418$	39.8	1 584	$-7.00(10^6)$	$-8.55(10^6)$
						$I_y = 39.02(10^6) \text{ mm}^4$
						say, $I_y = 39(10^6) \text{ mm}^4$

Solution b: About axis aa, from Fig. 10-26.

$$I_1 = \frac{bh^3}{12} = \frac{175(125^3)}{12} = 28.48(10^6) \text{ mm}^4$$

$$I_2 = -0.055r^4 = -0.055(50^4) = -0.34(10^6) \text{ mm}^4$$

$$I_3 = \frac{-\pi d^4}{64} = \frac{-\pi(75^4)}{64} = -1.55(10^6) \text{ mm}^4$$

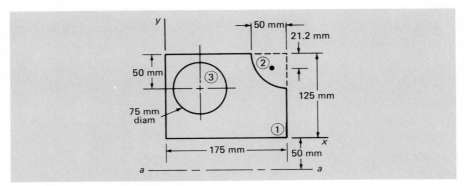

Fig. 10-26 Diagram for Sample Problem 14*b*.

Area	I	A	d	d^2	Ad^2	$I + Ad^2$
1	$28.48(10^6)$	21 880	112.5	12 660	$277(10^6)$	$305.48(10^6)$
2	$-0.34(10^6)$	$-1\ 963$	153.8	23 650	$-46.42(10^6)$	$-46.76(10^6)$
3	$-1.55(10^6)$	$-4\ 418$	125	15 620	$-69.01(10^6)$	$-70.56(10^6)$

$$I_{a\text{-}a} = 188.16(10^6) \text{ mm}^4$$
$$\text{say, } I_{a\text{-}a} = 188(10^6) \text{ mm}^4$$

PROBLEMS

10-1. Locate the centroid of the T section in Fig. Prob. 10-1 from point O.

10-2. Locate the centroid of Fig. Prob. 10-2.

*10-3. Locate the centroid of Fig. Prob. 10-3 with reference to axes through the lower left-hand corner of the figure.

10-4. Three unfinished planks are nailed to an unfinished plank, as shown in Fig. Prob. 10-4. How far is the centroid above the base?

10-5. In Fig. Prob. 10-5, $AB = 16$ ft, $BC = 6$ ft, and $AD = 12$ ft. Find the coordinates of the centroid. Use D as a reference point.

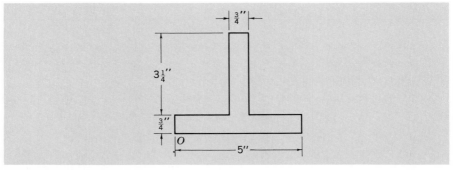

FIGURE PROBLEM 10-1

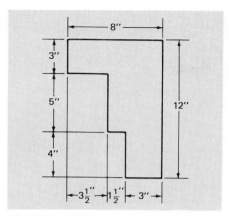

FIGURE PROBLEM 10-2

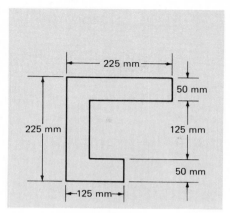

FIGURE PROBLEM 10-3

*10-6. Find the centroid of the area shown in Fig. Prob. 10-6.

10-7. Locate the centroid of the channel shown in Fig. Prob. 10-7.

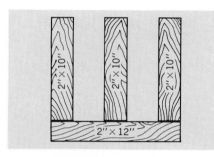

FIGURE PROBLEM 10-4

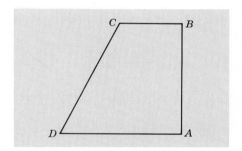

FIGURE PROBLEM 10-5

10-8. Find the centroid of each of three angle irons having the following dimensions: 4 by 3 by $\frac{1}{4}$, 6 by 6 by $\frac{3}{4}$, 8 by 6 by 1. Verify the results from Tables 7 and 8, App. B.

*10-9. Find the centroid of the piece of sheet iron shown in Fig. Prob. 10-9.

*10-10. Find the centroid of a plate 150 by 250 mm from which two circular holes have been cut as shown in Fig. Prob. 10-10.

10-11. Find the centroid of the section of a bridge member made of two 18- by $\frac{3}{4}$-in web plates, a 20- by $\frac{3}{8}$-in top plate, two top angles 4 by 3 by $\frac{1}{2}$ in, and two bottom angles 4 by 4 by $\frac{3}{8}$ in (Fig. Prob. 10-11).

10-12. A 6-in cube of wood weighing 40 lb/ft³ has a 1-in-diameter hole drilled through it parallel to one edge. The long axis of the hole is 2 in from the upper surface and 2 in from the left side of the cube.

    a. Locate the centroid.

    b. If this hole is filled with lead weighing 710 lb/ft³, where is the centroid?

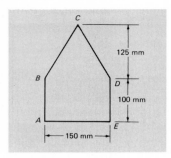

FIGURE PROBLEM 10-6

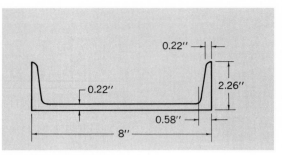

FIGURE PROBLEM 10-7

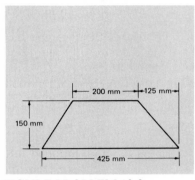

FIGURE PROBLEM 10-9

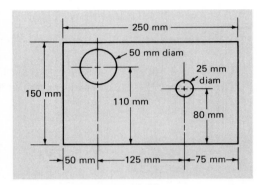

FIGURE PROBLEM 10-10

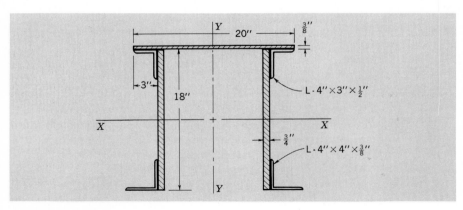

FIGURE PROBLEM 10-11

*10-13. In Fig. Prob. 10-13, the dam has a mass of 450 kg and $F = 3.6$ kN. If $AB = 4.8$ m, $BC = 1.8$ m, and $AD = 4.8$ m, find the maximum height above A where the force F can be applied without overturning the dam about point D. (*Note:* W acts as the centroid of $ABCD$.)

10-14. A dam stands 12 ft high. It is 4 ft wide at the top and 10 ft wide at the

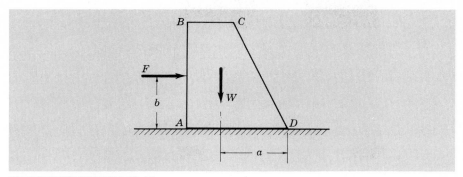

FIGURE PROBLEM 10-13

base. The upstream face is vertical. The coefficient of friction between the dam and its foundation is 0.6. What are the factors of safety
against sliding and overturning when the water has reached the top of
the dam? The dam is made of concrete (density = 150 lb/ft³).

*10-15. Find the volume of a cone by the method of revolving a right triangle
with a 75-mm base and 100-mm altitude about the vertical edge.

10-16. Find the volume of a sphere by revolving a semicircle about a diameter.

10-17. Find the volume of the rim of a flywheel 6 ft in diameter if the rim is
12 in wide and 8 in thick.

*10-18. Find the moment of inertia about the horizontal and vertical centroidal axes of the following rectangles:
a. 100 by 250 mm
b. 200 by 300 mm
c. 250 by 250 mm

10-19. Find the moment of inertia about the horizontal and vertical centroidal axes for the following figures:
a. Right triangle with horizontal base of 5 in and altitude of 7 in
b. $3\frac{1}{2}$-in-diameter circle
c. $1\frac{3}{4}$-in-radius semicircle with horizontal base
d. $1\frac{3}{4}$-in-radius quadrant (90° segment) with one side horizontal

10-20. Find I about the centroidal axis parallel to the base of the area shown
in Fig. Prob. 10-20.

10-21. Find the moments of inertia with respect to both principal (centroidal) axes of the T section shown in Fig. Prob. 10-1.

*10-22. In Fig. 10-8, find the moment of inertia with respect to the horizontal
centroidal axis and also with respect to the left edge of the figure.

10-23. What are the moments of inertia with respect to the principal (centroidal) axes of the planks shown in Fig. Prob. 10-4?

10-24. Find I_x and I_y for each of two angle irons, 4 by 4 by $\frac{1}{2}$ in and 4 by 3 by
$\frac{1}{2}$ in. Compare results with the values given in Tables 7 and 8, App. B.

10-25. Find I_x and I_y for a girder made with four 7- by 4- by $\frac{3}{8}$-in angles and
one 12- by $\frac{3}{8}$-in web plate (Fig. Prob. 10-25).

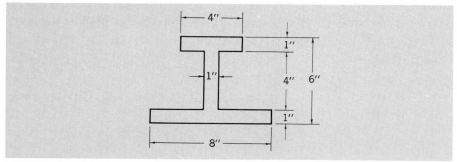

FIGURE PROBLEM 10-20

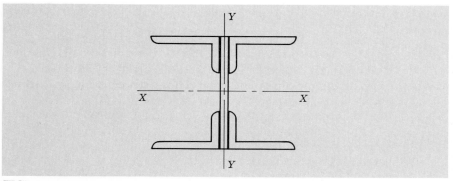

FIGURE PROBLEM 10-25

10-26. Find I_x and I_y for a girder built up of four 5- by $3\frac{1}{2}$- by $\frac{1}{2}$-in angles and a 10- by $\frac{1}{2}$-in web plate (Fig. Prob. 10-25).

10-27. What is the moment of inertia with respect to the horizontal centroidal axis of the column section shown in Fig. Prob. 10-27? The area of

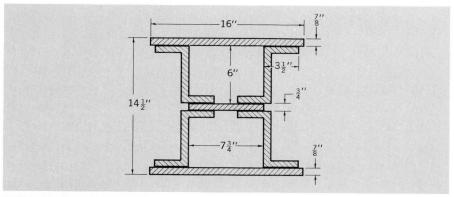

FIGURE PROBLEM 10-27

a Z bar is 8.63 in², and the moment of inertia of a Z bar with respect to its centroidal axis parallel to the shorter leg is 42.12 in⁴.

10-28. Find the moment of inertia of the section shown in Fig. Prob. 10-11 with respect to both principal (centroidal) axes.

10-29. A box girder is shown in Fig. Prob. 10-29. What is the moment of inertia with respect to the horizontal gravity axis if the bolt holes are deducted? Bolts are located in the center of the angle legs.

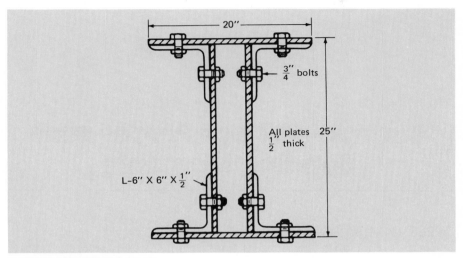

FIGURE PROBLEM 10-29

10-30. Find the moment of inertia about the horizontal and vertical centroidal axes of the plate shown in Fig. Prob. 10-10.

11

Beams — Shear Forces and Bending Moments

11-1 TYPES OF BEAMS

A beam is a member which resists transverse loads and forces by bending elastically. The beams discussed in this study will be confined to those which meet the following limitations.

1. All forces acting on the beam lie in the same plane; this (longitudinal) plane passes through the centroids of all the cross sections of the beam.
2. The elastic limit of the beam material is not exceeded.
3. The cross section of the beam is uniform in size and shape for the entire length.
4. The beam is thick enough to prevent localized buckling or wrinkling.
5. Vibration, shock, and impact loading do not occur.

Beams can be generally categorized as either statically determinate or statically indeterminate. A statically determinate beam is one whose unknown forces or moments can be evaluated by applying the conditions for static equilibrium, Eqs. (11-1).

$$\Sigma F_x = 0 \qquad \Sigma F_y = 0 \qquad \Sigma M = 0 \qquad (11\text{-}1)$$

Common beams of this type are shown in Fig. 11-1. Among them are a *simply supported beam* (Fig. 11-1a), an *overhanging beam* (Fig. 11-1b), and a *cantilever beam* (Fig. 11-1c). The loads, whether concentrated (F) or uniformly distributed (W), are assumed to be known; the reactions (R) of the supports and

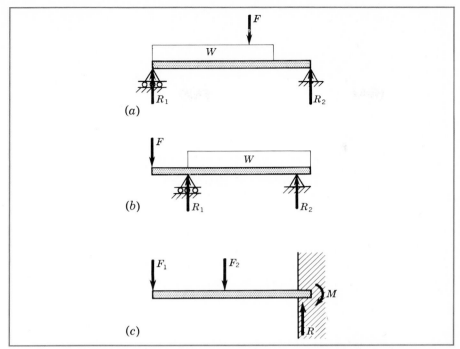

Fig. 11-1 Common statically determinate beams: (*a*) simply supported beam, R_1 and R_2 unknown; (*b*) overhanging beam, R_1 and R_2 unknown; (*c*) cantilever beam, R and M unknown.

the wall moment (M) are unknown. This chapter deals only with statically determinate beams.

The unknown forces acting on statically indeterminate beams cannot be found by Eqs. (11-1) only. Special procedures are used to solve problems of this type. Chapter 16 deals with statically indeterminate beams.

11-2 BEAM THEORY

Let Fig. 11-2 represent a simply supported beam carrying a load F at its center. The load produces a bending in the beam at all points. Any sections such as AB and CD are rotated into new positions by the bending of the beam. The length BD is shortened or compressed, and length AC is lengthened or stretched. The resistance offered to this shortening and lengthening of the fibers is called *internal fiber stress* or simply *stress.* The upper fibers are in compression, and the lower ones are in tension.

It was shown in Chap. 8 that the deformation of a body, if within the elastic limit, is proportional to the applied force. The diagram shows that the maximum internal stresses occur at the outer fibers AC and BD, because the deformation is greatest at those points and decreases to zero at the center G or H.

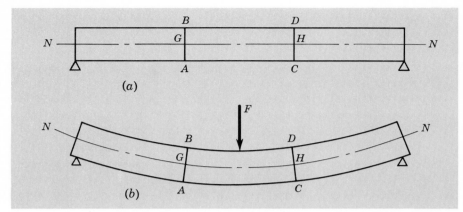

Fig. 11-2 Simply supported beam: (*a*) before load is applied; (*b*) after load is applied.

Axes *GG* and *HH* (Fig. 11-3) are called the *neutral axes of the sections,* or *axes of zero stress.* The neutral axis of a section of a beam passes through the centroid of the section.

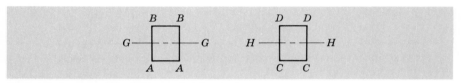

Fig. 11-3 Cross sections of beam in Fig. 11-2.

In addition to the bending of a beam, there is a tendency of one section of a beam to slip past the adjacent section. This tendency is called *shear,* and the shear forces must be resisted by the fibers of the beam. Shearing forces are parallel to the plane of the section (Fig. 11-4).

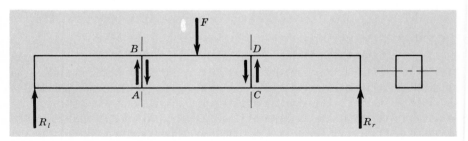

Fig. 11-4 Shear forces in simply supported beam.

Figure 11-5 shows the beam at section *AB.* The fibers resisting bending above the neutral axis *G* are subjected to compression forces, and those below *G* are subjected to tension forces. The compressive and tensile forces due to bending are indicated. Note that these forces increase as the distance from the neutral axis G increases. Consequently, we usually expect the most severe stresses to occur at the extreme fibers. The vertical shear force V at section AB also is shown. In Fig. 11-6 the same situation is represented, except that the compressive forces above the neutral axis (between B and G in Fig. 11-5) are replaced by their resultant C and, similarly, the tensile forces are replaced by resultant T. For the portion of the beam shown in Fig. 11-6, the only horizontal forces acting are C and T. Applying the condition for equilibrium that $\Sigma F_x = 0$, it is evident that $T = C$ in magnitude although their directions are opposite. These equal, opposite, parallel forces form a *couple,* the effect of which is to produce a *resisting moment.* This moment must be equal and opposite (direction of rotation) to the moment produced by external forces acting on the left portion of the beam so that the condition $\Sigma M = 0$ is satisfied. The reaction R_l at the left support is the only external force which produces a moment in Fig. 11-6 (the weight of the beam has been neglected). Its moment arm is the horizontal distance from the support to section AB.

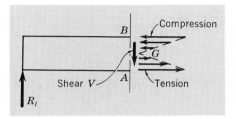

Fig. 11-5 Shear and bending forces at a beam section.

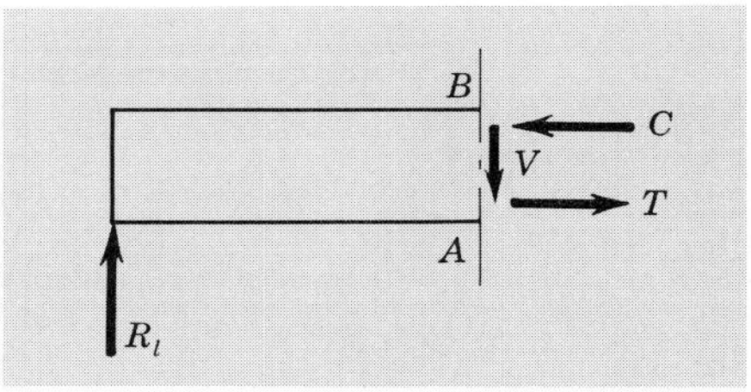

Fig. 11-6 Shear and resultant bending forces at a beam section.

It is customary to use the portion of the beam *to the left* of a section in solving problems. This practice will be followed in almost all cases.

For problems with statically determinate beams, it is usual to first calculate the unknown reactions. Types of loading which are symmetrically placed on simply supported beams enable the reactions to be determined by inspection. In Fig. 11-7 a simply supported beam with a concentrated load F at the center of the span is shown. Neglecting the weight of the beam, apply the condition $\Sigma F_y = 0$. It is seen that $R_l + R_r - F = 0$; and since the load F is centrally located on the beam, each reaction equals half the load, or $R_l = F/2$ and $R_r = F/2$. A similar analysis applies to the beam in Fig. 11-8, which carries a uniformly distributed load whose total weight is W. Again it is found that each reaction equals half the load, or $R_l = W/2$ and $R_r = W/2$. Uniform loads are often specified by the loads they apply either per linear foot or per linear meter of beam. The symbol w will be used for this purpose. In the U.S. customary system, w will be in pounds per foot; in the SI metric system, w will be in newtons per meter. Note that the total uniform load (W) equals the weight per foot (w) times the length of the load. For the load in Fig. 11-8, the length of the load equals the span length (L) of the beam; thus, $W = wL$.

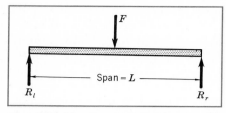

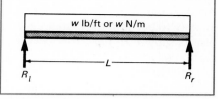

Fig. 11-7 Simply supported beam with load at midspan.

Fig. 11-8 Simply supported beam with uniformly distributed load.

For cantilever beams with vertical loading only, the unknown reaction R can be evaluated by applying the condition $\Sigma F_y = 0$. The beam shown in Fig. 11-9 is a cantilever with a concentrated load F at the unsupported end. The wall must provide a force (reaction) R to counteract the load F so that $\Sigma F_y = 0$ is satisfied. Therefore, $R - F = 0$, or $R = F$. Similarly, the cantilever beam shown in Fig. 11-10, carrying a uniform load $W = wL$, has a wall reaction R which balances the total load W. Thus, $R - W = 0$, or $R = W$. The weight of the beam has been neglected in both of the previous cases.

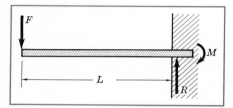

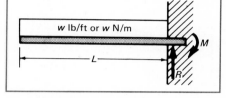

Fig. 11-9 Cantilever beam with load at free end.

Fig. 11-10 Cantilever beam with uniformly distributed load.

The unknown moment M at the wall for a cantilever beam can be calculated from $\Sigma M = 0$. Since there can be no unbalanced moment in static equilibrium, the moment due to the load on the beam must be counteracted by M at the wall. The moment arm of the load is always measured to the wall for this calculation. For the beam of Fig. 11-9, $M = FL = 0$, or $M = FL$. For the purpose of taking moments, the total weight of a uniform load is considered to act at the center of the load. Then, for the beam of Fig. 11-10, $M - W(L/2) = 0$, or $M = W(L/2)$.

Sample Problem 1[1]   Calculate the reactions for a simply supported beam with a 12-ft span which carries a concentrated load of 8000 lb placed 3 ft from the left support, if:

(a)   The weight of the beam is neglected.
(b)   The beam weighs 50 lb/ft.

---

[1] See also Sample Computer Program 16, Chap. 17.

Solution a: A sketch of the beam showing the loading and the reactions should be drawn as in Fig. 11-11.

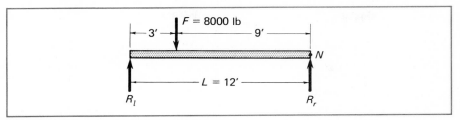

Fig. 11-11 Diagram for Sample Problem 1a.

To ensure static equilibrium, Eqs. (11-1) must be satisfied. Since there are no horizontal forces or forces with horizontal components, the condition $\Sigma F_x = 0$ is satisfied, but it gives no useful equation to help solve for the unknowns. The other conditions for equilibrium are $\Sigma F_y = 0$ and $\Sigma M = 0$. Each gives a useful equation which enables the solution of the unknowns R_l and R_r.

$\Sigma F_y = 0$

Up forces will be considered positive and down forces negative.

$$R_l + R_r - F = 0$$
$$R_l + R_r = 8000$$

$\Sigma M = 0$

This relation states that there can be no unbalanced moment about *any axis of rotation.* Hence, any point on or off the beam can be selected as an axis about which to take moments. The usual procedure, however, is to select an axis so that one (or more) of the unknowns is eliminated. Such an axis would be located on the line of action of one of the unknown reactions, since, if its line of action passes through the axis of rotation, the reaction will have a zero moment arm and thus a zero moment. Point N in Fig. 11-11 represents an axis passing through the centroid of the right end section of the beam in a direction perpendicular to the plane of the paper. Reaction R_r will have no moment about axis N. The common notation for this is $\Sigma M_r = 0$; that is, the sum of moments about R_r (actually axis N) is zero. Clockwise moments will be called positive, and counterclockwise moments will be taken as negative.

$\Sigma M_r = 0$ $\qquad\qquad R_l(12) - 8000(9) = 0$

$$R_l = \frac{8000(9)}{12} = 6000 \text{ lb}$$

Substituting this value in the equation from $\Sigma F_y = 0$ gives us

$$6000 + R_r = 8000$$
$$R_r = 2000 \text{ lb}$$

Note that the load F is supported by each reaction in proportion to the distance of F from the other reaction. The load is 3 ft from the left end of the 12-ft span; thus, $R_r = \frac{3}{12} F = 3(8000)/12 = 2000$ lb. F is 9 ft from the right end; thus, $R_l = \frac{9}{12} F = 9(8000)/12 = 6000$ lb. This method of division of loading between the reactions is called the *lever rule*.

Solution b: Figure 11-12 shows the beam with the loading and reactions. The weight of the beam is treated as a uniformly distributed load of $w = 50$ lb/ft. The total weight of the beam is $W = wL = 50(12) = 600$ lb. The procedure is similar to part a of this problem.

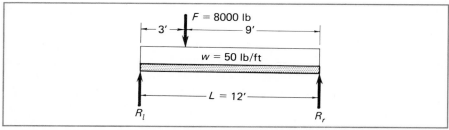

Fig. 11-12 Diagram for Sample Problem 1b.

$$\Sigma F_y = 0 \qquad\qquad R_l + R_r - F - W = 0$$
$$R_l + R_r = 8000 + 600$$
$$R_l + R_r = 8600$$

$$\Sigma M_r = 0$$
$$R_l(12) - 8000(9) - 600(6) = 0$$
$$R_l = \frac{8000(9) + 600(6)}{12} = \frac{72\,000 + 3600}{12}$$
$$= \frac{75\,600}{12} = 6300 \text{ lb}$$

Substituting, $\qquad\qquad\qquad 6300 + R_r = 8600$
$$R_r = 2300 \text{ lb}$$

Comparison of these results with those of part a indicates that the effect of the uniform load is divided equally between R_l and R_r.

*Sample Problem 2[1]   A 6-m-long beam is supported 0.5 m from the right end and 1.5 m from the left end. The beam carries a concentrated load of 4.5 kN at the left end of the beam and a uniform load of 3000 N/m extending from the left support to the right end of the beam. Calculate the reactions at the supports. Neglect the weight of the beam.

---

[1] See also Sample Computer Program 17, Chap. 17.

Solution: Sketch the beam as in Fig. 11-13.

$$W = 3000(4.5) = 13\ 500\ \text{N} = 13.5\ \text{kN}$$

$$\Sigma F_y = 0 \qquad\qquad R_l + R_r - F - W = 0$$

$$R_l + R_r - 4.5 - 13.5 = 0$$

$$R_l + R_r = 18$$

$$\Sigma M_r = 0 \qquad R_l(4) - 4.5(5.5) - 13.5(1.75) = 0$$

$$4R_l = 24.75 + 23.62 = 48.37$$

$$R_l = 12.1\ \text{kN}$$

Substituting, $\qquad\qquad 12.1 + R_r = 18$

$$R_r = 5.9\ \text{kN}$$

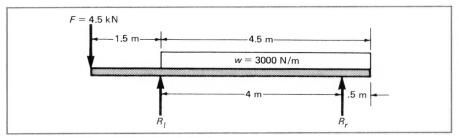

Fig. 11-13 Diagram for Sample Problem 2.

11-3   SHEAR FORCE DIAGRAM

It was indicated in Sec. 11-2 that a vertical force exists at a given section in a beam. It is useful to know what shear force a beam must resist at *every section*. The shear force at any section of a beam is the algebraic sum of all vertical forces acting on the beam to the left of that section. This information is conveniently represented in a shear force diagram, which is drawn in projection with the sketch of the beam it represents. The shear force diagram is a plot of the *net external shearing forces* which act at each beam cross section. The forces are caused by the loading on the beam, and the fibers of the beam material must resist them to maintain static equilibrium.

Several examples will demonstrate the procedure for obtaining a shear force diagram. In the examples, the shear force is calculated at sections 1 ft apart (0.5 m apart when metric units are used) starting from the left end. External up forces are considered positive, and external down forces are considered negative. All reaction forces must be known before the shear diagram can be sketched.

Sample Problem 3   Find the shear forces on a simply supported beam 10 ft long that are due to a concentrated load of 1000 lb at the center of the span. Neglect the weight of the beam. Sketch the shear force diagram for this beam.

Solution: From the beam shown in Fig. 11-14a it is apparent that the reactions are 500 lb each because of the symmetrical loading.

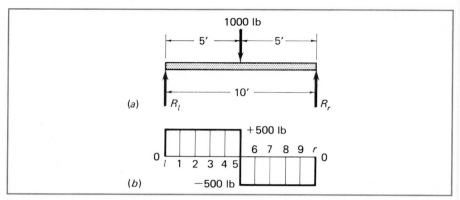

Fig. 11-14 (a) Beam diagram for Sample Problem 3. (b) Shear force diagram.

At the left end of the beam the only force acting is the reaction (500 lb up); therefore, the net shear force at the left end (actually, just slightly in from the end) is $V_l = +500$ lb. At a point 1 ft in from the reaction no new forces are encountered; thus $V_1 = +500$ lb. As we move from left to right along the beam and examine the net shear force at every foot, we find that

$$V_2 = +500 \text{ lb}$$
$$V_3 = +500 \text{ lb}$$
$$V_4 = +500 \text{ lb}$$

At a section just to the left of the center (say, at 4.99 ft from the left end), no new load has been encountered; thus $V_5 = +500$ lb. However, at the center of the beam a down load of 1000 lb appears. The net shear force is now $V_5' = +500 - 1000 = -500$ lb. No new loads are encountered for the remainder of the beam until the right reaction is reached; therefore,

$$V_6 = -500 \text{ lb}$$
$$V_7 = -500 \text{ lb}$$
$$V_8 = -500 \text{ lb}$$
$$V_9 = -500 \text{ lb}$$

Just before reaching the right reaction, the shear force is $V_r = -500$ lb. When the reaction is reached, the net shear force becomes $V_r' = -500 + 500 = 0$ lb.

These shear forces can be plotted (to a suitable scale) in projection with the beam. Such a plot is called the *shear force diagram*, as shown in Fig. 11-14b. Establish a horizontal line OO whose length is the same as that of the beam. Plot values of shear force V at the location along line OO corresponding to the beam section where V acts. Plot positive V values above line OO and negative values below line *OO*. Connect the plotted points with straight lines. The resulting diagram gives the net shear force at *all* sections across the beam.

The shear diagram shows a constant value of shear force between the left end of the beam and the concentrated load. Therefore, it is unnecessary to calculate shear forces for intermediate sections if no new forces are present in this range.

*Sample Problem 4   Determine the shear diagram for a simply supported beam that is 5 m long and carries a uniformly distributed load of 1400 N/m, including its own weight.

Solution: The total load is 1400 N/m $\times$ 5 m = 7000 N. Each reaction is one-half the total load, or 3500 N. Therefore, $V_0 = 3500$ N. Between the 0.5-m section and the left end of the beam there is 700 N of load. Hence, $V_{0.5} = 3500 - 700 = 2800$ N. The forces to the left of the 1-m section are 3500 N up and 1400 N of load down.

$$V_1 = 3500 - 1400 = 2100 \text{ N}$$

Similarly,
$$V_{1.5} = 1400 \text{ N}$$
$$V_2 = 700 \text{ N}$$
$$V_{2.5} = 0$$
$$V_3 = -700 \text{ N}$$
$$V_{3.5} = -1400 \text{ N}$$
$$V_4 = -2100 \text{ N}$$
$$V_{4.5} = -2800 \text{ N}$$
$$V_5 = -3500 \text{ N}$$

These values are plotted in projection with the beam, as shown in Fig. 11-15.

A uniform load on a beam results in a sloped straight line on the shear diagram. To construct the shear force diagram in this problem, it is necessary

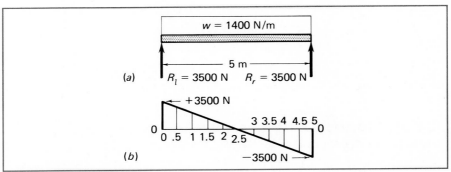

Fig. 11-15 (*a*) Beam diagram for Sample Problem 4. (*b*) Shear force diagram.

only to calculate and plot the shear force at each end of the uniform load V_0 and V_5 and to join the two points with a straight line.

*Sample Problem 5   Determine the shear diagram for a cantilever beam that is 3 m long and has a concentrated load of 70 kN at the free end. Neglect the weight of the beam.

Solution: Cantilever beams should always be sketched with the free end at the left so the signs (+ or −) given to forces and moments remain consistent.

At the left end of the beam, the shear force is 70 kN down because of the concentrated load. No other forces are encountered until the fixed end of the beam, where a reaction of $R = 70$ kN (up) must occur to ensure that $\Sigma F_y = 0$. Thus, the shear diagram can be sketched (Fig. 11-16) without intermediate calculations at other sections.

*Sample Problem 6   By using the same cantilever beam as in Sample Problem 5 and replacing the concentrated load with a uniform load of 3000 N/m, determine the shear diagram.

Solution: $W = 3000$ N/m $\times$ 3 m $= 9000$ N $= 9$ kN. The shear diagram for this beam is shown in Fig. 11-17.

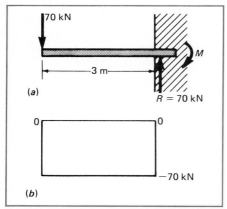

Fig. 11-16 (*a*) Beam diagram for Sample Problem 5. (*b*) Shear force diagram.

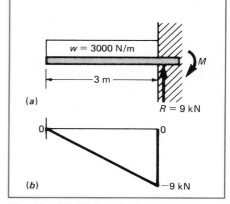

Fig. 11-17 (*a*) Beam diagram for Sample Problem 6. (*b*) Shear force diagram.

Sample Problem 7   Determine the shear force diagram for the overhanging beam shown in Fig. 11-18. The beam weighs 35 lb/ft.

Solution: Reactions R_l and R_r must be determined before the shear diagram can be sketched.

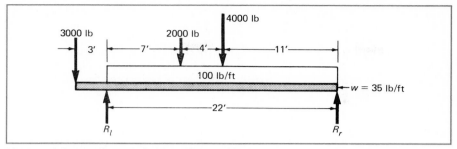

Fig. 11-18 Beam diagram for Sample Problem 7.

$\Sigma F_y = 0$

$$R_l + R_r - 3000 - 2000 - 4000 - 100(22) - 35(25) = 0$$
$$R_l + R_r = 12\ 075\ \text{lb}$$

$\Sigma M_r = 0$

$$R_l(22) - 3000(25) - 2000(15) - 4000(11) - 100(22)(11) - 35(25)(12.5) = 0$$
$$R_l = \frac{75\ 000 + 30\ 000 + 44\ 000 + 24\ 200 + 10\ 940}{22} = \frac{184\ 140}{22}$$
$$= 8370\ \text{lb}$$
$$R_r = 12\ 075 - 8370 = 3705\ \text{lb}$$

Start the shear diagram from the left end of the beam.

At the left end of the beam there is a down force of 3000 lb. The 3-ft section between the left end and R_l has a uniform load of 35 lb/ft that is due to the weight of the beam.

$$V_3 = -[3000 + 35(3)] = -[3000 + 105] = -3105\ \text{lb (down)}$$

The reaction force, $R_l = 8370$ lb, acts up.

$$V_3' = 8370 - 3105 = 5265\ \text{lb (up)}$$

The 7-ft section between R_l and the 2000-lb concentrated load has a uniform load of 135 lb/ft.

$$V_{10} = 5265 - 135(7) = 5265 - 945 = 4320\ \text{lb (up)}$$

The effect of the 2000-lb load is considered next.

$$V_{10}' = 4320 - 2000 = 2320\ \text{lb (up)}$$

The next 4-ft section has a uniform load of 135 lb/ft.

$$V_{14} = 2320 - 135(4) = 2320 - 540 = 1780\ \text{lb (up)}$$

The 4000-lb concentrated load will cause the net shear force to become negative.

$$V_{14}' = 1780 - 4000 = -2220\ \text{lb (down)}$$

The remaining 11-ft section increases the net down force by 135 lb/ft.

$$V_{25} = -2220 - 135(11) = -2220 - 1485 = -3705 \text{ lb (down)}$$

But the right reaction $R_r = 3705$ lb (up); thus the shear diagram returns to zero.

The shear diagram for this beam is shown in Fig. 11-19.

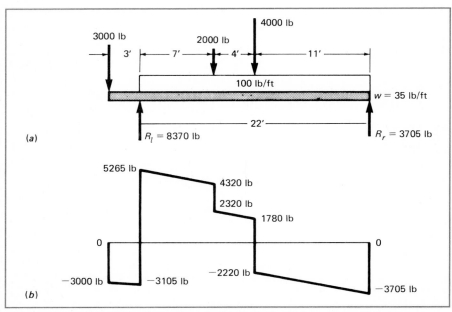

Fig. 11-19 (*a*) Beam diagram for Sample Problem 7. (*b*) Shear force diagram.

11-4 MOMENT DIAGRAM

In order that the designer can select a beam to carry a given load, he or she must know how the bending moment changes at sections from one end of the beam to the other. For this purpose, the bending moment is calculated about axes through sections 1 ft apart (0.5 m apart when metric units are used) and is plotted just as in the case of shear. The figure thus obtained is called the *bending moment diagram*. Bending moment diagrams are also used in determining the patterns of deflection in beams. The bending moment at any section of the beam is the algebraic sum of the moments due to forces acting to the left of that section. The bending moment diagram is a plot of the *net external moments* which act at the beam cross sections. These external moments, which are caused by external forces, must be resisted by the beam material to maintain equilibrium. As a result of resisting these moments, the beam material is stressed and bending occurs. For purposes of calculation, external *clockwise moments* are taken as *positive* and external *counterclockwise moments* as *negative*.

Sample Problem 8[1] Determine the bending moment diagram for a 10-ft-long simply supported beam with a concentrated load of 1000 lb at the center of the span. Neglect the weight of the beam. *Note:* Refer to Sample Problem 3 and Fig. 11-14. Also, refer to Fig. 11-20.

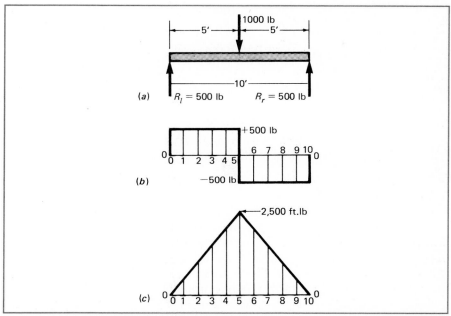

Fig. 11-20 (*a*) Beam diagram for Sample Problem 8. (*b*) Shear force diagram. (*c*) Moment diagram.

Solution: Calculate the bending moment at sections 1 ft apart along the length of the beam, starting from the left end.

At the left end of this beam the only external effect is the 500-lb reaction. The line of action of this force passes through the end section. Therefore, this force produces no moment at the end section. $M_0 = 0$. To calculate the moment about an axis through the section 1 ft from the left end, all external effects *to the left of that section* are considered. The reaction force of 500 lb has a moment arm of 1 ft. Therefore, $M_1 = 500(1) = 500$ ft·lb (clockwise). Note that M_1 is the net external moment at the 1-ft section. Equilibrium is maintained when the beam material successfully counterbalances M_1. By following a similar procedure, the moments at other sections are

$$M_2 = 500(2) = 1000 \text{ ft·lb}$$
$$M_3 = 500(3) = 1500 \text{ ft·lb}$$
$$M_4 = 500(4) = 2000 \text{ ft·lb}$$
$$M_5 = 500(5) = 2500 \text{ ft·lb}$$

[1] See also Sample Computer Program 18, Chap. 17.

At the 6-ft section, the counterclockwise moment effect of the 1000-lb force must be considered. Hence,

$$M_6 = 500(6) - 1000(1) = 3000 - 1000 = 2000 \text{ ft} \cdot \text{lb}$$

Similarly,
$$M_7 = 500(7) - 1000(2) = 1500 \text{ ft} \cdot \text{lb}$$
$$M_8 = 500(8) - 1000(3) = 1000 \text{ ft} \cdot \text{lb}$$
$$M_9 = 500(9) - 1000(4) = 500 \text{ ft} \cdot \text{lb}$$
$$M_{10} = 500(10) - 1000(5) = 0 \text{ ft} \cdot \text{lb}$$

The moment diagram shown in Fig. 11-20 is obtained by plotting the values of M calculated above.

*Sample Problem 9 Determine the bending moment diagram for a 5-m-long simply supported beam carrying a uniformly distributed load of 1400 N/m, including its own weight. *Note:* Refer to Sample Problem 4.

Solution: Calculate the bending moment at sections 0.5 m apart along the length of the beam, starting from the left end.

At the left end of this beam the only external effect is the 3500-N reaction. Since the line of action of this force passes through the end section, the moment $M_0 = 0$. For the section 0.5 m from the left end, consider only the external effects *to the left of that section.* There are two such effects: the 3500-N reaction force with a 0.5-m moment arm and the portion of the uniform load which lies to the left of the section with its moment arm. The portion of the uniform load in question is 0.5 m long; therefore, its *weight* is 1400 N/m × 0.5 m = 700 N. Since the weight of a uniform load can be considered concentrated at the center of gravity of that load for purposes of taking moments, the moment arm for this portion of the uniform load is 0.25 m. Thus, the moment at the 0.5-m section is

$$M_{0.5} = 3500(0.5) - 700(0.25) = 1750 - 175 = 1575 \text{ N} \cdot \text{m}$$

Similarly,
$$M_1 = 3500(1) - 1400(0.5) = 3500 - 700 = 2800 \text{ N} \cdot \text{m}$$
$$M_{1.5} = 3500(1.5) - 2100(0.75) = 5250 - 1575 = 3675 \text{ N} \cdot \text{m}$$
$$M_2 = 3500(2) - 2800(1) = 7000 - 2800 = 4200 \text{ N} \cdot \text{m}$$
$$M_{2.5} = 3500(2.5) - 3500(1.25) = 8750 - 4375 = 4375 \text{ N} \cdot \text{m}$$
$$M_3 = 3500(3) - 4200(1.5) = 10\,500 - 6300 = 4200 \text{ N} \cdot \text{m}$$
$$M_{3.5} = 3500(3.5) - 4900(1.75) = 12\,250 - 8575 = 3675 \text{ N} \cdot \text{m}$$
$$M_4 = 3500(4) - 5600(2) = 14\,000 - 11\,200 = 2800 \text{ N} \cdot \text{m}$$
$$M_{4.5} = 3500(4.5) - 6300(2.25) = 15\,750 - 14\,175 = 1575 \text{ N} \cdot \text{m}$$
$$M_5 = 3500(5) - 7000(2.5) = 17\,500 - 17\,500 = 0 \text{ N} \cdot \text{m}$$

Plotting these values to a suitable scale gives the moment diagram in Fig. 11-21.

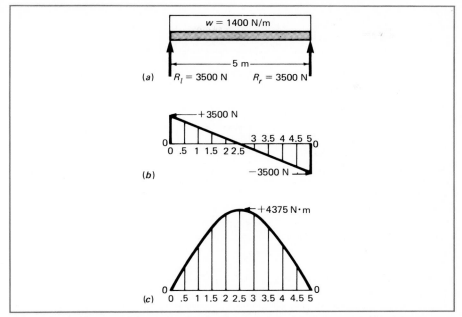

Fig. 11-21 (*a*) Beam diagram for Sample Problem 9. (*b*) Shear force diagram. (*c*) Moment diagram.

°Sample Problem 10   Determine the moment diagram for a 3-m-long cantilever beam with a concentrated load of 70 kN at the free end. Neglect the weight of the beam. *Note:* Refer to Sample Problem 5.

Solution: With the free end at the left, calculate the bending moment at sections 0.5 m apart along the length of the beam, starting from the free left end.

The load at the free end has no moment arm about the end section; thus, $M_0 = 0$ kN·m. At the section 0.5 m from the left end, the 70-kN load has a 0.5-m arm; thus, the counterclockwise moment

$$M_{0.5} = -70 \text{ kN}(0.5 \text{ m}) = -35 \text{ kN·m}$$

Similarly,
$$M_1 = -70 \text{ kN (1 m)} = -70 \text{ kN·m}$$
$$M_{1.5} = -105 \text{ kN·m}$$
$$M_2 = -140 \text{ kN·m}$$
$$M_{2.5} = -175 \text{ kN·m}$$
$$M_3 = -210 \text{ kN·m}$$

When these values are plotted to scale, a moment diagram as shown in Fig. 11-22 is obtained.

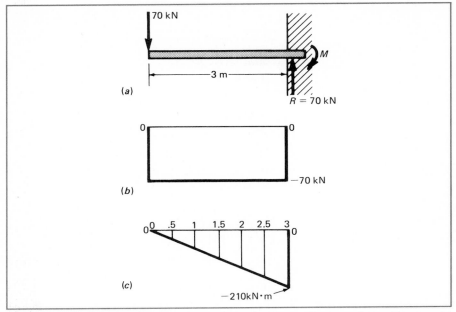

Fig. 11-22 (*a*) Beam diagram for Sample Problem 10. (*b*) Shear force diagram. (*c*) Moment diagram.

*Sample Problem 11   Determine the moment diagram for a 3-m-long cantilever beam with a uniform load of 3000 N/m (including the weight of the beam) for the entire length. *Note:* Refer to Sample Problem 6.

Solution $W = 3000$ N/m $\times$ 3 m $= 9000$ N $= 9$ kN

With the free end at the left, calculate the bending moment at sections 0.5 m apart along the length of the beam starting from the free left end.

Since the load at the free end is zero, $M_0 = 0$. The portion of the uniform load to the left of the 0.5-m section weighs 1500 N. The weight of a uniform load can be considered as acting at its center of gravity; therefore, the moment arm about the 0.5-m section is 0.25 m. Thus,

$$M_{0.5} = -1.5(0.25) = -0.375 \text{ kN} \cdot \text{m (counterclockwise)}$$

Similarly,
$$M_1 = -3(0.5) = -1.5 \text{ kN} \cdot \text{m}$$
$$M_{1.5} = -4.5(0.75) = -3.4 \text{ kN} \cdot \text{m}$$
$$M_2 = -6(1) = -6 \text{ kN} \cdot \text{m}$$
$$M_{2.5} = -7.5(1.25) = -9.4 \text{ kN} \cdot \text{m}$$
$$M_3 = -9(1.5) = -13.5 \text{ kN} \cdot \text{m}$$

This beam, with its shear and moment diagrams, is shown in Fig. 11-23.

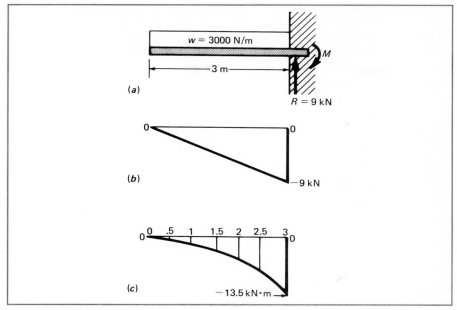

Fig. 11-23 (*a*) Beam diagram for Sample Problem 11. (*b*) Shear force diagram. (*c*) Moment diagram.

11-5 RELATIONS AMONG BEAM LOADING, SHEAR DIAGRAM, AND MOMENT DIAGRAM

Examination of the preceding examples (Sample Problems 8 to 11), and especially the sketches in those examples (Figs. 11-20 to 11-23), reveals that certain relations exist among beam loading, the shape of the shear force diagram, and the shape of the bending moment diagram. These relations can be summarized as follows:

Beam load	Shear diagram	Moment diagram
1. Concentrated load	Horizontal line	Sloped straight line
2. Uniform load	Sloped straight line	Curved line

Item 1 is demonstrated in Figs. 11-20 and 11-22; item 2 is shown in Figs. 11-21 and 11-23. The mathematical expressions which explain these relations will be discussed in Sec. 11-6.

In general, the *vertical height* (ordinate) at any point on a diagram will equal the *slope* of the next higher degree diagram at the same point. Thus, the vertical height at any point on the shear diagram equals the slope of the bending moment diagram at that same point. One important application of this relation is revealed by close inspection of how the shear force diagrams relate to the corresponding bending moment diagrams in Figs. 11-20 and 11-23. The relation can be stated as follows.

When the shear diagram passes through zero or equals zero for a particular beam section, the value on the moment diagram for that section will be either:

(a) A maximum (see Figs. 11-20 to 11-23)
(b) Zero (see Figs. 11-20 to 11-23)
(c) A relative maximum (see Fig. 11-25)

Sample Problem 12 Determine the moment diagram for the overhanging beam shown in Fig. 11-24. Neglect the weight of the beam.

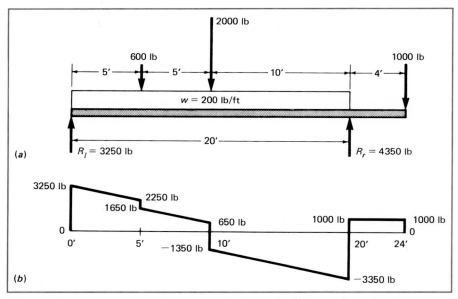

Fig. 11-24 (*a*) Beam diagram for Sample Problem 12. (*b*) Shear force diagram.

Solution: Application of the principles stated above can facilitate the determination of the moment diagram. Thus, the moment diagram will be curved for the portion of the beam between the reactions because of the uniform load. The overhanging portion, which has no uniform load, will have a straight sloped line for its moment diagram.

Furthermore, the shear diagram indicates two possible locations (10 and 20 ft from the left end) for maximum bending moment where the shear diagram goes through zero. Although the shear diagram is zero at both ends of the beam, the moments at the end sections are zero for a *free* or *simply supported end.*

If an accurate moment diagram is required, moments should be calculated for (1) each section where the shear diagram changes direction or jogs, (2) sections where the shear diagram crosses through or equals zero, and (3) several intermediate sections in regions where the moment diagram will curve.

Rather than calculate a moment for each 1-ft section, this solution will show only two or three intermediate calculations between important sections so that the shape of the diagram can be determined.

$M_0 = 0$ (by inspection)

$M_2 = 3250(2) - 200(2)(1) = 6500 - 400 = 6100 \text{ ft} \cdot \text{lb}$

$M_4 = 3250(4) - 200(4)(2) = 13\,000 - 1600 = 11\,400 \text{ ft} \cdot \text{lb}$

$M_5 = 3250(5) - 200(5)(2.5) = 16\,250 - 2500 = 13\,750 \text{ ft} \cdot \text{lb}$

$M_7 = 3250(7) - 600(2) - 200(7)(3.5) = 22\,750 - 1200 - 4900$
$$= 16\,650 \text{ ft} \cdot \text{lb}$$

$M_9 = 3250(9) - 600(4) - 200(9)(4.5) = 29\,250 - 2400 - 8100$
$$= 18\,750 \text{ ft} \cdot \text{lb}$$

$M_{10} = 3250(10) - 600(5) - 200(10)(5) = 32\,500 - 3000 - 10\,000$
$$= 19\,500 \text{ ft} \cdot \text{lb}$$

$M_{12} = 3250(12) - 600(7) - 2000(2) - 200(12)(6) = 39\,000 - 4200$
$$-4000 - 14\,400 = 16\,400 \text{ ft} \cdot \text{lb}$$

$M_{15} = 3250(15) - 600(10) - 2000(5) - 200(15)(7.5) = 48\,750$
$$-6000 - 10\,000 - 22\,500 = 10\,250 \text{ ft} \cdot \text{lb}$$

$M_{18} = 3250(18) - 600(13) - 2000(8) - 200(18)(9) = 58\,500$
$$-7800 - 16\,000 - 32\,400 = 2300 \text{ ft} \cdot \text{lb}$$

$M_{20} = 3250(20) - 600(15) - 2000(10) - 200(20)(10) = 65\,000$
$$-9000 - 20\,000 - 40\,000 = -4000 \text{ ft} \cdot \text{lb}$$

$M_{24} = 0$ (by inspection)

A plot of these moment values results in the moment diagram shown in Fig. 11-25. The calculations and the diagram reveal that the maximum bending moment occurs at the section 10 ft from the left end (note that the shear diagram passes through zero) and a relative maximum moment occurs at the 20-ft section (shear diagram passes through zero).

11-6   BENDING MOMENT FROM SHEAR DIAGRAM AREA

A mathematical relation exists between a bending moment at a given section in a statically determinate beam and the area of the shear diagram from the end of the beam to that section. The expression is

$$M_x = A_{0-x} \tag{11-2}$$

where A_{0-x} = shear diagram area between left end of beam and section x. Since the shear diagram is a plot of shear force V in pounds (or new-

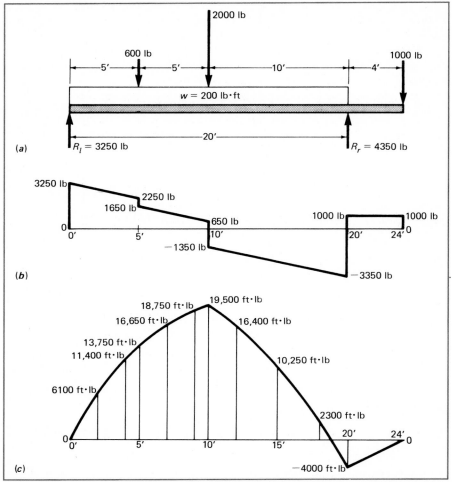

Fig. 11-25 (*a*) Beam diagram for Sample Problem 12. (*b*) Shear force diagram. (*c*) Moment diagram.

tons) and beam length in feet (or meters), areas on the diagram have units of foot-pounds (or newton-meters)

M_x = bending moment (ft·lb or N·m) at section x

Equation (11-2) is a specific application of a more general relation between successive diagrams. In general, the *area* between any two vertical heights (ordinates) on a diagram will equal the *difference in length* of the two corresponding vertical heights (ordinates) on the next higher degree diagram.

To demonstrate the use of Eq. (11-2), consider a simply supported weightless beam with a central concentrated load of 900 lb on a 12-ft span. Suppose the moment at the section 4 ft from the left end is required. Then the shaded area on the shear diagram of Fig. 11-26 will equal the value of the moment M_4.

$$M_4 = +450(4) = +1800 \text{ ft}\cdot\text{lb}$$

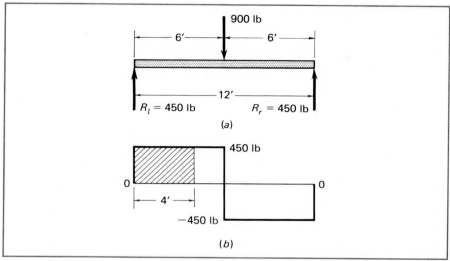

Fig. 11-26 (*a*) Beam diagram. (*b*) Shear force diagram. Cross-hatched area equals bending moment at 4-ft section.

Note that shear diagram areas above the zero line are positive and areas below the zero line are negative.

By the same method, the maximum bending moment at the center section (shear diagram crosses zero) is

$$M_6 = M_{\text{max}} = +450(6) = +2700 \text{ ft}\cdot\text{lb}$$

Suppose it is required to find the moment for the section 9 ft from the left end of this beam. From Eq. (11-2),

$$M_9 = A_{0-9}$$

Thus, the shear diagram area from the left end to the 9-ft section equals M_9. Figure 11-27 shows that some of the shaded area A_{0-9} is plus and some of it is minus. In totaling the areas, plus and minus signs must be considered; therefore,

$$M_9 = [+450(6)] + [-450(3)] = +2700 - 1350 = +1350 \text{ ft}\cdot\text{lb}$$

Although these examples are fairly simple, the method is similar regardless of the shape and complexity of the shear diagram. Bending moments determined by the shear diagram area method correctly represent the effect of the beam loading from which the shear diagram was sketched. However, if the beam is subjected to an external moment (for example, by a couple acting at the

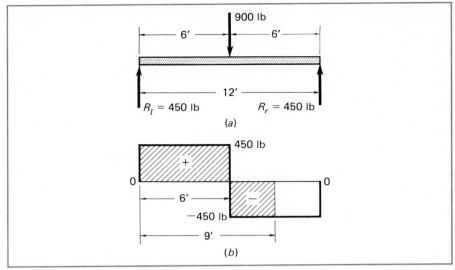

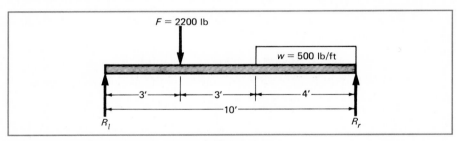

Fig. 11-27 (*a*) Beam diagram. (*b*) Shear force diagram. Algebraic sum of cross-hatched areas equals bending moment at 9-ft section.

end of the beam), then the true bending moment is determined by algebraically adding the external moment to the shear diagram area.

Sample Problem 13 Determine the shear force and bending moment diagrams for the beam shown in (*a*) Fig. 11-28 and (*b*) Fig. 11-29.

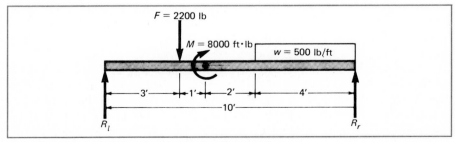

Fig. 11-28 Beam diagram for Sample Problem 13*a*.

Fig. 11-29 Beam diagram for Sample Problem 13*b*.

Solution a: This simply supported beam is solved in the same way as in Sample Problems 8, 9, and 12.

$$\Sigma M_r = 0 \quad R_l(10) - F(7) - w(4)(2) = 0$$

$$R_l = \frac{2200(7) + 500(4)(2)}{10} = 1940 \text{ lb (up)}$$

$$\Sigma F_y = 0 \quad R_l + R_r - F - w(4) = 0$$

$$R_r = 2200 + 500(4) - 1940 = 2260 \text{ lb (up)}$$

Solution b: The beam shown in Fig. 11-29 is identical to the one in Fig. 11-28 except that a couple is applied which adds an external moment at a section 4 ft from the left support. Note that both the shear and moment diagrams are affected by the addition of the couple; see Figs. 11-30 and 11-31.

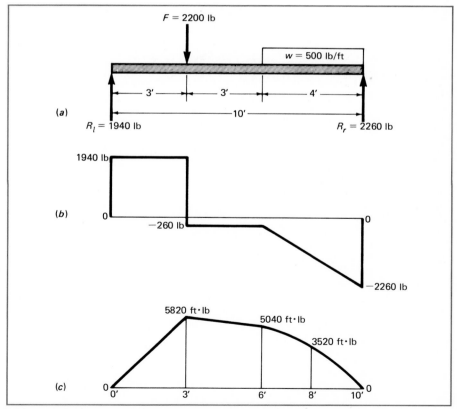

Fig. 11-30 (*a*) Beam diagram for Sample Problem 13*a*. (*b*) Shear force diagram. (*c*) Moment diagram.

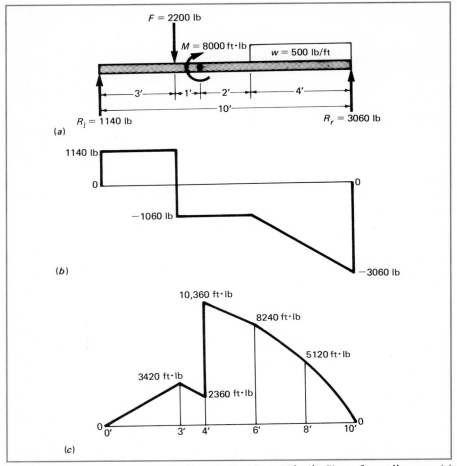

Fig. 11-31 (a) Beam diagram for Sample Problem 13b. (b) Shear force diagram. (c) Moment diagram.

$$\Sigma M_r = 0$$
$$R_l(10) - F(7) + M - w(4)(2) = 0$$
$$R_l(10) - 2200(7) + 8000 - 2000(2) = 0$$
$$R_l = \frac{2200(7) - 8000 + 2000(2)}{10} = 1140 \text{ lb (up)}$$

$$\Sigma F_y = 0$$
$$R_l + R_r - 2200 - 2000 = 0$$
$$R_r = 4200 - 1140 = 3060 \text{ lb (up)}$$

*Sample Problem 14   Determine the shear force and bending moment diagrams for the beam shown in (a) Fig. 11-32 and (b) Fig. 11-33.

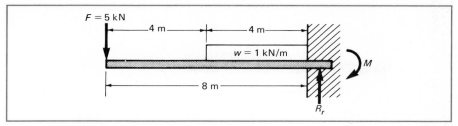

Fig. 11-32 Beam diagram for Sample Problem 14a.

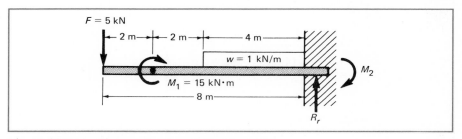

Fig. 11-33 Beam diagram for Sample Problem 14b.

Solution a: This cantilever beam is solved in the same way as in Sample Problems 10 and 11.

$$\Sigma M_r = 0 \quad -F(8) - w(4)(2) + M = 0$$
$$M = 5(8) + 1(4)(2) = 48 \text{ kN} \cdot \text{m (clockwise)}$$
$$\Sigma F_y = 0 \quad -F - w(4) + R_r = 0$$
$$R_r = 5 + 1(4) = 9 \text{ kN (up)}$$

Solution b: The beam shown in Fig. 11-33 is identical to the one shown in Fig. 11-32, except that a couple that is applied adds an external moment at a section 2 m from the left end. Note that since there is only one reaction R_r, its value remains the same in both cases and therefore the shear force diagram is unchanged. The moment diagram, however, is directly affected by the addition of the couple; see Figs. 11-34 and 11-35.

$$\Sigma M_r = 0$$
$$-F(8) + M_1 - w(4)(2) + M_2 = 0$$
$$M_2 = 5(8) - 15 + 1(4)(2)$$
$$M_2 = 40 - 15 + 8 = 33 \text{ kN} \cdot \text{m (clockwise)}$$
$$\Sigma F_y = 0$$
$$-F - w(4) + R_r = 0$$
$$R_r = 5 + 1(4) = 9 \text{ kN (up)}$$

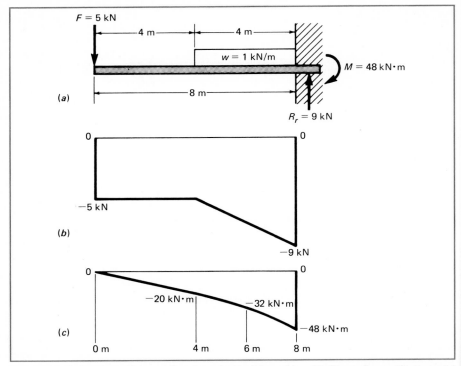

Fig. 11-34 (*a*) Beam diagram for Sample Problem 14*a*. (*b*) Shear force diagram. (*c*) Moment diagram.

11-7 MOVING LOADS

In calculating shear forces and bending moments, the assumption has been made that the loads on the beam were in a fixed position. There are many cases in which that is not true. A loaded truck or train passing over a bridge and a heavy machine being moved over a floor are examples of moving loads transmitted through wheels which are a fixed distance from each other. It is evident that the magnitude of the shear forces and bending moments will change as the load system moves across a beam. The problem, then, is to find the magnitude and location of the maximum shear force and bending moment. The following discussion will deal with simple beams only.

Maximum Shear: For simple beams, the maximum shear load V_{max} occurs at and is equal to the maximum reaction support force. Since the magnitude of the reactions will change as the loads move across the beam, it will be necessary to consider the loads at different locations in order to establish V_{max}. This is accomplished by alternately placing each load over a reaction and calculating the magnitude of that reaction. It should be kept in mind that the distance

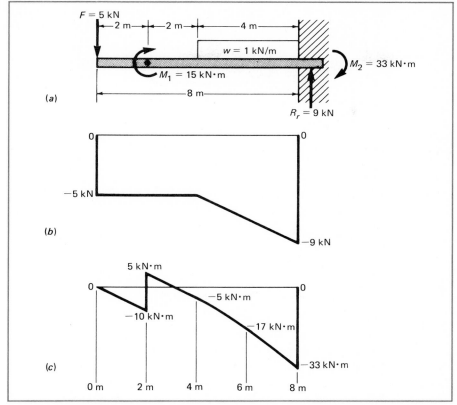

Fig. 11-35 (*a*) Beam diagram for Sample Problem 14*b*. (*b*) Shear force diagram. (*c*) Moment diagram.

between the loads must not change and that the loads can move in either direction.

Maximum Bending Moment: To establish the maximum bending moment M_{max} which develops because of the moving loads, the maximum moment under each load is calculated. The greatest of all the values so obtained is M_{max}. The following discussion will develop the theory of this problem; and although only three loads will be considered, the conclusions drawn will be general and can be applied to any number of loads.

Figure 11-36 shows a simple beam subjected to three moving loads: F_1, F_2, F_3. F is the resultant of the loads.

First, determine the reaction R_l.

$$\Sigma M_r = 0 \qquad R_l(L) - F(L - x - l) = 0$$

$$R_l = \frac{F(L - x - l)}{L}$$

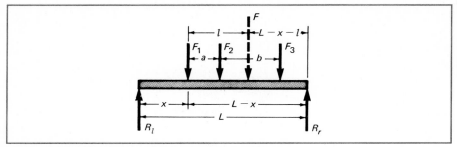

Fig. 11-36 Moving loads on a simply supported beam.

For bending moment under F_1, consider the section of the beam to the left of F_1.

$$M_x = R_l x$$
$$= \frac{F(Lx - x^2 - lx)}{L}$$
$$= -\frac{F}{L}[x^2 - (L - l)x]$$

But bending moment is a maximum when the expression in the brackets is greatest, since F is constant. Now, in algebra it is shown that the quadratic expression $Ax^2 + Bx + C$ is greatest or least when

$$x = -\frac{B}{2A}$$

In $x^2 - (L - l)x$, $A = 1$, $B = -(L - l)$, and $C = 0$
   Then $x = (L - l)/2$ gives a maximum or a minimum. Since the minimum moment at the supports is zero, the preceding value of x is for the maximum moment. The distance from R_l to F is

$$x + l = \frac{L - l}{2} + l = \frac{L + l}{2}$$

It is now evident that the distance from R_l to the resultant F is $(L + l)/2$. That is, the center of the beam is midway between the load F_1 and the resultant F of all the loads.
   The rule can be stated as follows: *The maximum bending moment under any load of a set of moving concentrated loads occurs when that load is as far on one side from the center of the beam as the resultant of all the loads on the beam is on the other side.* This rule applies to simple beams only.
   To determine the maximum bending moment for any possible placement of loads, the above calculation should be made for each of the moving loads

placed in its maximum bending moment position. The bending moment used for beam design will be the largest of the maximum bending moments.

*Sample Problem 15 Two loads of 40 and 20 kN that are 3 m apart roll over a beam 10 m long. Find (*a*) the maximum shear load (V_{max}) and (*b*) the maximum bending moment (M_{max}).

Solution a: Figures 11-37 and 11-38 show the positions of the loads to develop maximum shear at each of the reactions. V_{max} will occur at one of the two reactions.

From Fig. 11-37: $R_l = 40 + \dfrac{7}{10}(20) = 40 + 14 = 54$ kN

From Fig. 11-38: $R_r = 20 + \dfrac{7}{10}(40) = 20 + 28 = 48$ kN

Therefore, $V_{max} = 54$ kN

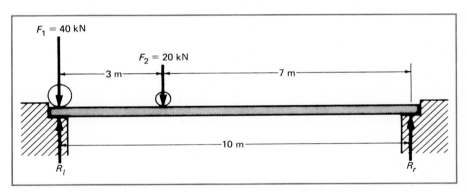

Fig. 11-37 Beam diagram for Sample Problem 15. Load positions arranged for maximum shear at R_l.

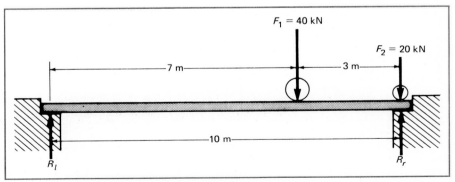

Fig. 11-38 Beam diagram for Sample Problem 15. Load positions arranged for maximum shear at R_r.

Solution b: Referring to Fig. 11-39, find the position of the resultant with reference to F_1 by taking moments about F_1.

$$F_2(3) - F(l) = 0$$
$$20(3) - 60(l) = 0$$
$$60 = 60l$$
$$l = 1 \text{ m}$$

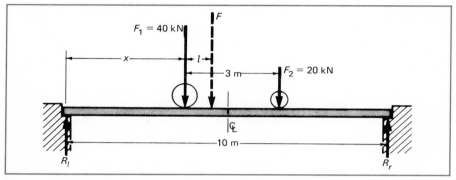

Fig. 11-39 Beam diagram for Sample Problem 15. Loads and resultant arranged in a general position.

Then, for maximum moment,

$$x = \frac{L - l}{2} = \frac{10 - 1}{2} = 4.5 \text{ m}$$

or F_1 is 0.5 m to the left of the center of the beam and the resultant F is 0.5 m to the right.

Figure 11-40 shows the location of the loads with F_1 in its maximum bending moment position.

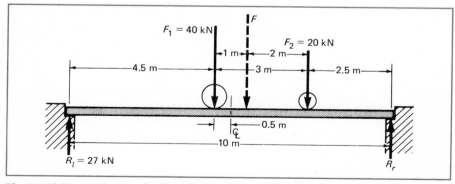

Fig. 11-40 Beam diagram for Sample Problem 15. Load positions arranged for maximum moment under F_1.

$$\Sigma M_r = 0 \qquad R_l(10) - 40(5.5) - 20(2.5) = 0$$
$$10R_l = 270$$
$$R_l = 27 \text{ kN}$$

Find the moment at the section under F_1 (40-kN load).

$$M_1 = 27(4.5) = 121.5 \text{ kN} \cdot \text{m}$$

M_1 is the maximum moment that is developed under the 40-kN load. Since F is 2 m from F_2, the maximum moment under the F_2 load occurs when F and F_2 are each 1 m from the center of the beam, as shown in Fig. 11-41.

$$\Sigma M_r = 0 \qquad R_l(10) - 40(7) - 20(4) = 0$$
$$10R_l = 360$$
$$R_l = 36 \text{ kN}$$

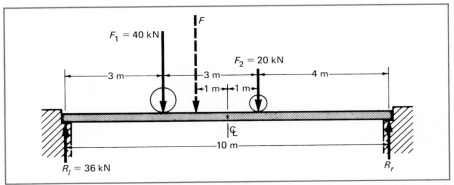

Fig. 11-41 Beam diagram for Sample Problem 15. Load positions arranged for maximum moment under F_2.

Next, take a section of the beam at F_2 (20-kN load) and find moments of the load to the left of that section.

$$M_2 = 36(6) - 40(3) = 216 - 120 = 96 \text{ kN} \cdot \text{m}$$

This is the maximum moment developed under the 20-kN load. Since the maximum moment under F_1 is greater than that under F_2, the maximum for the beam is 121.5 kN·m.

Sample Problem 16 Three loads of 6000, 9000, and 5000 lb are placed 6 and 10 ft apart, respectively (Fig. 11-42). They are moved across a bridge of span 40 ft. Find the maximum bending moment.

Solution: First, locate the resultant F with reference to F_1.

$$20\,000l = 6(9000) + 16(5000)$$
$$l = 6.7 \text{ ft}$$

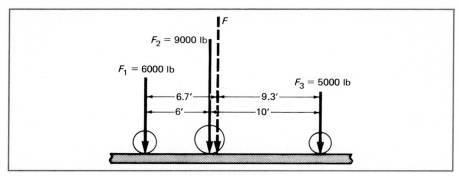

Fig. 11-42 Load diagram for Sample Problem 16.

A free-body diagram is shown for each load in the position for maximum moment (Figs. 11-43, 11-44, and 11-45), and results are given. The work of verifying results is left to the reader.

$$M_1 = 16.65(8320) = 138\ 500\ \text{ft}\cdot\text{lb}$$
$$M_2 = 19.65(9820) - 6(6000) = 157\ 000\ \text{ft}\cdot\text{lb}$$
$$M_3 = 24.65(12\ 320) - 16(6000) - 10(9000) = 117\ 700\ \text{ft}\cdot\text{lb}$$

Therefore, $M = 157\ 000$ ft $\cdot$ lb is the maximum bending moment.

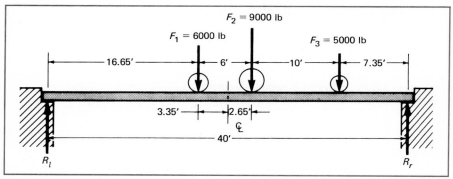

Fig. 11-43 Beam diagram for Sample Problem 16. Maximum moment at F_1.

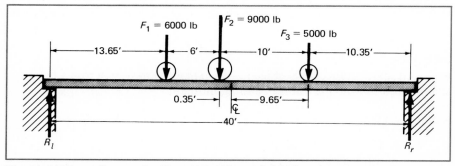

Fig. 11-44 Beam diagram for Sample Problem 16. Maximum moment at F_2.

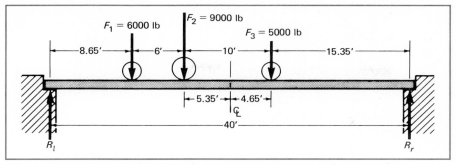

Fig. 11-45 Beam diagram for Sample Problem 16. Maximum moment at F_3.

PROBLEMS

Calculate the unknown reactions. Sketch and label the shear force diagrams for Probs. 11-1 to 11-10.

11-1. A simply supported beam with a 10-ft span carries a concentrated load of 2000 lb acting 4 ft from the left support. Neglect the weight of the beam.

*11-2. The beam in Fig. Prob. 11-2.

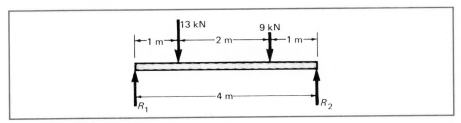

FIGURE PROBLEM 11-2

*11-3. A simply supported beam 5 m long carries a uniformly distributed load of 1.5 kN/m, including the weight of the beam and a concentrated load of 15 kN that is 1.5 m from the right end.

11-4. The beam in Fig. Prob. 11-4.

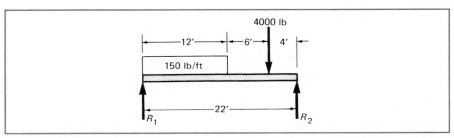

FIGURE PROBLEM 11-4

*11-5. A 3.5-m-long cantilever beam of negligible weight carries concentrated loads of 10 and 8 kN that act at 1.5 and 3.5 m, respectively, from the wall.

11-6. The beam in Fig. Prob. 11-6.

11-7. The beam in Fig. Prob. 11-7.

*11-8. The beam in Fig. Prob. 11-8.

*11-9. The beam in Fig. Prob. 11-9.

11-10. The beam in Fig. Prob. 11-10.

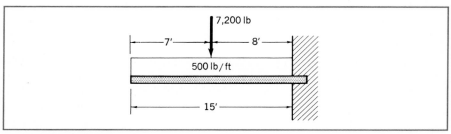

FIGURE PROBLEM 11-6

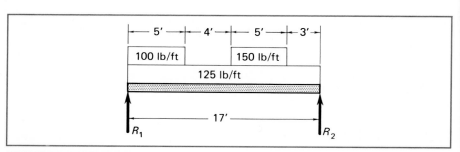

FIGURE PROBLEM 11-7

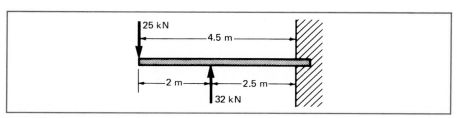

FIGURE PROBLEM 11-8

12

Beams — Design

12-1 STRESS DUE TO BENDING

In Sec. 11-2 it was shown that the fiber stresses over any cross section of a beam are a maximum at the extreme outer fibers and decrease to zero at the neutral axis.

The formula for the stress in a beam due to bending will be derived so that it will be applicable to the cross section of a beam of any shape. For simplicity, a rectangular section will be used. The result will be true, in general, within the elastic range, as proved in more advanced works on strength of materials.

Let Fig. 12-1 be the section of a rectangular beam. XX is the centroidal axis. Take a thin rectangle of area a whose centroid is at distance y_a from XX. Let s_a be the stress on area a. Then the force F_a on area a is $F_a = s_a a$. The moment of F_a about XX is

$$F_a y_a = s_a a y_a$$

But the fiber stress is 0 at the centroidal axis and increases in proportion to the distance y to a maximum value s at 1-2, the extreme fiber where $y = c$. Then,

$$\frac{s_a}{y_a} = \frac{s}{c}$$

or

$$s_a = \frac{s y_a}{c}$$

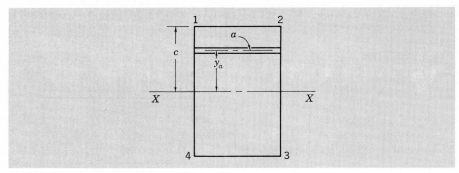

Fig. 12-1 Section of a rectangular beam.

By substituting in the equation for the moment, above, we obtain

$$F_a y_a = \frac{s y_a}{c} a y_a$$

$$= \frac{s}{c} y_a^2 a$$

If we divide the remaining cross-sectional area into a large number of very thin areas similar to a, Fig. 12-1, the bending moment for each can be represented by a formula as above. Since these moments are referred to a common axis XX, they can be added. Thus the total moment M is

$$M = \Sigma \left(\frac{s}{c} y^2 a \right)$$

But s/c is the same for every term in the summation; therefore, s/c can be factored, giving us

$$M = \frac{s}{c} \Sigma y^2 a$$

From Sec. 10-7, $\Sigma y^2 a = I$, the moment of inertia about axis XX. In Sec. 11-2 the moment of the resisting couple was shown to be equal to the bending moment of the external loads to the left of the section. Therefore,

$$M = \frac{s}{c} I$$

or
$$s = \frac{Mc}{I} \tag{12-1}$$

The expression I/c is called the *section modulus S*. Equation (12-1) can then be written

$$s = \frac{Mc}{I} = \frac{M}{I/c}$$

$$s = \frac{M}{S} \tag{12-2}$$

where s = maximum fiber stress due to bending, psi; N/mm^2 = MPa
   M = bending moment, in·lb; N·mm
   c = distance from neutral axis to the extreme fiber, in; mm
   I = moment of inertia about neutral axis, in^4; mm^4
   S = section modulus, in^3; mm^3

For a given cross section, the moment of inertia I and the distance c depend only on the size, shape, and axis location. Therefore, the section modulus $S = I/c$ also depends on those factors.

Expressions for I and S of a variety of common cross sections and structural shapes are given in App. B.

The equation for stress due to bending [Eq. (12-1)] applies only within the elastic range of the material, since proportionality of stress and strain was used in its derivation. An index of the rupture strength of the material in bending can be found by using the moment at the breaking point M_r in the flexure equation Eq. (12-1). Thus,

$$s_r = \frac{M_r c}{I} \qquad (12\text{-}3)$$

where s_r is a numerical index of rupture strength called *modulus of rupture*. Although s_r is useful for comparative purposes, it is not the true ultimate bending stress.

Sample Problem 1   An 8- by 10-in dressed-size timber beam carries a uniformly distributed load on a simply supported 14-ft span. Assume an allowable bending stress of 1100 psi. Find the maximum safe distributed load in pounds per foot.

Solution: The American Standard dressed size for a nominal 8 by 10 in is $7\frac{1}{2}$ by $9\frac{1}{2}$ in (see Table 11, App. B, for American Standard Timber Sizes).

Assume the cross section to be placed for maximum strength, that is, the maximum moment of inertia about the neutral axis. Figure 12-2a shows the cross section placed for maximum strength, and Fig. 12-2b shows the same cross section in a weaker position. Refer to Fig. 11-21, which shows that the maximum bending moment occurs at the center of the beam. The maximum stress occurs where the bending moment is maximum. Therefore, the maximum bending moment for a simply supported beam with a uniformly distributed load across the entire span, as shown in Fig. 12-3, is

$$M_{\text{max}} = R_1 \left(\frac{L}{2}\right) - \frac{W}{2}\left(\frac{L}{4}\right) = \frac{WL}{4} - \frac{WL}{8} = \frac{WL}{8} = \frac{wL^2}{8}$$

For the particular beam in Fig. 12-3,

$$M_{\text{max}} = \frac{W(14)}{8} = 1.75W \text{ ft·lb}$$

or
$$M_{\text{max}} = 1.75(12)W = 21W \text{ in·lb}$$

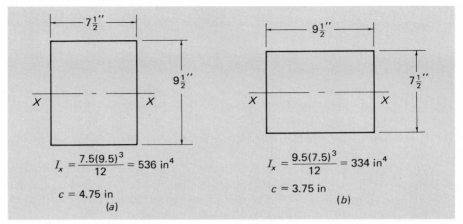

$$I_x = \frac{7.5(9.5)^3}{12} = 536 \text{ in}^4$$

$$c = 4.75 \text{ in}$$

(a)

$$I_x = \frac{9.5(7.5)^3}{12} = 334 \text{ in}^4$$

$$c = 3.75 \text{ in}$$

(b)

Fig. 12-2 Diagram for Sample Problem 1: (a) cross section placed for maximum bending strength; (b) cross section placed for minimum bending strength.

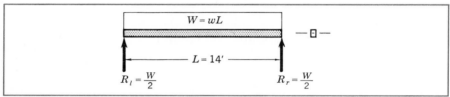

Fig. 12-3 Beam diagram for Sample Problem 1.

From Eq. (12-1),

$$s = \frac{Mc}{I} \qquad 1100 = \frac{21W(4.75)}{536}$$

$$W = \frac{1100(536)}{21(4.75)} = 5910 \text{ lb}$$

$$w = \frac{W}{L} = \frac{5910}{14} = 422 \text{ lb/ft (maximum safe load, including beam weight)}$$

Sample Problem 2   A run of 4-in schedule 40 seamless steel pipe (4.50-in *OD*, 0.237-in wall thickness) is to carry a $\frac{1}{4}$-ton-capacity chain hoist attached midway between pipe support hangers. The ultimate tensile strength of the steel pipe is 48 000 psi. A safety factor of 4 is specified. The pipe weighs 10 lb/ft. Assume no additional stress or load due to internal pressure. Treat the length of pipe between hangers as a simply supported beam. Find the maximum safe spacing of pipe support hangers.

Solution: Sketch the system as in Fig. 12-4.

$$\Sigma F_y = 0 \qquad R_l + R_r - 10L - 500 = 0$$

$$R_l = R_r = 5L + 250 \text{ (due to symmetry)}$$

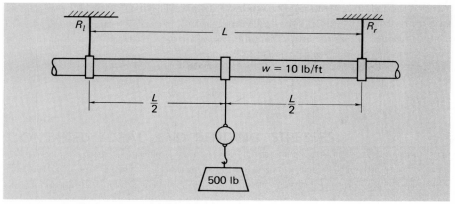

Fig. 12-4 Diagram for Sample Problem 2.

Find $M_{\max}$ at center of span.

$$M_{\max} = R_l\left(\frac{L}{2}\right) - \frac{10L}{2}\left(\frac{L}{4}\right)$$

$$= (5L + 250)\frac{L}{2} - \frac{5L^2}{4}$$

$$= \frac{5L^2}{2} + 125L - \frac{5L^2}{4} = \frac{5L^2}{4} + 125L \text{ ft} \cdot \text{lb}$$

$$= 15L^2 + 1500L \text{ in} \cdot \text{lb}$$

For a hollow circular section (Table 12, App. B),

$$I = \frac{\pi}{64}(d_0^4 - d_i^4) = \frac{\pi}{64}(4.50^4 - 4.03^4)$$

$$= \frac{\pi}{64}(410 - 264) = \frac{\pi}{64}(146) = 7.17 \text{ in}^4$$

$$c = \frac{4.50}{2} = 2.25 \text{ in}$$

Allowable stress $= 48\,000/4 = 12\,000$ psi. From Eq. (12-1),

$$s = \frac{Mc}{I} \qquad M = \frac{sI}{c}$$

$$M = \frac{12\,000(7.17)}{2.25} = 38\,200 \text{ in} \cdot \text{lb}$$

$$15L^2 + 1500L = 38\,200$$

$$L^2 + 100L - 2550 = 0$$

This quadratic equation can be solved by the quadratic formula.

$$L = \frac{-100 \pm \sqrt{100^2 - 4(1)(-2550)}}{2} = \frac{-100 \pm \sqrt{10\ 000 + 10\ 200}}{2}$$

$$= \frac{-100 \pm \sqrt{20\ 200}}{2} = \frac{-100 \pm 142}{2}$$

We select the positive result and obtain

$$L = \frac{-100 + 142}{2} = \frac{42}{2} = 21.0 \text{ ft (maximum safe spacing of supports)}$$

Sample Problem 3[1]   For the beam shown in Fig. 12-5, find the following.

(a)   The maximum compressive stress in the beam
(b)   The maximum tensile stress in the beam
(c)   The factors of safety in tension and compression based on the ultimate stresses if the material is AISI 1020 steel

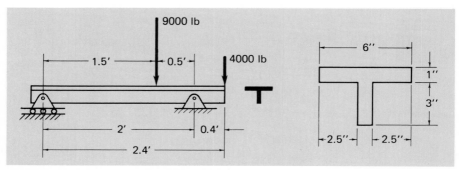

Fig. 12-5 Beam diagram for Sample Problem 3.

Note: In this problem, the weight of the beam will be neglected. As a rule of thumb, if the total beam weight is less than 10 percent of the total load, the weight of the beam may be neglected without serious error. The density of steel is about 0.30 lb/in³. The volume of this beam is

$$V = [1(6) + 1(3)](2.4)(12) = 9(2.4)(12)$$
$$V = 259.2 \text{ in}^3 \qquad \text{say, } V = 260 \text{ in}^3$$
$$W = 260(0.30) = 78 \text{ lb (weight of beam)}$$

But $9000 + 4000 = 13\ 000$ lb (total load)

Therefore, neglect the weight of the beam.

[1] See also Sample Computer Program 19, Chap. 17.

Solution a: To determine stress, the centroid and the moment of inertia of the T section must be found.

By symmetry, the centroid is located somewhere on the vertical centerline (see Fig. 12-6). The distance $\bar{y}$ from the base of the T is found.

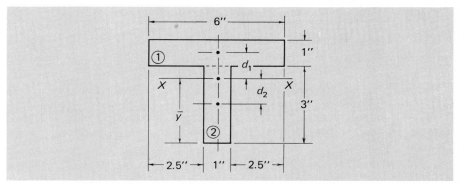

Fig. 12-6 Diagram for Sample Problem 3. Location of centroid of cross section.

Code	Dimen.	A	y	Ay
1	1×6	6	3.5	21
2	1×3	3	1.5	4.5
		$\Sigma A = 9$		$25.5 = \Sigma Ay$

$$\bar{y} = \frac{\Sigma Ay}{\Sigma A} = \frac{25.5}{9} = 2.83 \text{ in}$$

The moment of inertia about the horizontal centroidal axis is found as follows:

$$I_x = (I + Ad^2)_1 + (I + Ad^2)_2$$
$$d_1 = y_1 - \bar{y} = 3.5 - 2.83 = 0.67 \text{ in}$$
$$d_2 = \bar{y} - y_2 = 2.83 - 1.5 = 1.33 \text{ in}$$
$$I_x = \left[\frac{6(1^3)}{12} + 6(0.67^2)\right] + \left[\frac{1(3^3)}{12} + 3(1.33^2)\right]$$
$$= (0.5 + 2.68) + (2.25 + 5.33) = 3.18 + 7.58$$
$$= 10.76 \text{ in}^4$$

The distances to the extreme fibers are

$$c_1 = 4 - 2.83 = 1.17 \text{ in (to top)}$$
$$c_2 = 2.83 \text{ in (to base)}$$

To determine the maximum bending moment, calculate the reactions R_l and R_r and sketch the shear diagram as shown in Fig. 12-7.

$$\Sigma F_y = 0 \qquad R_l + R_r - 9000 - 4000 = 0$$
$$R_l + R_r = 13\,000 \text{ lb}$$
$$\Sigma M_l = 0 \quad -R_r(2) + 9000(1.5) + 4000(2.4) = 0$$
$$R_r = \frac{23\,100}{2} = 11\,550 \text{ lb}$$
$$R_l = 1450 \text{ lb}$$

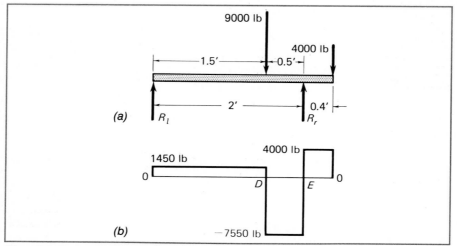

Fig. 12-7 (*a*) Beam diagram for Sample Problem 3. (*b*) Shear force diagram.

From the shear diagram, it is apparent that sections D and E must be investigated for maximum bending moment. Moments at D and E can be calculated either by taking moments of all forces to the left of the sections (see Sec. 11-4 and Sample Problem 12, Chap. 11) or by the shear diagram area method (see Sec. 11-6). By the shear diagram area method,

$$M_d = 1450(1.5) = 2180 \text{ ft·lb}$$
$$M_e = 1450(1.5) - 7550(0.5) = 2180 - 3780$$
$$= -1600 \text{ ft·lb}$$

Therefore, maximum bending moment $= M_d = 2180$ ft·lb.

From Eq. (12-1), the maximum compressive stress in the beam occurs at the top fiber of section D and is

$$s_c = \frac{Mc_1}{I} = \frac{2180(12)(1.17)}{10.76} = 2840 \text{ psi}$$

Solution b: The maximum tensile stress in the beam occurs at the bottom fiber of section D and is

$$S_t = \frac{Mc_2}{I} = \frac{2180(12)(2.83)}{10.76} = 6880 \text{ psi}$$

Solution c: From Table 1A, App. B, both the ultimate tensile and the compressive strengths for AISI 1020 steel are equal to 65 000 psi.

$$N_c = \frac{65\ 000}{2840} = 22.9$$

$$N_t = \frac{65\ 000}{6880} = 9.4$$

Sample Problem 4 A cantilever beam carries a uniform load of 400 lb/ft across the entire 9-ft length of span and a concentrated load of 3000 lb at the free end. For a maximum allowable stress of 22 000 psi, determine the most economical beam (lightest beam that satisfies the strength requirement), assuming that the section is:

(a) A W beam (a wide-flanged I-shaped beam)
(b) An S beam (a standard I-shaped beam)
(c) A standard channel

Solution: A cantilever beam has its maximum bending moment at the wall. Figure 12-8 shows this beam with its shear force diagram. The solution will omit the weight of beam until a selection is made, at which time the weight of beam will be compared to the total loading of 6600 lb for significance.

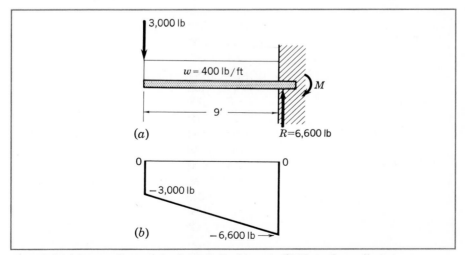

Fig. 12-8 (*a*) Beam diagram for Sample Problem 4. (*b*) Shear force diagram.

The maximum bending moment is

$$M = -3000(9) - 3600(4.5) = -27\,000 - 16\,200$$
$$= -43\,200 \text{ ft} \cdot \text{lb}$$

From Eq. (12-2), $s = \dfrac{M}{S}$

$$S = \frac{M}{s} = \frac{43\,200(12)}{22\,000} = 23.56 \text{ in}^3 \text{ (minimum required)}$$

For strength, the beam selected must have a section modulus greater than 23.56 in³. The most economical beam of a given type is selected for its light weight (therefore, low cost).

From the Section Modulus Table (Table 9, App. B):

(a)   W 10 × 26 (actual $S = 27.9$ in³)
(b)   S 10 × 25.4 (actual $S = 24.7$ in³)
(c)   C 12 × 25 (actual $S = 24.1$ in³)

Note that in each case the total beam weight is less than 5 percent of the load. If the beam weight were included in the calculations, it would increase both M and S only 2 percent.

12-2   HORIZONTAL AND VERTICAL SHEAR STRESSES

Shear has been defined as the tendency of two adjacent portions of a body to slide by each other like the two blades of a pair of scissors. The subject of vertical shear in beams has already been considered in Sec. 11-2. There is also a shearing tendency in a plane at right angles to the vertical shearing plane. This is called *horizontal shear.*

The reader will get the best idea of horizontal shear from a simple illustration. Place several 12- by 1-in boards on two supports, as shown in Fig. 12-9a. When the boards are of the same length, the ends will be practically straight and even. If now a load F is applied at the center of the pile, the boards will bend or sag in the middle and the result will be as shown in Fig. 12-9b. Each board will tend to slide on the one above or below it and in this way move the ends from their original positions. The horizontal motion of one board over the other is what causes horizontal shear.

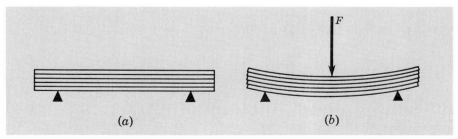

Fig. 12-9 (*a*) Boards before load is applied. (*b*) Boards after load is applied.

Now, if each board is glued securely to the ones next to it and a small load is applied, the ends of the pile of boards will remain square and even. There will still be a tendency to slide, but the motion will be prevented by the glue. This resistance measures the horizontal shear. Again, if the boards are securely bolted together and a load is applied, the boards cannot slip. The shearing resistance is offered by the bolts.

When the pile of boards is replaced by a single beam 12 by 6 in and a load F is applied, the beam will bend slightly. There is still the tendency of one horizontal portion to slide past the adjacent face. However, if the beam is not overloaded, the sliding motion is prevented by the internal fibers of the beam, just as it was resisted by the glue or bolts. The resistance that the fibers are capable of offering is called the *horizontal shearing strength.*

Some fibrous materials, such as wood and wrought iron, are more likely to fail from horizontal shear than materials that have no natural internal cleavage surfaces. Timber has a low shearing strength parallel to the grain; hence, a short beam tends to fail from horizontal shear rather than from bending.

Figure 12-10 represents a small cube cut from any part of a loaded beam. Let V_1 and V_2 be the vertical shearing forces and H_1 and H_2 the horizontal shearing forces. The dimension Δx is very small, so that the effect of the load so far as vertical shear on the two faces is concerned can be neglected and V_1 can be considered equal to V_2. Similarly,

$$H_1 = H_2$$

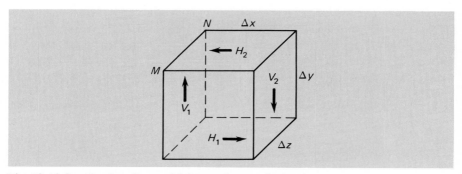

Fig. 12-10 Small cube of material from a beam with load.

Let s_v be the vertical shear stress and s_h the horizontal shear stress. Then

$$V_1 = \Delta y\, \Delta z\, s_v = V_2$$

and

$$H_1 = \Delta x\, \Delta z\, s_h = H_2$$

Since the block is in equilibrium under the shearing forces, moments about the line MN must equal zero, or

$$\Delta y\, H_1 = \Delta x\, V_2$$

$$\Delta x\, \Delta y\, \Delta z\, s_h = \Delta x\, \Delta y\, \Delta z\, s_v$$

Therefore,

$$s_h = s_v$$

In Chap. 7 it was assumed that the shearing stress was of the same intensity on each square inch of the cross section. In the derivations and the solutions that follow, it will be shown that the shearing stress in a beam of symmetrical section is a maximum at the neutral axis and decreases to zero at the outer fibers.

The relation between the stresses just developed can be stated as follows: *At any point in a member subjected to shearing forces, there exist equal shearing stresses in planes mutually at right angles to each other.* The subscripts can be dropped and the equal unit shearing stresses can be designated as s_s.

Imagine the segment of the beam in Fig. 12-11a between planes AB and CD to be removed as a free body. The enlarged view of this free body is shown in Fig. 12-11b with the distribution of stresses due to bending. The stresses on section CD are greater than the corresponding stresses on section AB because the bending moment at CD is greater than at AB.

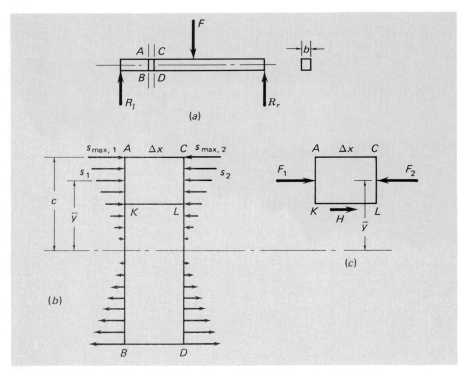

Fig. 12-11 Horizontal shear forces in a beam: (*a*) beam diagram; (*b*) bending stresses on a beam segment; (*c*) bending and horizontal shear forces on a portion of a beam segment.

In order to develop an equation for the shear stress on any horizontal plane, let us consider a plane such as KL in Fig. 12-11b. In Sec. 11-4 it was seen that, in general, bending moments vary from one end of the beam to the other. Let F_1 be the force due to the compressive stresses on surface AK (whose thickness is b) and F_2 that on surface CL, as shown in Fig. 12-11c.

Since F_1 and F_2 are different, they must be balanced by a force H on the surface KL. This is the horizontal shear force on that surface. Since F_2 is greater than F_1, for equilibrium,

$$H = F_2 - F_1$$

Let s_1 and s_2 be the fiber stresses at the centroids of the faces AK and CL, respectively, and a the area of each of the faces.

$$a = b(CL) = b(AK)$$

Then $$F_1 = s_1 a \quad \text{and} \quad F_2 = s_2 a$$

From Sec. 12-1, $$\frac{s_1}{\bar{y}} = \frac{s_{max,1}}{c} \quad \text{and} \quad \frac{s_2}{\bar{y}} = \frac{s_{max,2}}{c}$$

where $\bar{y}$ is the distance from the neutral axis to the centroid of area a, c is the distance from the neutral axis to the outside fibers, and $s_{max,1}$ and $s_{max,2}$ are the stresses in the outside fibers. Then

$$F_1 = s_1 a = \frac{s_{max,1}}{c}\,\bar{y}a$$

$$F_2 = s_2 a = \frac{s_{max,2}}{c}\,\bar{y}a$$

$$H = \frac{s_{max,2}}{c}\,a\bar{y} - \frac{s_{max,1}}{c}\,a\bar{y} = \frac{a\bar{y}}{c}(s_{max,2} - s_{max,1})$$

The horizontal shear force H on plane KL, whose area is $b\,\Delta x$, is also given by

$$H = (b\,\Delta x)s_h$$

Equating the two expressions for H gives us

$$b\,\Delta x\, s_h = \frac{a\bar{y}}{c}(s_{max,2} - s_{max,1})$$

$$s_h = \frac{a\bar{y}}{bc\,\Delta x}(s_{max,2} - s_{max,1})$$

If M_1 and M_2 are the bending moments on the respective faces, we have, by Eq. (12-1),

$$s_{max,1} = \frac{M_1 c}{I} \quad \text{and} \quad s_{max,2} = \frac{M_2 c}{I}$$

Then $$s_h = \frac{a\bar{y}}{bc\,\Delta x}\left(\frac{M_2 c}{I} - \frac{M_1 c}{I}\right)$$

$$= \frac{a\bar{y}}{Ib}\frac{M_2 - M_1}{\Delta x}$$

When Δx is small, as in this discussion,

$$\frac{M_2 - M_1}{\Delta x} = V$$

where V is the vertical shear force at sections AB and CD (since Δx is small). The value of V is taken from the shear force diagram at the section to be analyzed. By substituting, we obtain the final form for horizontal shear stress.

$$s_h = \frac{Va\bar{y}}{Ib} \qquad (12\text{-}4)$$

where s_h = horizontal shear stress, psi; N/mm² = MPa
  V = vertical shear force at section being considered, lb; N
  a = area of cross section of beam between fiber being considered and nearest extreme fiber, in²; mm²
  $\bar{y}$ = distance from neutral axis of entire cross section to centroid of area a, in; mm
  I = moment of inertia of entire cross section of beam, in⁴; mm⁴
  b = width of cross section at fiber being considered, in; mm

Sample Problem 5 A 10-ft-long, simply supported, 4- by 8-in rough-cut beam carries a uniform load of 100 lb/ft (Fig. 12-12*a*). Find the maximum horizontal shear stresses on each of the following horizontal planes.

(a) At the neutral axis
(b) 1 in above the neutral axis
(c) 2 in above the neutral axis
(d) 3 in above the neutral axis
(e) 4 in above the neutral axis

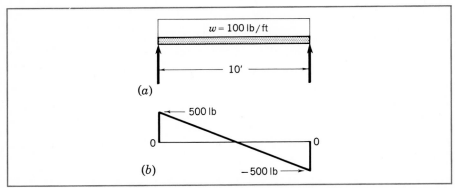

Fig. 12-12 (*a*) Beam diagram for Sample Problem 5. (*b*) Shear force diagram.

Solution: Equation (12-4) shows that s_h varies directly as V. Therefore, maximum s_h will occur at the section where V has its largest value. In Fig. 12-12*b* the shear force diagram shows that V is a maximum (500 lb) at both reactions

$$V = 500 \text{ lb} \qquad b = 4 \text{ in} \qquad h = 8 \text{ in}$$
$$I = \frac{bh^3}{12} = \frac{4(8^3)}{12} = 170.7 \text{ in}^4$$

(a) At the neutral axis XX (area a is shown shaded in Fig. 12-13a):

$$a = 16 \text{ in}^2 \qquad \bar{y} = 2 \text{ in}$$

$$s_h = \frac{Va\bar{y}}{Ib} = \frac{500(16)(2)}{170.7(4)} = 23.4 \text{ psi}$$

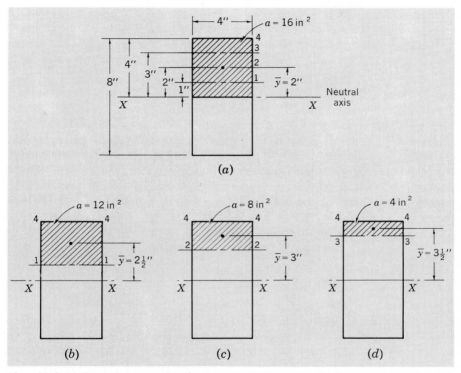

(a)

(b)   (c)   (d)

Fig. 12-13 (*a*) Beam cross section for Sample Problem 5*a*. (*b*) Beam cross section for Sample Problem 5*b*. *(c)* Beam cross section for Sample Problem 5*c*. (*d*) Beam cross section for Sample Problem 5*d*.

(b) At horizontal plane 1-1 (see Fig. 12-13b):

$$a = 12 \text{ in}^2 \qquad \bar{y} = 2.5 \text{ in}$$

$$s_h = \frac{500(12)(2.5)}{170.7(4)} = 21.9 \text{ psi}$$

(c) At horizontal plane 2-2 (see Fig. 12-13c):

$$a = 8 \text{ in}^2 \qquad \bar{y} = 3 \text{ in}$$

$$s_h = \frac{500(8)(3)}{170.7(4)} = 17.5 \text{ psi}$$

(d)   At horizontal plane 3-3 (see Fig. 12-13d):

$$a = 4 \text{ in}^2 \qquad \bar{y} = 3.5 \text{ in}$$

$$s_h = \frac{500(4)(3.5)}{170.7(4)} = 10.3 \text{ psi}$$

(e)   At horizontal plane 4-4 (the area a represents the area above the horizontal plane in question; thus the area above plane 4-4 is zero):

$$a = 0 \qquad \text{therefore} \qquad s_h = 0$$

The values of s_h for horizontal planes below the neutral axis are determined in the same way, except that area a is the area below the horizontal plane. This problem illustrates the principle that s_h is a maximum at the neutral axis and zero at the outer fibers.

To study the manner in which the shearing stresses vary, the stresses occurring at different distances from the neutral axis can be laid off to scale as shown in Fig. 12-14. The vectors representing the stresses are of varying length, as can be seen from the values computed in the foregoing example. If the ends of the vectors are joined by means of a smooth curve, any horizontal length between the vertical axis and the curve will represent the shearing stress at the corresponding point in the beam. The curve so determined is found to be a parabola.

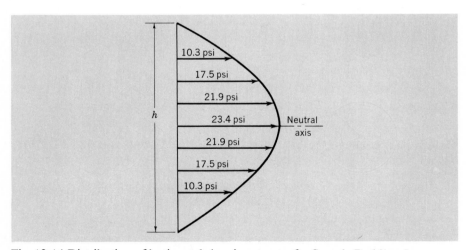

Fig. 12-14 Distribution of horizontal shearing stresses for Sample Problem 5.

12-3   MAXIMUM HORIZONTAL SHEAR STRESS FOR COMMON CROSS SECTIONS

For the special case of horizontal shear stress at the neutral axis when the cross section is a rectangle, a simplified formula is available.

For a rectangular cross section b in [or mm] wide by h in [or mm] high,

$$a = \frac{bh}{2} \qquad \bar{y} = \frac{h}{4} \qquad I = \frac{bh^3}{12}$$

Substituting this information into Eq. (12-4) gives us

$$s_h = \frac{Va\bar{y}}{Ib} = \frac{V\left(\dfrac{bh}{2}\right)\left(\dfrac{h}{4}\right)}{\dfrac{bh^3}{12}} \quad (b)$$

$$s_h = \frac{3V}{2A} \qquad\qquad (12\text{-}5)$$

where s_h = horizontal shear stress at the neutral axis for a rectangular cross section, psi; N/mm² = MPa
  V = vertical shear force at section being considered, lb; N
  $A = bh$, the area of the entire cross section, in²; mm²

Similarly, for a beam of circular cross section whose diameter is d in [or mm] (see Fig. 12-15),

$$a = \frac{\pi d^2}{8} \qquad \bar{y} = \frac{4r}{3\pi} = \frac{2d}{3\pi} \qquad b = d \qquad I = \frac{\pi d^4}{64}$$

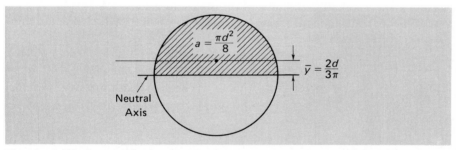

Fig. 12-15 Circular cross section for determination of maximum horizontal shear stress.

Substituting into Eq. (12-4) gives us

$$s_h = \frac{V\left(\dfrac{\pi d^2}{8}\right)\left(\dfrac{2d}{3\pi}\right)}{(\pi d^4/64)d}$$

$$s_h = \frac{4V}{3A} \qquad\qquad (12\text{-}6)$$

where s_h = horizontal shear stress at the neutral axis for a circular cross section, psi; N/mm² = MPa
  V = vertical shear force at section being considered, lb; N
  $A = \pi d^2/4$, the area of the entire cross section, in²; mm²

The reader is cautioned that Eqs. (12-5) and (12-6) are limited to calculations at the neutral axis for rectangular and circular cross sections, respectively.

12-4 MAXIMUM VERTICAL SHEAR STRESS IN S-SHAPE AND W-SHAPE BEAMS

The assumption is usually made in the case of S- and W-shape beams that the flange carries all the bending stress and the web, extending through the entire depth, takes the vertical shear. Although the assumption is not quite true, it is on the side of safety. The vertical shear stress in the webs of S- and W-shape beams is given by the equation

$$s_v = \frac{V}{td} \tag{12-7}$$

where s_v = maximum vertical shear stress, psi; N/mm^2 = MPa
V = maximum vertical shear force, lb; N
t = thickness of the web, in; mm
d = depth of the beam, in; mm

From Sec. 12-2, $s_v = s_h$.
In the design of a beam, tests should always be made for shearing strength as well as bending strength.

The code frequently used for the design of structural beams is the American Institute of Steel Construction (AISC) code. Table 12-1 gives the allowable stresses for the structural steels in bending and shear.

Sample Problem 6 An S 15 × 50 carries a total uniformly distributed load of 40 000 lb on a 20-ft simple span. Find the maximum bending stress and the shear stress in the web.

TABLE 12-1 AISC ALLOWABLE STRESSES FOR BEAMS

Material Specification (ASTM Designation)	Allowable Stress, psi	
	Bending (s_b)	Shear ($s_v = s_h$)
A 36, structural steel ($s_y = 36\ 000$ psi)	24 000	14 500
A 529, structural steel ($s_y = 42\ 000$ psi)	27 700	16 800
A 572, structural steel		
Grade 42 ($s_y = 42\ 000$ psi)	27 700	16 800
Grade 50 ($s_y = 50\ 000$ psi)	33 000	20 000
Grade 60 ($s_y = 60\ 000$ psi)	39 600	24 000
Grade 65 ($s_y = 65\ 000$ psi)	42 900	26 000
A 242, structural steel ($s_y = 50\ 000$ psi)	33 000	20 000
A 441, structural steel ($s_y = 46\ 000$ psi)	30 400	18 400
A 588, structural steel ($s_y = 50\ 000$ psi)	33 000	20 000

Solution: From Table 5, App. B, for an S 15 × 50, $S = 64.8$ in³, depth $d = 15$ in, and web thickness $t = 0.550$ in. For a simply supported beam with a uniform load,

$$M_{max} = \frac{WL}{8} = \frac{40\,000(20)}{8} = 100\,000 \text{ ft} \cdot \text{lb}$$

In bending,

$$s = \frac{M}{S} = \frac{100\,000(12)}{64.8} = 18\,500 \text{ psi}$$

Note that the weight of beam has been neglected.

From Fig. 12-16, $V_{max} = 20\,000$ lb. From Eq. (12-7) for web shear,

$$s_v = s_h = \frac{V}{td} = \frac{20\,000}{0.550(15)} = 2420 \text{ psi}$$

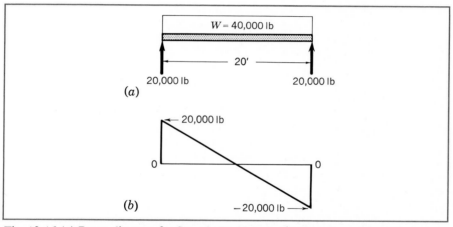

Fig. 12-16 (*a*) Beam diagram for Sample Problem 6. (*b*) Shear force diagram.

12-5 DISCUSSION OF BEAM DEFLECTION

When a load is placed on a beam, the beam tends to sag or deflect as shown in Fig. 12-17.

Deflection plays a very important part in the design of structures and machines. If floor beams, or joists, deflect too far, the plaster on the ceiling under them may crack. Although no damage to the structure may result, the appearance of the ceiling may be ruined. Also, a floor supported by such beams may be so out of level that its usefulness for machinery may be impaired.

Under load, the neutral axis becomes a curved line and is called the *elastic curve*. The deflection *y* is the vertical distance between a point on the elastic curve and the unloaded neutral axis.

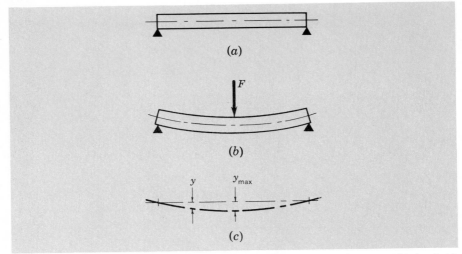

Fig. 12-17 Beam deflection due to load: (*a*) beam without load; (*b*) beam with load; (*c*) comparison of loaded and unloaded neutral axes.

12-6 RADIUS OF CURVATURE

The radius of curvature at any point on a curve is the radius of the circle which would match the shape of the curve at and in the immediate neighborhood of the point. In general, the elastic curve is not a circle, but for a very short length, such as Δl, it can be considered to be the arc of a circle (Fig. 12-18). Before the beam is loaded, sections AB and GH are parallel to each other (Fig. 12-18*a*). After loading, sections AB and GH occupy new positions $A'B'$ and $G'H'$ owing to deflection. These new positions are no longer parallel to each other; instead, both are perpendicular to the elastic curve (neutral axis) as shown in Fig. 12-18*b*. From Fig. 12-18*c* it is apparent that fiber AG has been shortened (compression) to length $A'G'$ and fiber BH has been stretched (tension) to $B'H'$, whereas the portion of the neutral axis XX has not changed in length. The elongation δ of fiber BH is the change in its length; therefore,

$$\delta = B'B + HH' = 2B'B \text{ (since } B'B = HH')$$

Also, $\Delta l = XX = BH$

The strain ϵ at the extreme fiber is

$$\epsilon = \frac{\delta}{\Delta l}$$

But the stress at this fiber is

$$s = \frac{Mc}{I}$$

and since the modulus of elasticity $E = s/\epsilon$,

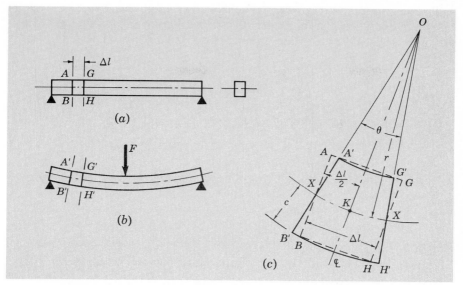

Fig. 12-18 Radius of curvature of a deflected beam: (*a*) beam before load is applied; (*b*) beam after load is applied; (*c*) effect of deflection on a beam segment.

$$E = \frac{Mc\,\Delta l}{\delta I}$$

By observing that triangle *OKX* is similar to triangle *XBB'*, we find that

$$\frac{OX}{XB'} = \frac{XK}{B'B}$$

But $\qquad\qquad OX = r$ (the radius of curvature)

$$XB' = c$$

$$XK = \frac{\Delta l}{2}$$

$$B'B = \frac{\delta}{2}$$

Substituting, $\qquad\qquad \dfrac{r}{c} = \dfrac{\Delta l/2}{\delta/2} = \dfrac{\Delta l}{\delta}$

From which $\qquad\qquad r = \dfrac{c\,\Delta l}{\delta}$

By combining the equation for *r* with the equation for *E*, we obtain

$$E = \frac{Mr}{I}$$

or $\qquad\qquad r = \dfrac{EI}{M} \qquad\qquad\qquad (12\text{-}8)$

It is evident that the greater the deflection at a point in a beam, the smaller will be the radius of curvature. That is, the deflection varies inversely with r. From Eq. (12-8), the deflection is found to depend directly on M and inversely on EI.

12-7   METHODS OF DETERMINING DEFLECTION FORMULAS

Several different methods are available for deriving beam deflection formulas. They are as follows.

1.   Moment-area method, which will be used in this book to determine maximum deflections.
2.   Double-integration method, a direct procedure involving higher mathematics[1]

The moment-area method makes use of relations between various diagrams. In Secs. 11-5 and 11-6, two important principles are stated as follows.

The area between any two vertical heights (ordinates) on a diagram will equal the difference in length of the two corresponding ordinates on the next higher degree diagram.

The vertical height (ordinate) at any point on a diagram will equal the slope of the next higher degree diagram at the same point.

The first of these statements was used in determining bending moments from the shear force diagram areas. The second statement was used to help predict the shape of successive diagrams and, most important, to pinpoint the possible locations of maximum bending moments from shear diagram zero locations.

There are two more successive diagrams which can be added to those we have used thus far. They will be introduced briefly merely to emphasize the relations. The next higher degree diagram ($EI\theta$ diagram) after the bending moment diagram is one which gives the *slope θ* of the deflected neutral axis multiplied by EI. The next higher degree diagram (EIy diagram) gives the *deflection y* of the neutral axis from the normal horizontal unloaded position multiplied by EI. Thus, areas on the moment diagram relate to differences in ordinate heights on the $EI\theta$ diagram, which in turn influence the slopes of the *elastic curve* (neutral axis) in the EIy diagram. These interrelations provide the basis for the moment-area method of determining deflection.

The moment-area method for a simply supported beam is developed as follows: Consider a simply supported beam which, under load, deflects so that its neutral axis takes the shape of the curve $ABCD$ in Fig. 12-19a. A very small length along the curve between points B and C has a radius of curvature r and

---

[1] E. F. Byars, R. D. Snyder, H. L. Plants, *Engineering Mechanics of Deformable Bodies,* 4th ed., Harper & Row, New York, 1983.

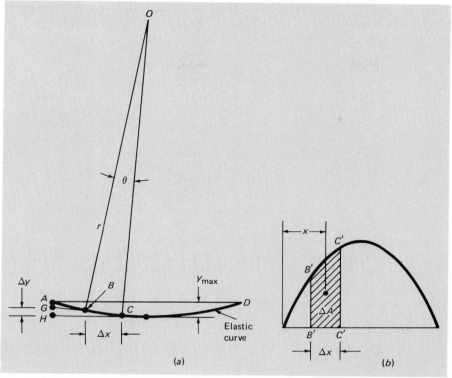

Fig. 12-19 (*a*) Elastic curve of a deflected beam. (*b*) Moment diagram of a deflected beam.

subtends angle $BOC = \theta$. Now

$$\tan \theta = \frac{BC}{r}$$

But

$$BC \approx \Delta x$$

since the length is short. Thus,

$$\tan \theta = \frac{\Delta x}{r}$$

By Eq. (12-8),

$$r = \frac{EI}{M}$$

where M is the bending moment at both $B'B'$ and $C'C'$ of Fig. 12-19*b*, since length Δx is short. Combining these equations gives us

$$\tan \theta = \frac{M \, \Delta x}{EI}$$

From the moment diagram (Fig. 12-19b) the shaded area ΔA is approximately rectangular, so that

$$\Delta A = M \, \Delta x$$

Then
$$\tan \theta = \frac{\Delta A}{EI}$$

If tangents to the elastic curve such as BG and CH are drawn at points B and C (Fig. 12-19a), angles OBG and OCH are right angles and the angle between BG and CH is θ. The vertical distance $\Delta y = GH$ along a perpendicular through point A represents a *portion* of the maximum deflection in this case. Since BG and CH are almost horizontal, the average of their lengths can be taken as x, the distance from the left reaction to the centroid of area ΔA.

From this discussion,

$$\tan \theta = \frac{\Delta y}{(BG + CH)/2} = \frac{\Delta y}{x}$$

Setting the two expressions for $\tan \theta$ equal gives us

$$\frac{\Delta y}{x} = \frac{\Delta A}{EI}$$

or
$$\Delta y = \frac{x \, \Delta A}{EI}$$

This equation indicates that Δy depends upon the *moment of an area* on the bending moment diagram.

For several systems, Δy is a portion of the maximum deflection y_{max}. If all the Δy values are added for all segments along the elastic curve between the left end A (Fig. 12-19a) and the point of maximum deflection, their total will be y_{max}. That is true for simple beams with symmetrical loading and for cantilever beams. When it is not true, a more general method is necessary. Derivation and detailed discussion of such methods are beyond the scope of this book.

Proceeding with the case at hand, we can find an expression for the maximum deflection by adding all the Δy components as follows.

$$
\begin{aligned}
y_{\text{max}} &= \Sigma \, \Delta y \\
&= \Delta y_1 + \Delta y_2 + \Delta y_3 + \cdots + \Delta y_n \\
&= \frac{x_1 \, \Delta A_1}{EI} + \frac{x_2 \, \Delta A_2}{EI} + \frac{x_3 \, \Delta A_3}{EI} + \cdots + \frac{x_n \, \Delta A_n}{EI} \\
&= \frac{\displaystyle\sum_{i=1}^{n} (x_i \, \Delta A_i)}{EI}
\end{aligned}
$$

Equation (10-2) can be written in the following form:

$$\bar{x} = \frac{\sum\limits_{i=1}^{n} (x_i \, \Delta A_i)}{\sum\limits_{i=1}^{n} \Delta A_i}$$

$$\sum_{i=1}^{n} (x_i \, \Delta A_i) = \bar{x} \sum_{i=1}^{n} \Delta A_i = \bar{x} A$$

where $\bar{x}$ is the distance from the centroid of shaded area A (Fig. 12-20) to the left reaction. Therefore,

$$y_{\text{max}} = \frac{\bar{x} A}{EI} \qquad (12\text{-}9)$$

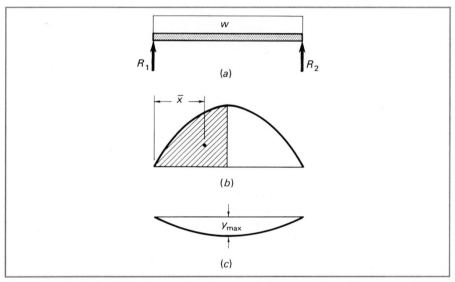

Fig. 12-20 (*a*) Beam diagram of a simply supported beam with uniform load. (*b*) Moment diagram. (*c*) Elastic curve.

Note that deflection is a function of a moment divided by EI, as discussed toward the end of Sec. 12-6.

Equation (12-9) can also be used to determine the deflection at any point in a cantilever beam, which will be demonstrated later in this chapter.

To determine the deflection at any point on a simply supported beam with symmetrical loading, it is necessary to find the maximum deflection and subtract from it an amount which represents the difference between y_{max} and the desired deflection.

The equation for deflection at any point on a simply supported beam with symmetrical loading (Fig. 12-21) is

$$y = \frac{\bar{x}_1 A_1}{EI} - \frac{\bar{x}_2 A_2}{EI} \qquad (12\text{-}10)$$

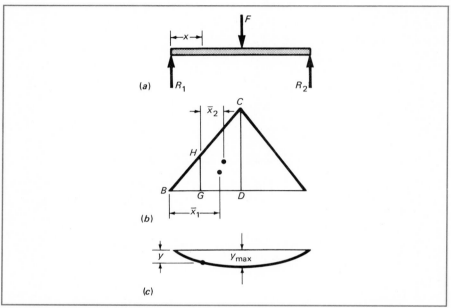

Fig. 12-21 (*a*) Beam diagram of a simply supported beam with concentrated load at midspan. (*b*) Moment diagram. (*c*) Elastic curve.

where $\dfrac{\bar{x}_1 A_1}{EI} = y_{\text{max}} \qquad \dfrac{\bar{x}_2 A_2}{EI} = y_{\text{max}} - y$

y = deflection at any point on a simply supported beam
A_1 = area of the bending moment diagram between the left reaction and the point of maximum deflection
$\bar{x}_1$ = horizontal distance from the centroid of area A_1 to the left reaction
A_2 = area of the bending moment diagram between the point where y is to be found and the point of maximum deflection
$\bar{x}_2$ = horizontal distance from the centroid of area A_2 to an axis through the point where deflection is to be determined
E = modulus of elasticity of the beam material
I = moment of inertia of the beam cross section

Note that for the beam in Fig. 12-21,

A_1 = area BCD of the bending moment diagram
$\bar{x}_1$ = moment arm of centroid of A_1 about the left end of the beam
A_2 = area $GHCD$ of the bending moment diagram
$\bar{x}_2$ = moment arm of centroid of A_2 about point x

12-8 DEFLECTION OF A SIMPLY SUPPORTED BEAM (CONCENTRATED LOAD AT CENTER)

Maximum deflection. By Eq. (12-9),

$$y_{max} = \frac{\bar{x}A}{EI}$$

From Fig. 12-22b,

$$A = \frac{1}{2}\left(\frac{FL}{4}\frac{L}{2}\right) = \frac{FL^2}{16}$$

$$\bar{x} = \frac{2}{3}\left(\frac{L}{2}\right) = \frac{L}{3}$$

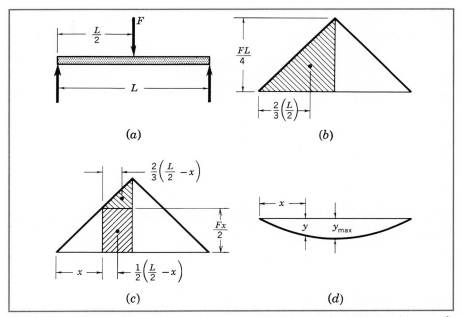

Fig. 12-22 (a) Beam diagram of a simply supported beam with load at midspan. (b) Moment diagram for derivation of maximum deflection equation. (c) Moment diagram for derivation of equation for deflection at any point. (d) Elastic curve.

Therefore,

$$y_{max} = \frac{\dfrac{L}{3}\left(\dfrac{FL^2}{16}\right)}{EI}$$

or
$$y_{max} = \frac{FL^3}{48EI}$$
(12-11)

Deflection at any point. By Eq. (12-10),

$$y = \frac{\bar{x}_1 A_1}{EI} - \frac{\bar{x}_2 A_2}{EI}$$

where $\bar{x}_1 A_1/EI = y_{max}$ (Sec. 12-7).

$\bar{x}_2 A_2$ is the moment of the entire shaded area in Fig. 12-22c. For convenience, the entire shaded area is divided into a rectangle and a triangle. The sum of the moment of the rectangular and triangular areas equals $\bar{x}_2 A_2$. Then

$$\bar{x}_2 A_2 = \left[\frac{1}{2}\left(\frac{L}{2} - x\right) \frac{Fx}{2} \left(\frac{L}{2} - x\right) \right]_{rectangle}$$
$$+ \left[\frac{2}{3}\left(\frac{L}{2} - x\right) \left(\frac{FL}{4} - \frac{Fx}{2}\right) \left(\frac{L}{2} - x\right) \frac{1}{2} \right]_{triangle}$$

from which
$$\frac{\bar{x}_2 A_2}{EI} = \frac{F}{48EI}(L^3 - 3L^2 x + 4x^3)$$

Substituting into Eq. (12-10) gives us

$$y = \frac{FL^3}{48EI} - \frac{F}{48EI}(L^3 - 3L^2 x + 4x^3)$$

which reduces to

$$y = \frac{Fx}{48EI}(3L^2 - 4x^2)$$
(12-12)

12-9 DEFLECTION OF A SIMPLY SUPPORTED BEAM (UNIFORM LOAD)

Maximum deflection. By Eq. (12-9),

$$y_{max} = \frac{\bar{x}A}{EI}$$

From Fig. 12-23b, $A = \frac{2}{3}$ of the enclosing rectangle.

$$A = \frac{2}{3}\left(\frac{WL}{8}\right)\left(\frac{L}{2}\right) = \frac{WL^2}{24}$$

$$\bar{x} = \frac{5}{8}\left(\frac{L}{2}\right) = \frac{5L}{16}$$

Substituting,
$$y_{max} = \frac{\frac{5L}{16}\left(\frac{WL^2}{24}\right)}{EI}$$

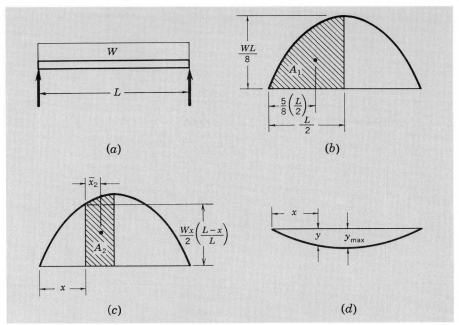

Fig. 12-23 (*a*) Beam diagram of a simply supported beam with uniform load. (*b*) Moment diagram for derivation of maximum deflection equation. (*c*) Moment diagram for derivation of equation for deflection at any point. (*d*) Elastic curve.

or

$$y_{\text{max}} = \frac{5WL^3}{384EI} \tag{12-13}$$

Deflection at any point. By applying a similar procedure as in Sec. 12-8 to Fig. 12-23*c*, we find that

$$y = \frac{Wx}{24EI} \left(\frac{L^3 - 2Lx^2 + x^3}{L} \right) \tag{12-14}$$

12-10 DEFLECTION OF A CANTILEVER BEAM (GENERAL)

The maximum deflection for a cantilever beam is given by

$$y_{\text{max}} = \frac{\bar{x}A}{EI} \tag{12-15}$$

where A is the entire area of the moment diagram and $\bar{x}$ is the distance from the centroid to the free end, as shown in Fig. 12-24*b*.

Equation (12-15) for a cantilever beam can be developed by using a procedure similar to that used in Sec. 12-7.

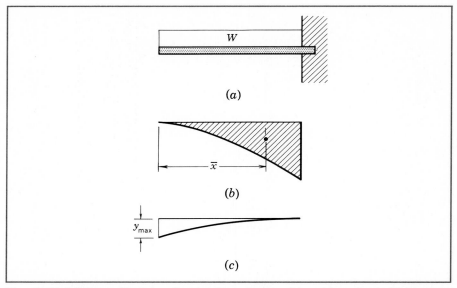

Fig. 12-24 (*a*) Beam diagram of a cantilever beam with uniform load. (*b*) Moment diagram. (*c*) Elastic curve.

12-11 DEFLECTION OF A CANTILEVER BEAM (CONCENTRATED LOAD AT FREE END)[1]

Maximum deflection. By Eq. (12-15),

$$y_{max} = \frac{\bar{x}A}{EI}$$

From Fig. 12-25*b*,

$$A = \frac{1}{2}(L)FL = \frac{FL^2}{2}$$

$$\bar{x} = \frac{2L}{3}$$

Substituting,

$$y_{max} = \frac{\dfrac{2L}{3}\left(\dfrac{FL^2}{2}\right)}{EI}$$

$$y_{max} = \frac{FL^3}{3EI} \qquad (12\text{-}16)$$

---

[1] For cases not shown here, refer to *Manual of Steel Construction,* American Institute of Steel Construction.

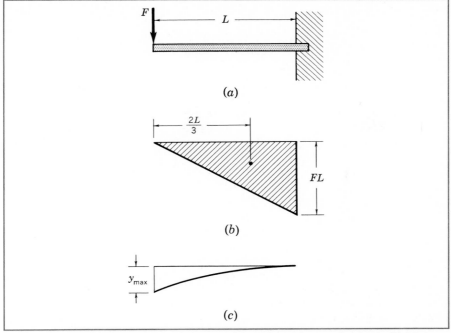

Fig. 12-25 (*a*) Beam diagram of a cantilever beam with load at free end. (*b*) Moment diagram. (*c*) Elastic curve.

-12 DEFLECTION OF A CANTILEVER BEAM (CONCENTRATED LOAD AT ANY POINT)

Maximum deflection. By Eq. (12-15),

$$y_{max} = \frac{\bar{x}A}{EI}$$

From Fig. 12-26*b*,

$$A = \frac{1}{2}(L-a)F(L-a) = \frac{F}{2}(L-a)^2$$

$$\bar{x} = \frac{2}{3}(L-a) + a = \frac{2}{3}L + \frac{a}{3}$$

Substituting, $$y_{max} = \frac{\left(\dfrac{2L}{3} + \dfrac{a}{3}\right)\left(\dfrac{F}{2}\right)(L-a)^2}{EI}$$

which reduces to

$$y_{max} = \frac{F}{6EI}(2L^3 - 3aL^2 + a^3) \tag{12-17}$$

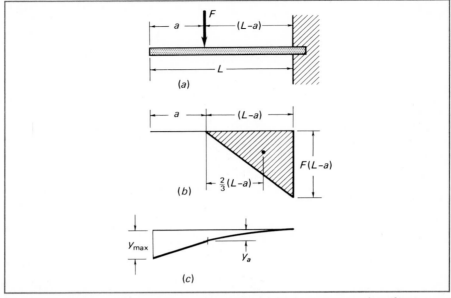

Fig. 12-26 (*a*) Beam diagram of a cantilever beam with load at any point. (*b*) Moment diagram. (*c*) Elastic curve.

Deflection under the load. The equation for the deflection y_a under load F (Fig. 12-26*c*) can be determined by considering a cantilever beam of length $L - a$ with a concentrated load at the free end. This deflection is given by an equation of the form of Eq. (12-16) with a span of $L - a$.

$$y_a = \frac{F(L - a)^3}{3EI} \qquad (12\text{-}18)$$

12-13   DEFLECTION OF A CANTILEVER BEAM (UNIFORM LOAD)

Maximum deflection. By Eq. (12-15),

$$y_{\text{max}} = \frac{\bar{x}A}{EI}$$

From Fig. 12-27*b*, $A = \frac{1}{3}$ of the enclosing rectangle.

$$A = \frac{1}{3} L \frac{WL}{2} = \frac{WL^2}{6}$$

$$\bar{x} = \frac{3L}{4}$$

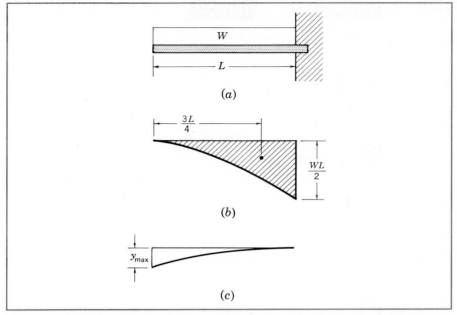

Fig. 12-27 (*a*) Beam diagram of a cantilever beam with uniform load. (*b*) Moment diagram. (*c*) Elastic curve.

Substituting,
$$y_{max} = \frac{\dfrac{3L}{4}\left(\dfrac{WL^2}{6}\right)}{EI}$$

$$y_{max} = \frac{WL^3}{8EI} \tag{12-19}$$

-14 DEFLECTION OF BEAMS WITH COMBINED LOADS

When a beam carries more than one load (or type of load), the deflection at a given point is determined by calculating the deflection at that point that is due to each load and adding the results. Table 12-2 gives maximum deflections for various beam loadings.

To demonstrate this principle, consider the beam shown in Fig. 12-28. The load F causes the beam to deflect as shown in Fig. 12-29*a*. The effect of the uniform load W is shown in Fig. 12-29*b*. The actual beam deflection due to both loads is obtained by adding the individual effects, as in Fig. 12-29*c*.

Sample Problem 7 An S12 × 31.8 simply supported beam that is 12 ft long carries a total uniform load of 8000 lb and a concentrated load of 14 000 lb at midspan (see Fig. 12-30).

TABLE 12-2 SELECTED MAXIMUM DEFLECTION FORMULAS

Beam and Loading	Maximum Deflection	Location
	$\dfrac{FL^3}{48EI}$	Midspan
	$\dfrac{5WL^3}{384EI}$	Midspan
	$\dfrac{23FL^3}{648EI}$	Midspan
	$\dfrac{F}{24EI}(3L^2a - 4a^3)$	Midspan
	$\dfrac{19FL^3}{384EI}$	Midspan
	$\dfrac{FL^3}{3EI}$	Free end
	$\dfrac{F}{6EI}(2L^3 - 3aL^2 + a^3)$	Free end
	$\dfrac{WL^3}{8EI}$	Free end

(a) Find the maximum deflection of the beam.
(b) If permissible deflection is limited to $\frac{1}{360}$ of the span (AISC code), is this beam acceptable based on deflection?

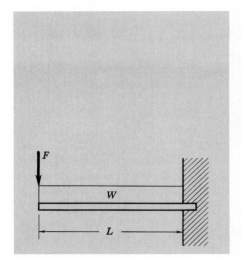

Fig. 12-28 Cantilever beam with load at free end and uniform load.

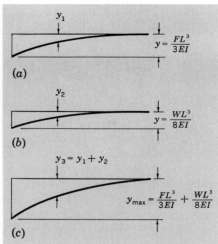

Fig. 12-29 (*a*) Elastic curve—deflection due to *F*. (*b*) Elastic curve—deflection due to *W*. (*c*) Elastic curve—combined effect.

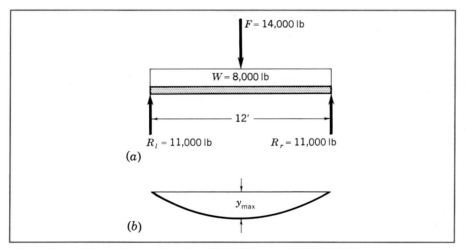

Fig. 12-30 (*a*) Beam diagram for Sample Problem 7. (*b*) Elastic curve.

Solution a: The deflection at the center of the beam due to load *F* is given by Eq. (12-11) or Table 12-2.

$$y_F = \frac{FL^3}{48EI}$$

In this case,

$$F = 14\,000 \text{ lb}$$
$$L = 12 \text{ ft} = 144 \text{ in}$$

and for an S12 × 31.8,

$$I_x = 218 \text{ in}^4 \text{ (Table 5, App. B)}$$
$$E = 30 \times 10^6 \text{ psi}$$

Substituting into Eq. (12-11),

$$y_F = \frac{14\,000(144^3)}{48(30)(10^6)(218)} = 0.133 \text{ in}$$

The deflection at midspan due to W is given by Eq. (12-13) or Table 12-2.

$$y_W = \frac{5WL^3}{384EI}$$

Where $W = 8000$ lb,

$$y_W = \frac{5(8000)(144^3)}{384(30(10^6)(218)} = 0.048 \text{ in}$$

The maximum deflection due to both loads is

$$y_{\text{max}} = y_F + y_W = 0.133 + 0.048 = 0.181 \text{ in}$$

Solution b: The maximum permissible deflection is

$$y = \frac{L}{360} = \frac{12(12)}{360} = \frac{144}{360} = 0.40 \text{ in}$$

Since actual $y_{\text{max}} = 0.181$ in, the beam is acceptable, based on deflection.

12-15   DESIGN OF A BEAM

The fundamental factors upon which beam analysis depend are:

1. Loading — arrangement and magnitude
2. Span
3. Type of beam support (simple, cantilever, etc.)
4. Allowable stresses (bending, shear)
5. Permissible deflection
6. Shape of beam cross section
7. Size of beam

The most common design procedure involves determining the size of beam. A suitable beam must have the required strength in bending and in shear. A further limitation may be imposed with respect to maximum deflection. For example, the AISC code limits maximum deflection for beams carrying plastered ceilings to $\frac{1}{360}$ of the span length. The following sample problem, which includes a limitation on deflection, will help to clarify this procedure.

Sample Problem 8[1]   Design a W-shape, A36 structural steel cantilever beam with a span of 12 ft carrying a uniform load of 670 lb/ft and a concentrated load of 16 500 lb at its midpoint. The allowable stresses, as given by the AISC code, are 24 000 psi in bending and 14 500 psi in web shear. The permissible deflection is $\frac{1}{360}$ of span.

Solution:

Bending   (*See Fig.* 12-31*a*):

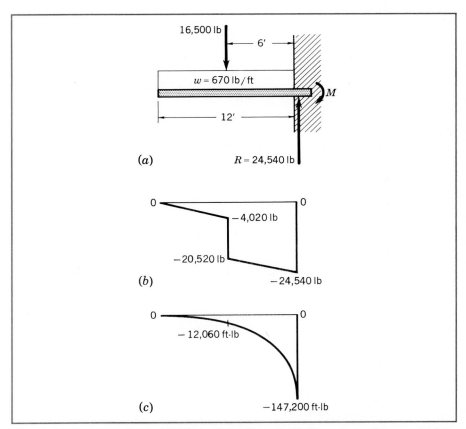

Fig. 12-31 (*a*) Beam diagram for Sample Problem 8. (*b*) Shear force diagram. (*c*) Moment diagram.

[1] See also Sample Computer Program 20, Chap. 17.

Allowable $s = 24\,000$ psi

$$M_{max} = 670(12)(6) + 16\,500(6) = 48\,200 + 99\,000$$
$$= 147\,200 \text{ ft·lb} \quad \text{(at wall; see Fig. 12-31}c)$$

$$s = \frac{M}{S}$$

$$S = \frac{M}{s} = \frac{147\,200(12)}{24\,000} = 73.6 \text{ in}^3 \text{ (minimum)}$$

By selecting on the basis of economy from the Section Modulus Table in App. B (Table 9), we find that a W16 × 50 (S = 81.0 in³) beam meets the bending requirement.

Deflection:

$$\text{Allowable deflection} = \frac{12(12)}{360} = 0.40 \text{ in}$$

$$y_{max} = \frac{WL^3}{8EI} + \frac{F}{6EI}(2L^3 - 3aL^2 + a^3)$$

$W = 670(12) = 8040$ lb $a = 6(12) = 72$ in $L = 12(12) = 144$ in

$$0.40 = \frac{8040(144^3)}{8(30)(10^6)I} + \frac{16\,500}{6(30)(10^6)I}[2(144^3) - 3(72)(144^2) + (72^3)]$$

$$= \frac{100}{I} + \frac{171}{I} = \frac{271}{I}$$

$$I = \frac{271}{0.40} = 678 \text{ in}^4 \text{ (minimum)}$$

The moment of inertia I for the W16 × 50 is 659 in⁴, which is not sufficient for the deflection requirement. However, a W18 × 50 beam will satisfy both bending and deflection requirements ($S = 88.9$ in³ and $I = 800$ in⁴). It should be noted that other beams would satisfy the bending and deflection requirements, such as W16 × 57, W18 × 55, and W18 × 60, but they are more expensive because their weights exceed the 50 lb/ft of the W18 × 50.

Web Shear: Check the W18 × 50 in web shear.

$$\text{Depth} = d = 17.99 \text{ in}$$
$$\text{Web thickness} = t = 0.355 \text{ in.}$$
$$V_{max} = 24\,540 \text{ lb} \quad \text{(see Fig. 12-31}b)$$

$$s_v = \frac{V}{td} = \frac{24\,540}{0.355(17.99)} = 3840 \text{ psi}$$

Allowable $s_v = 14\,500$ psi (therefore OK)

Use W18 × 50.

-16 LATERAL BUCKLING

Beams with long spans tend to buckle laterally, wrinkle locally, or twist under heavy loading. Intermediate lateral (side) supports are used to reduce such failures. If lateral supports are not used, the AISC recommends[1] reducing the allowable bending stress s (use only for W, S, HP, and channel shapes) to a value given by

$$s = \frac{12\ 000\ 000\ C_b}{ld/A_f} \tag{12-20}$$

where l = unsupported length of simple beam or twice the span of cantilever beam, in; mm
   d = depth of beam, in; mm
   A_f = area of flange in compression, in²; mm²
   C_b = 1, for simple and cantilever beams

The AISC code further specifies that the reduced allowable bending stress shall not exceed $0.60s_y$. Equation (12-20) is used only when ld/A_f exceeds

555 for A36 structural steel
475 for A529 structural steel
475 for A572, Grade 42 structural steel
400 for A572, Grade 50 structural steel
335 for A572, Grade 60 structural steel
310 for A572, Grade 65 structural steel
400 for A242, A441, A588 structural steel

PROBLEMS

 *12-1. A simple beam that is 150 by 200 mm in rough-sawn cross section and 3 m long carries a concentrated load of 36 kN at the center. Find the largest bending stresses developed at sections 0.9, 1.5, and 2.4 m from the left end.

 12-2. A wooden beam 12 ft long is supported at the ends. If the rough-sawn cross section is 6 by 8 in and the allowable bending stress is 1100 psi, what total uniform load can the beam support?

 12-3. In Prob. 12-2, if the allowable bending stress is 1200 psi and the rough-sawn cross section of the beam is 6 by 9 in, what is the maximum length of the span if the total uniform load is 6000 lb?

 *12-4. What should be the diameter of the hollow axle for the railway car shown in Fig. Prob. 12-4? The safe bending stress is 70 N/mm². The outside diameter of the axle is twice the inside diameter.

 12-5. What is the factor of safety based on the ultimate strength of the Class 60 cast-iron beam shown in Fig. Prob. 12-5? Consider the weight of the beam as a uniform load.

[1] For additional information, see *Manual of Steel Construction,* 8th ed., AISC, Specifications, Section 1.5.1.4.5.

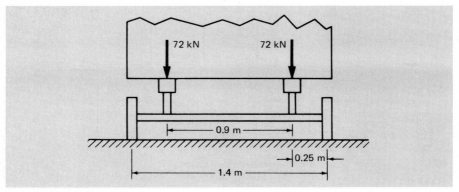

FIGURE PROBLEM 12-4

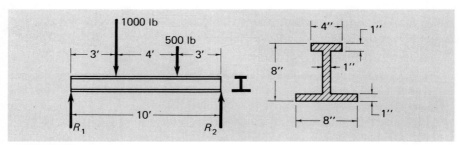

FIGURE PROBLEM 12-5

12-6. A total uniform load of 16 000 lb is carried by a beam of Douglas fir. If the span is 14 ft, select a dressed size whose nominal width is 8 in.

12-7. How far apart can the supports for an A36 structural steel W8 × 35 beam be placed if the beam supports a load of 8000 lb at the middle of the span? Use the AISC code.

12-8. What is the maximum bending stress in an A36 structural steel S 15 × 50 beam of 24-ft span if the beam supports a uniform load of 800 lb in addition to its own weight and loads of 6000 and 9000 lb at 6 and 16 ft, respectively, from the right support?

12-9. What should be the size of A36 structural steel S-shape beams spaced 4 ft on centers to carry a balcony load of 225 psf? The balcony is constructed as a cantilever and extends 8 ft beyond the wall. Use the AISC code. (*Hint:* Find the load on one beam.)

12-10. Select an A36 structural steel W-shape beam that is 24 ft long, overhangs each support 4 ft, and is to carry loads of 16 000 lb at the left end, 12 000 lb at the center, and 8000 lb at the right end. Use the AISC code.

12-11. A floor is supported by A529 structural steel S12 × 35 beams spaced 8 ft center to center. What uniformly distributed floor load (psf) will they support with a span of 18 ft? Use the AISC code.

*12-12. What should be the rough-sawn depth of a simply supported mountain hemlock beam that has a full width of 100 mm, is 3.6 m long, and is to carry equal concentrated loads of 4.5 kN at the third points?

12-13. What must be the diameter of a steel shaft 10 ft long between the bearings if the allowable bending stress is 10 000 psi? A pulley located 3 ft from one end is subjected to a total downward pull of 1600 lb. (Steel weighs 490 lb/ft³.)

12-14. What will be the maximum stress in the box girder of Fig. Prob. 12-14? The girder is made up by connecting four 2- by 12-in (rough-sawn) planks as shown. The span is 12 ft, and the uniformly distributed load is 1250 lb per linear foot.

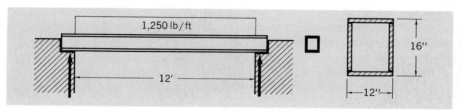

FIGURE PROBLEM 12-14

12-15. An AISI 1020 steel pin in a pin-connected truss is to be considered as a simple beam of 8-in span. It resists a variable force of 8000 lb at its center. What should be the diameter of this pin?

*12-16. The maximum force between a connecting rod and the crank-pin is 540 kN. This force may be assumed to be uniformly distributed against the pin, which is 200 mm long. What must be the diameter of the AISI 1045 steel pin if shock loading is assumed?

*12-17. In a scaffolding used in construction, a plank 50 by 300 mm (rough sawn) is used to support the workmen. The plank is held up by means of brackets 2.4 m apart. If a person having a mass of 91 kg stands at the middle and a person having a mass of 68 kg stands 0.3 m from the left end, could eastern white pine be used for the plank? $\rho = 705$ kg/m³ for eastern white pine.

*12-18. What will be the value of equal loads placed at the quarter points of a rough-sawn Douglas fir beam 150 mm wide, 250 mm deep, and 3.6 m long to develop the maximum allowable bending stress?

12-19. Select a dressed-size ponderosa pine beam 16 ft long to carry a uniform load of 200 lb/ft and concentrated loads of 1200 lb at 6 ft from the right end and 800 lb at 4 ft from the right end.

12-20. A wooden beam, dressed size 6 by 8 in, 12 ft long, and supported at the ends, carries a load of 2000 lb at the middle of the span.
   a. What is the unit shearing stress at the neutral axis?
   b. What is the unit shearing stress at a point 3 ft from the end and 2 in from the neutral axis?
   c. What is the maximum bending stress in the beam?

12-21. A simple beam, dressed size 6 by 6 in and 6 ft long, carries a concentrated load at the middle of the span. What is the load if it is determined by the allowable shearing stress of 120 psi?

12-22. A rectangular Sitka spruce beam, dressed size 8 by 12 in and with a 12-ft simple span, carries a uniform load of 1600 lb/ft. Find the unit horizontal shearing stress at the following points:
  a. At the neutral axis and at the end of the beam
  b. Four inches from the neutral axis and 1 ft from the support
  c. Two inches from the neutral axis and at a quarter point

12-23. **a.** What will be the total uniform load that can be placed on a mountain hemlock beam, dressed size 10 by 14 in, 8 ft long, and supported at the ends, if the allowable shear stress alone is considered?
  b. What will be the stress due to bending as a result of this load?

12-24. What should be the proper length of a dressed-size 10- by 12-in oak beam carrying a uniform load of 1800 lb per linear foot? Solve for both a simple span and a cantilever. Consider both horizontal shear and bending. Use an allowable shear stress of 120 psi and an allowable bending stress of 1450 psi.

12-25. What should be the spacing of $\frac{1}{2}$-in bolts in a beam consisting of a 4- by 6-in (rough-sawn) fir timber bolted to a 6- by 6-in (rough-sawn) timber? The pieces will be so placed that the cross section of the beam will be 6 by 10 in. The span and loading are shown in Fig. Prob. 12-25. The allowable shear stress in the bolts is 10 000 psi.

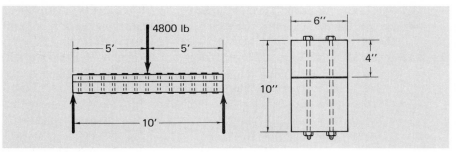

FIGURE PROBLEM 12-25

12-26. A shape is built up by bolting together three rough-sawn 50-mm by 200-mm planks 3.6 m long as shown in Fig. Prob. 12-26. What should be the spacing of 10-mm bolts to resist the shearing stresses? The bolts are made of AISI 1020 steel, and a safety factor of 5 is specified.

12-27. What is the maximum shearing stress developed in an A36 structural steel S10 × 35 beam that carries a uniform load over a simple span of 12 ft? Use the AISC code. Determine load from allowable bending stress.

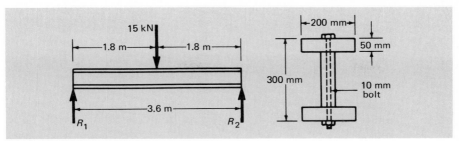

FIGURE PROBLEM 12-26

12-28. A plate girder used as a simple beam 36 ft long must support a concentrated load of 180 000 lb at the middle of the span and 45 000 lb at each quarter point (Fig. Prob. 12-28). The girder is composed of a 48- by $\frac{3}{8}$-in web plate and four 6- by 6- by $\frac{3}{4}$-in flange angles, all made of A36 structural steel. What should be the bolt spacing if $\frac{7}{8}$-in A307 bolts are used? Use the AISC code. Assume bearing-type connections.

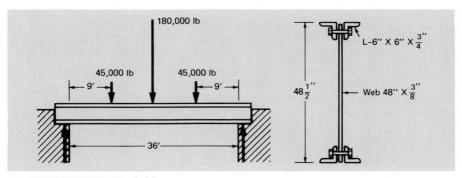

FIGURE PROBLEM 12-28

12-29. A W12 × 72 A36 structural steel beam is used for a simple span of 12 ft and carries a uniformly distributed load. What is the maximum shearing stress at the base of the upper flange? Consider the flange as a rectangle. Use the AISC code. Determine load based on allowable bending stress.

12-30. A Douglas fir beam 10- by 12-in in dressed size and 14 ft long is supported at the ends. The beam carries a total uniform load of 9000 lb and a concentrated load of 3600 lb at the middle of the span.
    a. What will be the maximum bending stress, shearing stress, and deflection?
    b. Are the values within safe limits?

12-31. What safe uniformly distributed load will a 6- by 10-in yellow pine beam carry if the span is 10 ft? Design for both dressed and full sizes.

Use an allowable bending stress of 1750 psi and an allowable shear stress of 120 psi.

12-32. For what length of beam will the total uniform load be determined by shear only and for what length by bending? Assume a ponderosa pine beam and 8- by 10-in dressed size cross section.

12-33. A W12 × 79 A36 structural steel beam carries a uniformly distributed load of 48 000 lb on a simple span of 24 ft. Compute the maximum bending stress, shearing stress, and deflection. Use the AISC code.

12-34. A mountain hemlock beam, dressed size 4 by 8 in and 7 ft long, is used as a cantilever to support an end load of 900 lb. What are the maximum bending stress, shearing stress, and deflection?

12-35. A floor carrying a load of 140 psf is supported by A441 structural steel S12 × 35 beams spaced 6 ft from center to center. The span of the beams is 21 ft. Use the AISC code.
   a. Does the deflection exceed the allowable value?
   b. What are the maximum bending and shearing stresses?

*12-36. A steel shaft 250 mm in diameter resting in bearings 1.05 m apart carries a flywheel having a mass of 18.14×10^3 kg midway between the bearings. Treat the shaft as a simple beam and find the maximum values of bending stress, shearing stress, and deflection.

12-37. A 4-in steel shaft 10 ft long and supported at the ends carries a 200-lb pulley midway between the supports. A belt pull of 500 lb acts on the pulley in a downward direction.
   a. What will be the resultant deflection?
   b. Determine the horizontal shear stress and the bending stress developed.

12-38. **a.** Select the most economical A36 structural steel S-shape beam for the loading shown in Fig. Prob. 12-38. Use the AISC code.
   b. Is this beam safe in vertical web shear?
   c. What is the maximum deflection?

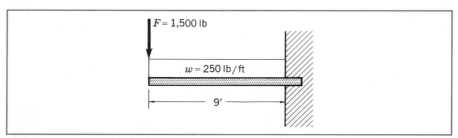

FIGURE PROBLEM 12-38

12-39. Select the most economical A36 structural steel S-shape beam for the loading shown in Fig. Prob. 12-39. Use the AISC code.

*12-40. A Class 20 hollow circular cast-iron post with an outside diameter of

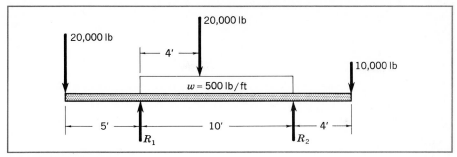

FIGURE PROBLEM 12-39

75 mm and a wall thickness of 12.5 mm is proposed for the loading shown in Fig. Prob. 12-40. Does this post satisfy a required safety factor of 7? Neglect the weight of the post.

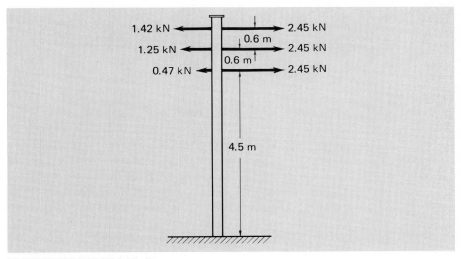

FIGURE PROBLEM 12-40

13

Torsion, Shafts, Shaft Couplings, and Keys

13-1 TORSION

In Fig. 13-1, F_1 and F_2 are the tensions on a belt, with $F_2 > F_1$. The difference in tensions tends to rotate the shaft clockwise. A cable with an attached weight W is wound around the shaft as shown. This weight tends to rotate the shaft in a counterclockwise direction. When a shaft is acted upon by two equal and opposite twisting moments in parallel planes, it is said to be in *torsion*. When such a condition exists, it may be either stationary or rotating uniformly. The twisting moment, or torque, is generally expressed in inch-pounds or newton-meters.

If, in Fig. 13-1, $F_2 = 250$ lb, $F_1 = 100$ lb, and the pulley radius $r = 6$ in, the torque $T = (F_2 - F_1)r = (250 - 100)(6) = 900$ in·lb.

If the radius of the shaft is 1 in and the moments are equal, W must equal 900 lb. This assumes negligible bearing friction.

In the preceding illustration, the twisting moment, or torque, is constant for the length of shaft between the pulley and the cable. However, if a shaft has several pulleys attached to it, the torque will be different in different sections of the shaft.

In Fig. 13-2, pulley B drives the shaft in clockwise rotation and provides power for pulleys A and C. Pulley B transmits a torque $T_b = (F_3 - F_4)r_b = (1500 - 500)(4) = 4000$ in·lb, which is available to pulleys A and C. Thus, the torque at pulley A, transmitted along the shaft between A and B, is

$$T_a = (F_5 - F_6)r_a = (300 - 100)(6) = 1200 \text{ in·lb}$$

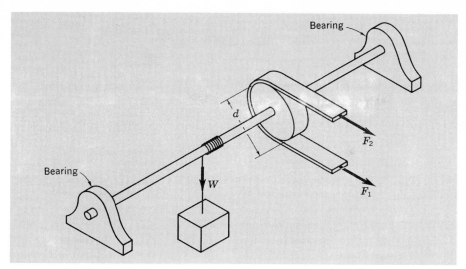

Fig. 13-1 Torsional system.

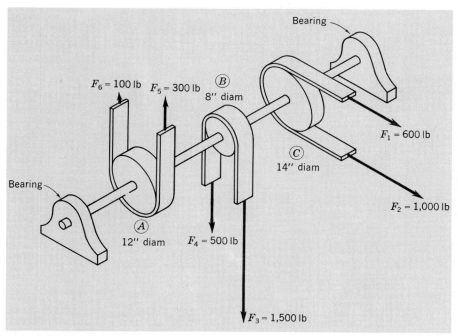

Fig. 13-2 Torque transmission.

Similarly, the shaft from B to C transmits the torque

$$T_c = (F_2 - F_1)r_c = (1000 - 600)(7) = 2800 \text{ in} \cdot \text{lb}$$

Since bearing friction has been assumed to be negligible, no torque exists in the shaft between the right bearing and pulley C or between the left bearing and pulley A.

13-2 TORSIONAL SHEARING STRESS

When a shaft is in torsion, any cross section of it tends to slip by or shear across the adjacent face. The resistance of the fibers per unit area of the shaft is called the *torsional shearing stress*. When a couple is applied to the lever shown in Fig. 13-3a, an element AB, originally straight, is twisted into the position AC. Radius OB of the end section is rotated into the position OC. Any portion of material originally at B' is rotated to C'. Now, the shear stress resisting that change is proportional to the strain, since $G = s_s/\epsilon_s$. But the elongation $B'C'$ is proportional to the distance OB' from the center of the shaft. Then, within the elastic limit of the material, the torsional shearing stress at any point of the cross section of a shaft is proportional to the distance of that point from the center of the section.

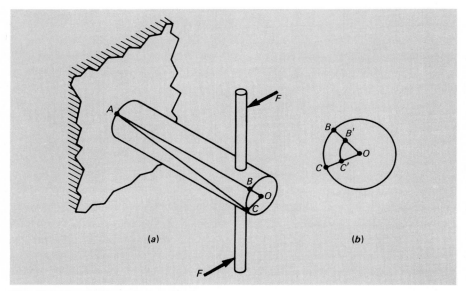

Fig. 13-3 Torsional strain in a shaft.

Since the center of the section has no shear stress, the axis of the shaft is a neutral axis. If the radius OB is 3 in, the shear stress at B is 6000 psi, and B' is 1.5 in from the center, the stress at B' is 3000 psi (see Fig. 13-3b).

An enlarged cross section of the shaft is shown in Fig. 13-4. Consider a thin ring of material (cross-hatched) whose area is a at an average radius r. The shearing force acting on area a is $F_r = a s_{sr}$, where s_{sr} represents the shear stress at radius r. The torque associated with this shearing force is

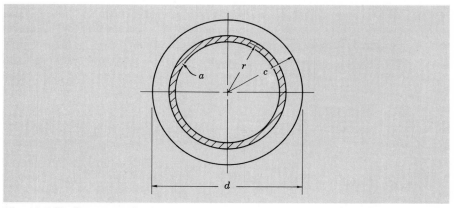

Fig. 13-4 Shaft cross section.

$$T_r = F_r r = s_{sr} a r$$

Now let s_s be the maximum shearing stress that occurs on the extreme outer fiber at radius c. Since shear stress is proportional to radial distance,

$$\frac{s_{sr}}{r} = \frac{s_s}{c} \quad \text{or} \quad s_{sr} = \frac{s_s r}{c}$$

Then, by substitution,

$$T_r = \frac{s_s (a r^2)}{c}$$

The total shearing torque on the entire cross section is the sum of the torques on all rings such as a. Thus,

$$T = \frac{s_s}{c} [\Sigma a r^2]$$

The term in brackets is the moment of inertia of the area of the circle about a centroidal axis *perpendicular to the plane of the circle*. This is called the *polar moment of inertia,* and it will be represented by J. Then

$$T = \frac{s_s J}{c}$$

or

$$s_s = \frac{T c}{J} \tag{13-1}$$

where s_s = maximum shearing stress due to torsion at the outer fiber of a uniform, solid, or hollow circular member,[1] psi; N/mm^2 = MPa

T = torque, in·lb; N·mm

c = distance from neutral axis to the extreme fiber, in; mm

$J = I_x + I_y$ = polar moment of inertia about a centroidal axis perpendicular to cross section, in^4; mm^4

The reader should note the similarity between the torsion formula, Eq. (13-1), and the flexure formula, Eq. (12-1). In the same manner as for bending, it is convenient to group the moment of inertia and the c-distance terms to form $S' = J/c$, where S' is the polar section modulus. Thus, Eq. (13-1) may be written as

$$s_s = \frac{T}{S'} \tag{13-2}$$

For a solid shaft of *circular cross section,*

$$J = I_x + I_y = \frac{\pi d^4}{64} + \frac{\pi d^4}{64} = \frac{\pi d^4}{32}$$

$$c = \frac{d}{2}$$

$$S' = \frac{J}{c} = \frac{\pi d^4/32}{d/2} = \frac{\pi d^3}{16}$$

Then

$$s_s = \frac{T}{S'} = \frac{T}{\pi d^3/16} = \frac{16T}{\pi d^3} \tag{13-3}$$

For a *hollow circular shaft* of inside diameter d_i and outside diameter d_o,

$$J = \frac{\pi d_o^4}{32} - \frac{\pi d_i^4}{32} = \frac{\pi(d_o^4 - d_i^4)}{32}$$

$$c = \frac{d_o}{2}$$

$$S' = \frac{J}{c} = \frac{\pi(d_o^4 - d_i^4)/32}{d_o/2} = \frac{\pi}{16}\left(\frac{d_o^4 - d_i^4}{d_o}\right)$$

Then

$$s_s = \frac{T}{S'} = \frac{T}{\dfrac{\pi}{16}\left(\dfrac{d_o^4 - d_i^4}{d_o}\right)} = \frac{16 T d_o}{\pi(d_o^4 - d_i^4)} \tag{13-4}$$

The preceding equations apply to pure torsional systems *within the elastic limit* of the material. The torsional strength of a shaft depends on the maximum safe torque the shaft can transmit. The strengths of two shafts of the same

---

[1] Equations for other shapes are to be found in R. J. Roark and W. C. Young, *Formulas for Stress and Strain,* 6th ed., McGraw-Hill Book Company, New York, 1982.

material can be compared by finding the ratio of the maximum safe torques of the shafts.

*Sample Problem 1   Find the diameter of a solid AISI 1045 steel shaft which is to transmit a torque of 18 kN·m if the factor of safety based on the ultimate shearing strength is 10.

Solution: From Table 1A, App. B, the ultimate $s_s = 480$ MPa for AISI 1045. Therefore, the allowable $s_s = 480/10 = 48$ MPa. From Eq. (13-3) for a circular cross section,

$$s_s = \frac{16T}{\pi d^3} \quad \text{or} \quad d^3 = \frac{16T}{\pi s_s}$$

$$d^3 = \frac{16(18)(10^3)(10^3)}{\pi(48)} = 1.91 \times 10^6$$

$$d = 124 \text{ mm}$$

Sample Problem 2   A solid shaft with a diameter of 3 in has the same cross-sectional area as a hollow shaft of $d_o = 5$ in and $d_i = 4$ in. If the shafts are made of the same material, which shaft is stronger (torsionally) and by what factor?

Solution: It is unnecessary to know the allowable shear stress because only the ratio of the torques is required, not the actual values of the torques. For the solid shaft [Eq. (13-3)],

$$s_s = \frac{16T_1}{\pi d^3} \quad \text{or} \quad T_1 = \frac{\pi d^3 s_s}{16}$$

$$T_1 = \frac{\pi(3^3)s_s}{16} = 5.3s_s$$

For the hollow shaft [Eq. (13-4)],

$$s_s = \frac{16T_2 d_o}{\pi(d_o^4 - d_i^4)} \quad \text{or} \quad T_2 = \frac{\pi(d_o^4 - d_i^4)s_s}{16d_o}$$

$$T_2 = \frac{\pi(5^4 - 4^4)s_s}{16(5)} = \frac{\pi(625 - 256)s_s}{16(5)} = 14.5s_s$$

$$\frac{T_2}{T_1} = \frac{14.5s_s}{5.3s_s} = 2.73$$

The hollow shaft is 2.73 times as strong in torsion as the solid shaft.

Sample Problem 3   What is the maximum safe torque, considering torsional shear only, that can be applied in tightening a 1″-8 UNC bolt whose allowable shear stress is 8800 psi?

Solution: Root diameter of threads $= 0.8647$ in (Table 3A, App. B). From Eq. (13-3),

$$S_s = \frac{16T}{\pi d^3} \quad \text{or} \quad T = \frac{\pi d^3 s_s}{16}$$

$$T = \frac{\pi(0.8647^3)(8800)}{16} = 1120 \text{ in} \cdot \text{lb (max safe torque)}$$

13-3   ANGLE OF TWIST

In some types of machinery it is desirable to know the amount of angle of twist in a shaft due to the torque. Let Fig. 13-5 represent a part of a shaft subjected to a uniform torque throughout. AB represents an element of the untwisted shaft,

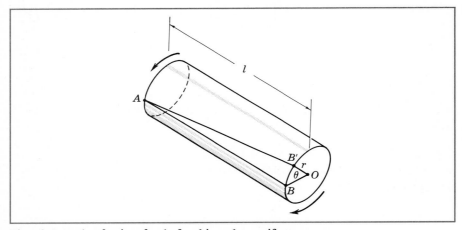

Fig. 13-5 Angle of twist of a shaft subjected to uniform torque.

and AB' is the same element after torque has been applied. Owing to the torque, the radius OB assumes a position OB'. The angle BOB' is the angle of twist θ. The displacement from B to B' is a shearing displacement. For the fiber represented by AB, the deformation in the length l is BB'. The unit deformation (strain) is

$$\epsilon_s = \frac{BB'}{l} \quad \text{and} \quad BB' = r\theta$$

Then,
$$\epsilon_s = \frac{r\theta}{l} \quad \text{(refer to Fig. 8-3c)}$$

where θ is the angle BOB' expressed in radians. Also, from the relations from Hooke's law,

$$\epsilon_s = \frac{S_s}{G}$$

where G is the modulus of rigidity (modulus of elasticity in shear). Then

$$\frac{r\theta}{l} = \frac{s_s}{G}$$

Note that $r = c$. Therefore,

$$\theta = \frac{s_s l}{Gc} \qquad (13\text{-}5)$$

The above expression gives the angle of twist in terms of the fiber stress. However, the design of shafts usually involves the angle of twist expressed in terms of torque; thus, from Eq. (13-1),

$$s_s = \frac{Tc}{J}$$

Substituting for s_s in Eq. (13-5) gives us

$$\theta = \frac{Tcl}{JGc} = \frac{Tl}{JG} \qquad (13\text{-}6)[1]$$

where θ = angle of twist, rad
  T = torque, in·lb; N·mm
  l = length of shaft subjected to torque, in; mm
  J = polar moment of inertia, in⁴; mm⁴
  G = modulus of rigidity, psi; N/mm² = MPa

Attention is called to the fact that the angle of twist increases with the length of the shaft and decreases as the radius increases.

13-4 POWER TRANSMISSION

When a force F that is constant in magnitude and direction acts on a body through a distance L in the direction in which the force acts, the work done is the product of the force and distance, or

$$\text{Work} = F(L)$$

When F is in pounds and L in feet, work is in foot-pounds. For example, if $F = 20$ lb and $L = 10$ ft,

$$\text{Work} = 20(10) = 200 \text{ ft·lb} = 2400 \text{ in·lb}$$

In Fig. 13-6, the driving force is $P_2 - P_1 = F$ on the pulley of radius r. Suppose the shaft is making n revolutions per minute (rpm). In one revolution, the force F acts through the distance $2\pi r$ (the circumference of the pulley). The work per revolution is $2\pi Fr$. But Fr is the moment, or torque, T. Then, in n revolutions, the total work done is $2\pi Tn$.

---

[1] Equation (13-6) has the same limitation as Eq. (13-1).

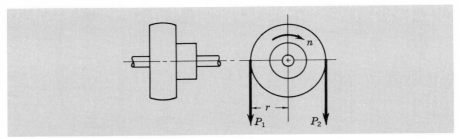

Fig. 13-6 Forces on rotating pulley.

The total work done can also be found from the relation that the work done is the product of torque and the angle, in radians, through which the torque acts, or

$$\text{Work} = T\theta$$

For n revolutions,

$$\text{Work} = T \times \left(n \text{ rev.} \times \frac{2\pi \text{ rad}}{1 \text{ rev.}} \right) = 2\pi T n$$

since

$$\text{Power} = \frac{\text{work}}{\text{time}}$$

$$\text{Power} = \frac{2\pi T n}{\text{time}}$$

For U.S. customary units, the usual power unit is horsepower (hp). One horsepower is defined as 33 000 ft·lb/min, or 33 000 × 12 in·lb/min. Thus,

$$\text{hp} = \frac{2\pi T n}{33\,000(12)} = \frac{Tn}{63\,000} \qquad (13\text{-}7a)$$

where T = torque, in·lb
$\quad\quad n$ = rpm

For SI metric units, the power unit is the watt (W). One watt is equal to 1 N·m/s. Thus,

$$\text{W} = \frac{2\pi T n}{60} = \frac{Tn}{9.55} \qquad (13\text{-}7b)$$

where W = watts (usually expressed in kilowatts)
$\quad\quad T$ = torque, N·m
$\quad\quad n$ = rpm

*Sample Problem 4[1]   A 50-mm-diameter shaft of AISI 4130 steel is to be driven at 1800 rpm. A safety factor of 10, based on the ultimate torsional shearing strength, is specified.

[1] See also Sample Computer Program 21, Chap. 17.

(a) What maximum power may be safely transmitted?
(b) What will be the angle of twist in a 1-m (20-diameter) length of the shaft when the maximum safe power is transmitted?

Solution: From Table 1A, App. B, for AISI 4130 steel,

$$\text{Ultimate } s_s = 825 \text{ MPa}$$

$$\text{Allowable } s_s = \frac{825}{10} = 82.5 \text{ MPa}$$

$$d = 50 \text{ mm}$$

$$n = 1800 \text{ rpm}$$

(a) From Eq. (13-3),

$$s_s = \frac{16T}{\pi d^3} \quad \text{or} \quad T = \pi \frac{d^3 s_s}{16}$$

$$T = \frac{\pi(50^3)(82.5)}{16} = 2.025(10^6) \text{ N} \cdot \text{mm} = 2.025(10^3) \text{ N} \cdot \text{m}$$

From Eq. (13-7*b*),

$$W = \frac{Tn}{9.55} = \frac{2.025(10^3)(1800)}{9.55} = 381(10^3)W$$

$$= 381 \text{ kW (maximum safe power)}$$

(b) $l = 1$ m, $G = 80(10^3)$ MPa (Table 1A, App. B), $c = 25$ mm

Using Eq. (13-5),

$$\theta = \frac{s_s l}{Gc} = \frac{82.5(1)(10^3)}{80(10^3)(25)} = 0.0413 \text{ rad}$$

or, since π rad $= 180°$,

$$\theta = \frac{180}{\pi}(0.0413) = 57.3(0.0413) = 2.37°$$

Note that a limitation of 1° (0.0175 rad) twist in 20 diameters of length is often specified. If this specification were enforced in this problem, the shaft would be stressed to only 35 MPa and would transmit 162 kW at a torque of 860 N·m.

Sample Problem 5 A solid shaft 8 in in diameter has the same cross-sectional area as a hollow shaft of the same material with inside diameter of 6 in.

(a) Compare the horsepower transmission of these shafts at the same revolutions per minute.
(b) Compare the angle of twist in equal lengths of these shafts when stressed to the same intensity.

Solution: Find the outside diameter d_o of the hollow shaft. Since the cross-sectional areas are equal,

$$\frac{\pi d^2}{4} = \frac{\pi}{4}(d_o{}^2 - d_i{}^2)$$

$$\frac{\pi(8^2)}{4} = \frac{\pi}{4}(d_o{}^2 - 6^2)$$

$$8^2 = d_o{}^2 - 6^2$$

$$d_o{}^2 = 64 + 36 = 100$$

$$d_o = 10 \text{ in}$$

(a) From Eq. (13-3) for the solid shaft,

$$S_s = \frac{16T}{\pi d^3} \quad \text{or} \quad T = \frac{\pi d^3 s_s}{16}$$

$$T = \frac{\pi(8^3)s_s}{16}$$

From Eq. (13-7a) for the solid shaft,

$$\text{hp}_{\text{solid}} = \frac{Tn}{63\,000} = \frac{\pi(8^3)s_s n}{63\,000(16)} = \frac{\pi s_s n}{63\,000(16)}(512)$$

From Eq. (13-4) for the hollow shaft,

$$S_s = \frac{16Td_o}{\pi(d_o{}^4 - d_i{}^4)} \quad \text{or} \quad T = \frac{\pi(d_o{}^4 - d_i{}^4)s_s}{16d_o}$$

$$T = \frac{\pi(10^4 - 6^4)s_s}{16(10)}$$

From Eq. (13-7a) for the hollow shaft,

$$\text{hp}_{\text{hollow}} = \frac{Tn}{63\,000} = \frac{\pi(10^4 - 6^4)s_s n}{63\,000(16)(10)} = \frac{\pi s_s n}{63\,000(16)}\left(\frac{10\,000 - 1300}{10}\right)$$

$$= \frac{\pi s_s n}{63\,000(16)}(870)$$

$$\frac{\text{hp}_{\text{hollow}}}{\text{hp}_{\text{solid}}} = \frac{870}{512} = 1.7$$

The hollow shaft can transmit 1.7 times the horsepower that the solid shaft can transmit.

(b) For the solid shaft, $c = d/2 = \frac{8}{2} = 4$ in. From Eq. (13-5),

$$\theta_{\text{solid}} = \frac{s_s l}{Gc} = \frac{s_s l}{G(4)}$$

For the hollow shaft, $c = d_o/2 = \frac{10}{2} = 5$ in. Using Eq. (13-5),

$$\theta_{\text{hollow}} = \frac{s_s l}{Gc} = \frac{s_s l}{G(5)}$$

$$\frac{\theta_{\text{hollow}}}{\theta_{\text{solid}}} = \frac{1/5}{1/4} = \frac{4}{5} = 0.80$$

The twist in the hollow shaft is only 80 percent of the twist in the solid shaft.

13-5 SHAFT COUPLINGS

A number of devices are used to connect sections of coaxial shafting. The simplest and most widely used of these devices is the bolted flanged coupling. This type of coupling provides a clear application of torsional shearing stress induced by power transmission. Figure 13-7 represents such a flange coupling. The torque is transmitted from one shaft to the other as follows:

1. From shaft A to key B
2. From key B to the left side of the coupling C
3. From the left side of the coupling C to the six bolts D
4. From the bolts D to the right side of the coupling E
5. From the right side of the coupling E to the key F
6. From key F to the other shaft G

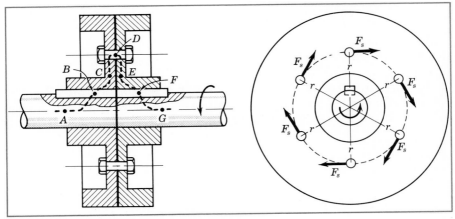

Fig. 13-7 Torque transmission in flange coupling.

The transmitted torque will develop a resisting torque in the bolts equal and opposite to it. The resisting torque in each bolt will be

$$T_r = \frac{T}{n_b} \qquad \text{(assuming that each bolt will take an equal share)}$$

where T_r = resisting torque in each bolt
T = total transmitted torque
n_b = number of bolts

But torque equals force times radius. Therefore,

$$T_r = Fr$$

where F = resisting force in each bolt
r = bolt-circle radius

Shear, Bolts: The shear stress developed will be

$$s_s = \frac{F}{A_s}$$

where s_s = shear stress, psi; N/mm² = MPa
F = load causing shear, lb; N
A_s = cross-sectional area of bolt, in²; mm²

Bearing, Bolts: In addition to causing shear in the bolts, it should be noted that the force F will also cause bearing between the bolts and the web of the flange coupling. The compressive stress in the bolt will be

$$s_c = \frac{F}{A_c}$$

where s_c = compressive stress, psi; N/mm² = MPa

F = load causing bearing, lb; N

A_c = projected area of bolt in the web, in²; mm²

Shear, Hub: Further analysis of the coupling should include the consideration of possible failure due to shearing the hub from the web. Such failure might occur at the section where the web joins the hub, shown in Fig. 13-8a. The area resisting shear is the cylindrical area (striped area) in Fig. 13-8b, which is t in [mm] wide and d_{hub} in [mm] in diameter. The shear force on this area is

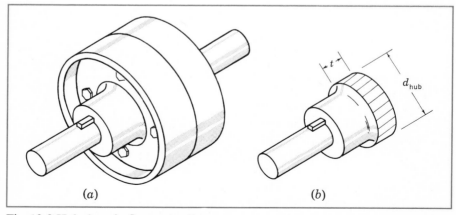

Fig. 13-8 Hub shear in flange coupling.

$F = T/r_{hub}$. Therefore, the shear stress is

$$s_s = \frac{F}{A_s}$$

where s_s = shear stress, psi; N/mm² = MPa

$F = T/r_{hub}$ = shearing force on this area, lb; N

$A_s = \pi d_{hub} t$, in²; mm² (see Fig. 13-8*b*)

3-6   KEYS

When torque is transmitted by means of couplings, gears, or pulleys, some method must be used to fasten those devices to their shafts. One commonly used device is the key. Figure 13-9 shows a rectangular key fastening a spur gear to a shaft. A clockwise torque is transmitted from the shaft through the key to the gear.

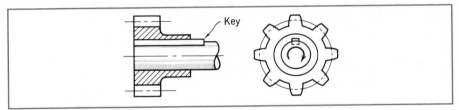

Fig. 13-9 Spur gear keyed to a shaft.

Shear, Key: There is a tendency for the key to shear on the rectangular area (cross-hatched in Fig. 13-10*a*) at the surface of the shaft. The force F which produces shear at section AA (Fig. 13-10*b*) is equal to T/r_{shaft}. The shear stress

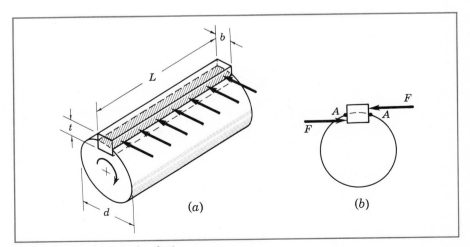

Fig. 13-10 Shear analysis of a key.

developed in the key will then be equal to

$$s_s = \frac{F}{A_s}$$

where s_s = shear stress, psi; N/mm² = MPa

$F = \dfrac{T}{r_{\text{shaft}}}$ = force causing shear in key, lb; N

$A_s = bL$ = shear area, in²; mm²

Bearing, Key: The analysis of the key can be carried further to determine the bearing stress. The force F will tend to crush the key at the cross-hatched area (Fig. 13-11a). This force can be taken as the same force in the shear analysis of the key. Therefore, the bearing stress developed in the key is

$$s_c = \frac{F}{A_c}$$

where s_c = bearing stress, psi; N/mm² = MPa

$F = \dfrac{T}{r_{\text{shaft}}}$ = force causing bearing, lb; N

$A_c = \dfrac{tL}{2}$ = bearing area, in²; mm²

ˌSample Problem 6[1]   A flange coupling with four bolts transmits 19 kW at 200 rpm. The bolt circle diameter is 150 mm. The thickness of the flange web is 20 mm, and the flange-hub diameter is 110 mm. The shafts joined by the

[1] See also Sample Computer Program 22, Chap. 17.

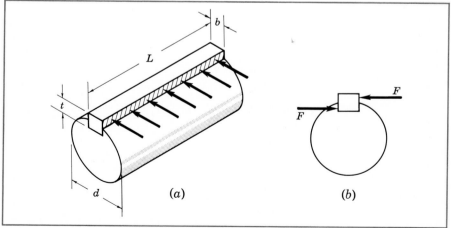

Fig. 13-11 Bearing analysis of a key.

coupling are 60 mm in diameter. The bolts and coupling are made of AISI 1020 steel. A safety factor of 4, based on ultimate, is specified.

(a)   Determine the required diameter of the bolts.
(b)   Is the hub safe in shear where it joins the web?

Solution:

$$T = \frac{9.55 \text{ W}}{n} = \frac{9.55(19)(10^3)}{200} = 907 \text{ N} \cdot \text{m} = 907 \times 10^3 \text{ N} \cdot \text{mm}$$

$$F_{\text{bolt}} = \frac{T}{n_b r_{bc}} = \frac{907(10^3)}{4(75)} = 3.02 \times 10^3 \text{ N}$$

From Table 1A, App. B,

$$\text{Ultimate } s_s = 345 \text{ MPa}$$
$$\text{Ultimate } s_c = 450 \text{ MPa}$$

then,

$$\text{Allowable } s_s = \frac{345}{4} = 86.2 \text{ MPa}$$

$$\text{Allowable } s_c = \frac{450}{4} = 112 \text{ MPa}$$

(a) Shear:

$$A_s = \frac{F_{\text{bolt}}}{s_s} = \frac{3.02(10^3)}{86.2} = 35 \text{ mm}^2$$

$$A_s = 0.785 d^2$$

$$d^2 = \frac{35}{0.785} = 44.6$$

$$d = 6.7 \text{ mm (minimum)}$$

(b) Bearing:

$$A_c = \frac{F_{\text{bolt}}}{s_c} = \frac{3.02(10^3)}{112} = 27 \text{ mm}^2$$

$$A_c = dt$$

$$d = \frac{27}{20} = 1.35 \text{ mm (minimum)}$$

From these results, the minimum safe bolt diameter is 6.7 mm. Use the next largest available standard size, $d = 8$ mm.[1]

$$F_{\text{hub}} = \frac{T}{r_{\text{hub}}} = \frac{907(10^3)}{55} = 16.5 \times 10^3 \text{ N}$$

[1] Refer to Table 3B, App. B.

Find the shear stress in the hub where it joins the web, and compare that with the allowable stress.

$$A_s = \pi d_{hub} t = \pi(110)(20) = 6.91 \times 10^3 \text{ mm}^2$$

$$S_s = \frac{F_{hub}}{A_s} = \frac{16.5(10^3)}{6.91(10^3)} = 2.39 \text{ MPa}$$

Since the actual s_s is below the allowable s_s of 86.2 MPa, the hub is safe in shear.

*Sample Problem 7[1]   In the above problem, a 16- by 10-mm flat key of AISI 1020 steel is used to fasten each side of the coupling to its shaft. Use a safety factor of 4 based on ultimate. What length of key is required?

Solution:

$$F_{key} = \frac{T}{r_{shaft}} = \frac{907(10^3)}{30} = 30.2 \times 10^3 \text{ N}$$

Shear:

$$A_s = \frac{F_{key}}{S_s} = \frac{30.2(10^3)}{86.2} = 350 \text{ mm}^2$$

$$bL = 350$$

$$b = 16 \text{ mm}$$

$$L = \frac{350}{16} = 21.9 \text{ mm (minimum)}$$

Bearing:

$$A_c = \frac{F_{key}}{S_c} = \frac{30.2(10^3)}{112} = 270 \text{ mm}^2$$

$$A_c = \frac{t}{2} L$$

$$t = 10 \text{ mm}$$

$$L = \frac{2(270)}{10} = 54 \text{ mm (minimum)}$$

The minimum safe length of key is 54 mm.

Sample Problem 8   Two hollow shafts are joined by a flange coupling according to the following specifications.

Shaft: 302 stainless steel
        5 in OD
        $2\frac{1}{2}$ in ID
Key: AISI 1045 steel
        1 by 1 by 8 in

[1] See also Sample Computer Program 22, Chap. 17.

Coupling: Class 60 cast iron
Hub diameter = 8 in
Web thickness = $1\frac{1}{2}$ in
Bolt circle diameter = 11 in
Bolt diameter = $\frac{3}{4}$ in
Number of bolts = 6
Material of bolts—AISI 1045 steel
Factor of safety on all parts = 6 (based on ultimate)

What maximum horsepower can this arrangement transmit at 350 rpm?

Solution: In order to determine the maximum horsepower for this power transmission arrangement, it is necessary to calculate the maximum safe torque which the combination can transmit. The maximum safe torque will be the torque which the weakest component can transmit. To determine this torque, the following calculations will be made.

1. Shaft in torsional shear
2. Key in shear
3. Key in bearing
4. Hub in shear at the web
5. Bolts in shear
6. Bolts in bearing with web

1. 302 Stainless Shaft — Torsion:

$$\text{Ultimate } s_s = 110\ 000 \text{ psi}$$

$$\text{Allowable } s_s = \frac{110\ 000}{6} = 18\ 330 \text{ psi}$$

$$T = s_s S'$$

$$S' = \frac{J}{c} = \frac{\pi}{16}\left(\frac{d_o^4 - d_i^4}{d_o}\right) \qquad \text{(see Sec. 13-2)}$$

$$= \frac{\pi}{16}\left(\frac{5^4 - 2.5^4}{5}\right) = \frac{\pi}{16}\left(\frac{625 - 39.1}{5}\right)$$

$$= \frac{\pi}{16}\left(\frac{585.9}{5}\right)$$

$$= 23 \text{ in}$$

$$T_{\text{shaft}} = 18\ 330(23) = 421\ 000 \text{ in} \cdot \text{lb}$$

2. AISI 1045 Key in Shear:

$$\text{Ultimate } s_s = 70\ 000 \text{ psi}$$

$$\text{Allowable } s_c = \frac{70\,000}{6} = 11\,670 \text{ psi}$$

$$T = F_{\text{key}}(r_{\text{shaft}})$$

$$F_{\text{key}} = A_s s_s$$

$$A_s = bL = 1(8) = 8 \text{ in}^2$$

$$F_{\text{key}} = 8(11\,670) = 93\,300 \text{ lb}$$

$$T = 93\,300(2.5) = 233\,000 \text{ in} \cdot \text{lb}$$

3. AISI 1045 Key in Bearing:

$$\text{Ultimate } s_c = 95\,000 \text{ psi}$$

$$\text{Allowable } s_c = \frac{95\,000}{6} = 15\,830 \text{ psi}$$

$$T = F_{\text{key}}(r_{\text{shaft}})$$

$$F_{\text{key}} = A_c s_c$$

$$A_c = \frac{tL}{2} = \frac{1}{2}(8) = 4 \text{ in}^2$$

$$F_{\text{key}} = 4(15\,830) = 63\,300 \text{ lb}$$

$$T = 63\,300(2.5) = 158\,000 \text{ in} \cdot \text{lb}$$

From calculations 2 and 3, $T_{\text{key}} = 158\,000$ in·lb

4. Class 60 Cast-Iron Hub in Shear:

$$\text{Ultimate } s_s = 65\,000 \text{ psi}$$

$$\text{Allowable } s_s = \frac{65\,000}{6} = 10\,830 \text{ psi}$$

$$T = F_{\text{hub}}(r_{\text{hub}})$$

$$F_{\text{hub}} = A_s s_s$$

$$A_s = \pi d_{\text{hub}} t = \pi(8)(1.5) = 37.7 \text{ in}^2$$

$$F_{\text{hub}} = 37.7(10\,830) = 408\,000 \text{ lb}$$

$$T_{\text{hub}} = 408\,000(4) = 1\,632\,000 \text{ in} \cdot \text{lb}$$

5. AISI 1045 Bolts in Shear:

$$\text{Allowable } s_s = 11\,670 \text{ psi} \qquad \text{(from calculation 2)}$$

$$T = n_b F_{\text{bolt}} r_{bc}$$

$$F_{\text{bolt}} = A_s s_s$$

$$A_s = \frac{\pi(d_{\text{bolt}}^2)}{4} = \frac{\pi}{4}(0.75^2) = 0.442 \text{ in}^2$$

$$F_{\text{bolt}} = 0.442(11\,670) = 5160 \text{ lb}$$

$$n_b = 6 \qquad r_{bc} = 5\tfrac{1}{2} \text{ in}$$

$$T = 6(5160)(5.5) = 170\,000 \text{ in} \cdot \text{lb}$$

6. AISI 1045 Bolts in Bearing against Class 60 Cast-iron Web: When bearing occurs between materials with different compressive strengths, the permissible stress is that of the weaker material.

AISI 1045 bolts   Ultimate $s_c = 95\,000$ psi

Class 60 CI web   Ultimate $s_c = 170\,000$ psi

Therefore, use allowable stress of bolts.

Allowable $s_c = 15\,830$ psi   (see calculation 3)

$$T = n_b F_{\text{bolt}} r_{bc}$$
$$F_{\text{bolt}} = A_c s_c$$
$$A_c = d_{\text{bolt}} t = 0.75(1.5) = 1.125 \text{ in}^2$$
$$F_{\text{bolt}} = 1.125(15\,830) = 17\,800 \text{ lb}$$
$$n_b = 6 \quad r_{bc} = 5\tfrac{1}{2} \text{ in}$$
$$T = 6(17\,800)(5.5) = 587\,000 \text{ in} \cdot \text{lb}$$

From calculations 5 and 6, $T_{\text{bolt}} = 170\,000$ in·lb.

SUMMARY OF TORQUE CALCULATIONS

	Allowable torque, in·lb
Shaft	421 000
Key	158 000
Hub	1 632 000
Bolts	170 000

The maximum safe torque for this power transmission arrangement is 158 000 in·lb. The maximum allowable horsepower is

$$\text{hp} = \frac{Tn}{63\,000} = \frac{158\,000(350)}{63\,000} = 878$$

PROBLEMS

*13-1.   What torque can a 75-mm-diameter solid Monel shaft transmit safely under steady load conditions?

13-2.   A $1\tfrac{3}{4}$-in-diameter solid shaft of 302 stainless steel transmits a torque of 31 000 in·lb. Assuming varying load conditions, is this shaft satisfactory?

*13-3.   An AISI 1045 steel shaft transmits a torque of 1.1 kN·m under shock loading.

     a.   What diameter solid shaft is required?

     b.   What size hollow shaft is required if the outside diameter is twice the inside diameter?

13-4. What is the length of a $\frac{1}{2}$-in AISI 1095 steel rod that can be twisted through one-half revolution without exceeding a shearing stress of 22 000 psi?

*13-5. An AISI 1045 steel rod 20 mm in diameter and 300 mm long is subjected to a torque of 180 kN·mm. What will be the angle of twist?

13-6. **a.** Find the horsepower which a 3-in shaft can transmit at 200 rpm if the torsional shearing stress is not to exceed 7000 psi.

      b. What material would you recommend for the shaft? (Assume the stress specified in part **a** includes shock loading conditions.)

*13-7. Find the diameter of a solid AISI 1020 steel shaft to transmit 110 kW at 2200 rpm under shock loading.

*13-8. Design a hollow shaft with $d_o = 1.8d_i$ for the conditions in Prob. 13-7.

13-9. At what revolutions per minute would you operate a 12-in-diameter AISI 1020 steel shaft to transmit 6000 hp at a maximum shear stress of 5000 psi?

13-10. A 302 stainless-steel shaft has the following specifications.

Outside diameter = 0.500 in
Inside diameter = 0.375 in
Length = 3.25 in
Loading = varying
rpm = 10 000

Find *a*. Maximum safe horsepower. *b*. Angle of twist (degrees).

*13-11. A gear transmitting 18 kW at 130 rpm is fastened to a 75-mm-diameter AISI 1045 steel shaft by means of an AISI 1045 steel flat key 20 mm wide by 12 mm thick. If the permissible shear and bearing stresses are 55 and 110 MPa, respectively, what length of key is required?

13-12. A torque of 90 000 in·lb is transmitted from a 6-in AISI 1095 steel shaft to a pulley by means of two AISI 1095 steel square keys, each of which is 6 in long. If $s_s = 10\ 000$ psi and $s_c = 18\ 000$ psi, what should be the dimensions of the keys?

13-13. A $\frac{3}{4}$-in-diameter AISI 1045 steel shaft transmits 10 hp at 500 rpm to a pulley which is keyed to the shaft by an AISI 1045 steel flat key $\frac{3}{16}$ in wide by $\frac{1}{8}$ in thick and 4 in long. What are the safety factors of the key based on the allowable stresses of 8000 psi in shear and 16 000 psi in bearing?

13-14. What torque can be transmitted by the eight steel bolts shown in the coupling of Fig. Prob. 13-14? The bolts, $\frac{7}{8}$ in in diameter, are in bearing against the $\frac{1}{2}$-in web of the coupling. The bolt circle diameter is 11 in. Allowable stresses are as in Prob. 13-12.

*13-15. An AISI 1045 steel shaft transmits 373 kW at 200 rpm. The flange couplings use 10 AISI 1020 steel bolts 20 mm in diameter arranged in a circle 300 mm in diameter. The web of the coupling is 25 mm thick. What are the stresses in the bolts?

13-16. An AISI 1095 steel shaft 1 in in diameter is twisted through an angle of 2° in a length of 10 ft. What will be the shearing stress?

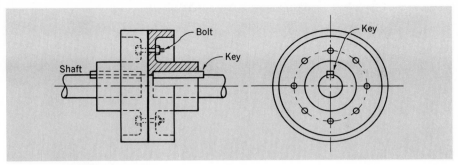

FIGURE PROBLEM 13-14

*13-17. An AISI 1045 steel shaft 3 m long is subjected to a torque of 200 kN·mm. If the shearing stress is not to exceed 83 MPa and the angle of twist may not exceed 0.035 rad, what diameter is needed?

13-18. A hollow AISI 1045 steel shaft of 5 in OD and 4 in ID transmits 180 hp at a speed of 240 rpm. The length of shaft subjected to the torque is 10 ft.

    a. What is the angle of twist?

    b. What is the shearing stress?

14

Combined Stresses

14-1 PRINCIPLE OF SUPERPOSITION

It has been shown that either direct axial loading (Chap. 7) or bending (Chap. 12) can induce tensile (or compressive) stresses. Furthermore, we have seen that a shearing stress may result from either direct shearing forces (Chap. 7) or torsion (Chap. 13).

Stresses of the *same* kind which act simultaneously on a given area (or point) can be added to get their combined effect or resultant. If such stresses are collinear (act along the same line), they can be added algebraically. If the stresses are not collinear, they must be added vectorially. These procedures resemble the methods used in Chap. 2 to resolve concurrent force systems into resultants.

This principle of superposition will be applied in this chapter to the following situations.

1. Combined tension or compression
   (a) Combined axial and bending stresses in beams
   (b) Eccentrically loaded short columns
   (c) Eccentrically loaded machine members
2. Combined direct and torsional shear
   (a) Eccentrically loaded bolted joints

Members which are simultaneously loaded in tension (or compression) and shear are designed by considering the combined effect of these *different* kinds of stress. Of the several methods available for such combinations, two are demonstrated in this chapter: the maximum principal stress method and the maximum shear stress method. These methods will be applied to:

3. Combined tension and shear
    (a) Combined bending and torsional shearing stresses which occur in shafts

4-2 COMBINED AXIAL AND BENDING STRESSES

Consider a simply supported beam with a uniformly distributed load which is subjected to an axial tensile force F, as shown in Fig. 14-1. The problem of determining the combined stress at any point in the beam can be subdivided by examining the effect of the axial load and the effect of the uniform load separately and then superimposing those effects to get the combined result.

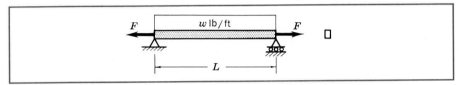

Fig. 14-1 Beam subjected to combined axial and bending stresses.

The axial force induces a tensile stress on all cross sections of the beam given by

$$s_1 = \frac{F}{A}$$

where A is the cross-sectional area of the beam. This stress is assumed to act equally at all points of a given cross section, as shown in Fig. 14-2.

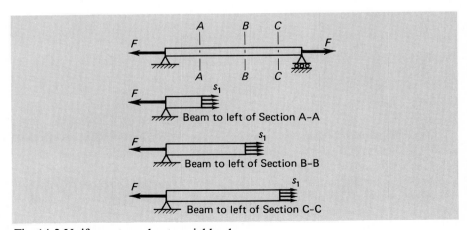

Fig. 14-2 Uniform stress due to axial load.

The uniform load w causes the beam to bend and induces a stress at each cross section which varies from maximum tension at the bottom fibers through zero at the neutral axis to maximum compression at the top fibers. The bending stresses can be determined from

$$s_2 = \frac{Mc}{I} \qquad \text{(see Chap. 12)}$$

Since a design is usually concerned with critical conditions, the extreme fiber values of s_2 are most often required. Furthermore, the largest of these extreme fiber values tends to occur at the cross section where M (bending moment) is maximum. For the case of a uniformly distributed load on a simple span, M_{max} occurs at the middle cross section of the beam. The bending stress distribution at this cross section is shown in Fig. 14-3.

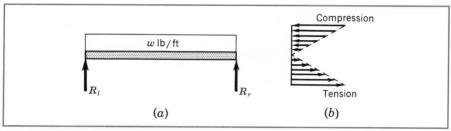

Fig. 14-3 Bending stress distribution.

The resultant combined stresses are determined by adding the individual axial and bending stresses algebraically,[1] as in Fig. 14-4. Then

$$s = s_1 \pm s_2$$

or
$$s = \frac{F}{A} \pm \frac{Mc}{I} \qquad (14\text{-}1)$$

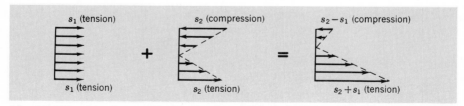

Fig. 14-4 Combined stress distribution by addition of axial and bending stresses.

In the case of axial tension and bending of a simply supported beam, when the bottom fibers are considered, both F/A and Mc/I are tensile stresses; thus the plus sign is proper. When the axial stress is tension and Mc/I is compression in

[1] Valid for small deflections.

the top fibers, the values must be subtracted. Figure 14-5 shows a plot of combined stresses for top and bottom fibers at all sections. The axial stress F/A could be either tension or compression, depending upon the nature of the load F; and we have seen that Mc/I can be either tension or compression, depending upon which fiber of the beam is considered. In general, when both F/A and Mc/I are tension (or both are compression), they should be added; but when one is tension and the other is compression, they should be subtracted. The resulting combined value should always be identified by stating whether it is tension or compression. These procedures will be clarified in the sample problems which follow.

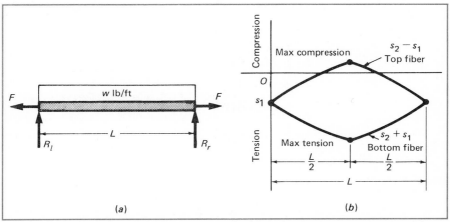

Fig. 14-5 (*a*) Simply supported beam with uniform load and axial load. (*b*) Distribution of stresses in top and bottom fibers of the beam.

Sample Problem 1
A simply supported W 12 × 65 beam that is 16 ft long carries a concentrated load of 6000 lb at each quarter point and is subjected to an axial tensile force of 25 000 lb applied at the end sections.

(a) Find maximum combined tensile stress and maximum combined compressive stress.
(b) If it were necessary to make a 1½-in hole in the web of this beam at the center cross section so that a water pipe could be accommodated, where on the cross section would you recommend that the hole center be located?

Solution a: The weight of the beam is 65(16) = 1040 lb. The total vertical load on the beam is 6000 + 6000 + 6000 = 18 000 lb. Since the weight of the beam is only (1040/18 000)(100) = 5.8 percent of the total vertical load, it can be neglected without excessive error. (Approximately 4 percent error occurs here.)

Figure 14-6 shows the beam with its shear force and bending moment diagrams. Note that the 25 000-lb axial force does not affect these diagrams. For the W 12 × 65, $A = 19.1$ in², $S = 87.9$ in³, and $d = 12.12$ in (Table 4, App. B.),

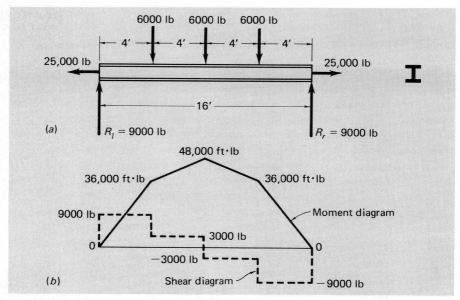

Fig. 14-6 (*a*) Beam diagram for Sample Problem 1. (*b*) Shear force and bending moment diagrams.

Direct Stress:

$$s_1 = \frac{F}{A} = \frac{25\,000}{19.1} = 1310 \text{ psi (tension)}$$

Bending Stress:

$$s_2 = \frac{Mc}{I} = \frac{M}{S}$$

The shear diagram indicates that the center cross section is the location of the maximum moment. A free-body diagram of the left half of the beam facilitates calculating M_{max} (Fig. 14-7).

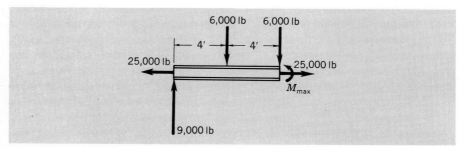

Fig. 14-7 Free-body diagram of left half of beam for Sample Problem 1.

$\Sigma M = 0$ about an axis through the center cross section gives us

$M_{max} = 9000(8) - 6000(4) = 72\ 000 - 24\ 000 = 48\ 000$ ft·lb

$$s_2 = \frac{M}{S} = \frac{48\ 000(12)}{87.9} = 6550 \text{ psi}$$

$s_2 = 6550 \text{ psi} \begin{cases} \text{tension at bottom fiber} \\ \text{compression at top fiber} \end{cases}$

Therefore, from Eq. (14-1), at center cross section,

Top fiber $s = s_2 - s_1 = 6550 - 1310$

Top fiber $s = 5240$ psi (maximum compression)

Bottom fiber $s = s_2 + s_1 = 6550 + 1310$

Bottom fiber $s = 7860$ psi (maximum tension)

Solution b: The distribution of combined stress at the center cross section is shown in Fig. 14-8.

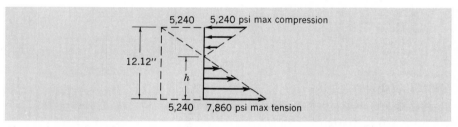

Fig. 14-8 Location of point of zero stress from distribution of combined stress at center cross section.

The location of zero combined stress would be the most preferable position for a hole in the web. The level of zero stress can be found from Fig. 14-8 by similar triangles.

$$h = \frac{7860}{7860 + 5240}(12.12) = \frac{7860}{13\ 100}(12.12) = 7.27 \text{ in}$$

The center of the $1\frac{1}{2}$-in hole should be located 7.27 in above the bottom of the lower flange.

*Sample Problem 2** A 0.45-m cantilever member is subjected to a load $F = 27$ kN, as shown in Fig. 14-9a. Find the maximum tensile and compressive stresses. The centroid of the area is 50 mm from the top.

Solution: Force F can be resolved into vertical and horizontal components F_y and F_x.

$F_y = F \sin 30° = 27(0.5) = 13.5 \text{ kN} = 13.5 \times 10^3 \text{ N}$

$F_x = F \cos 30° = 27(0.866) = 23.4 \text{ kN} = 23.4 \times 10^3 \text{ N}$

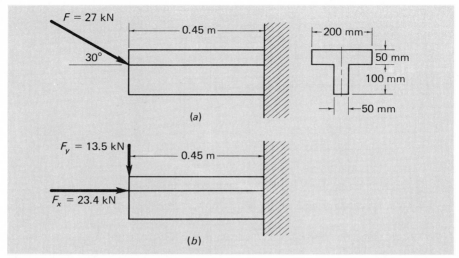

Fig. 14-9 (a) Beam diagram for Sample Problem 2. (b) Resolution of force F into horizontal and vertical components.

The beam shown in Fig. 14-9b can now be treated as a combined-stress problem.

Direct Stress:

F_x causes uniform compression at all sections.

$$A = 200(50) + 50(100) = 15\ 000\ \text{mm}^2 = 15 \times 10^3\ \text{mm}^2$$

$$s_1 = \frac{F_x}{A} = \frac{23.4(10^3)}{15(10^3)} = 1.56\ \text{N/mm}^2\ \text{(compression)}$$

Bending Stress:

$$s_2 = \frac{Mc}{I} \qquad M_{\text{max}} = F_y(L) = 13.5(10^3)(0.45)$$

$$= 6.075(10^3)\ \text{N}\cdot\text{m} = 6.075 \times 10^6\ \text{N}\cdot\text{mm}\ \text{(at wall)}$$

$$I_x = \frac{200(50^3)}{12} + 10\ 000(25^2) + \frac{50(100^3)}{12} + 5000(50^2)$$

$$I_x = 2.08(10^6) + 6.25(10^6) + 4.17(10^6) + 12.5(10^6)$$
$$= 25(10^6)\ \text{mm}^4$$

$$\text{Top fiber } s_2 = \frac{6.075(10^6)(50)}{25(10^6)} = 12.15\ \text{N/mm}^2\ \text{(tension)}$$

$$\text{Bottom fiber } s_2 = \frac{6.075(10^6)(100)}{25(10^6)} = 24.3\ \text{N/mm}^2\ \text{(compression)}$$

Combined Stresses:

Top fiber $s = s_2 - s_1$

$= 12.15 - 1.56$

$s = 10.59$ N/mm² $\qquad$ say, $s = 10.6$ N/mm² (tension)

Bottom fiber $s = s_2 + s_1$

$= 24.3 + 1.56$

$s = 25.86$ N/mm² $\qquad$ say, $s = 25.9$ N/mm² (compression)

14-3 ECCENTRICALLY LOADED SHORT COMPRESSION MEMBERS

The reader will recall from Sec. 2-8 that a member under compression which does not tend to buckle or bend is called a short compression member.

Let us examine a short compression member that has a load F applied eccentric to the centroid of its cross section, Fig. 14-10a. Now imagine two

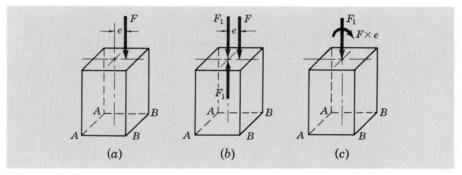

Fig. 14-10 (a) Eccentrically loaded short compression member. (b) Introduction of equal and opposite forces at centroid. (c) Equivalent system for eccentrically loaded short compression member.

equal and opposite forces F_1 equal to the original load F, applied at the centroid, as indicated in Fig. 14-10b. The loading has not changed, since the sum of the additional loads equals zero. We have effectively replaced the eccentric force system of Fig. 14-10a by a downward force F_1 at the centroid acting to compress the member and the upward force F_1 and the force F, which together form a couple Fe which subjects the member to a clockwise moment, Fig. 14-10c. The distance e, called the *eccentricity*, is the perpendicular distance from the centroid to the line of action of the force F. This problem can now be solved by applying the principles of combined axial and bending stresses, as in Sec. 14-2. The compressive stress due to the compressive load $F_1 = F$ is

$$s_1 = \frac{F}{A}$$

The flexural stress due to the moment Fe is

$$s_2 = \frac{Mc}{I} = \frac{Fec}{I}$$

Then the combined stress at the base is

$$s = s_1 \pm s_2$$

or

$$s = \frac{F}{A} \pm \frac{Fec}{I} \tag{14-2}$$

At edge AA the negative sign will be used because the direct stress due to force F is compression and the bending stress due to the moment is tension. For edge BB the stresses will be added, since both the direct and bending stresses are compressive.

Sample Problem 3 A short compression member 6 by 8 in in cross section and 10 in high has a load of 25 000 lb applied $1\frac{1}{2}$ in from the centroid of the cross section, as indicated in Fig. 14-11. Determine the stresses at sides AB and CD.

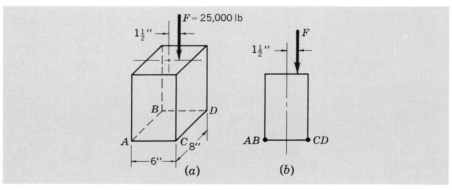

Fig. 14-11 Diagram for Sample Problems 3 and 4.

Solution: The direct compressive effect of the 25 000-lb force will cause a compressive stress on the area $ABCD$.

$$s_1 = \frac{F}{A}$$

$$s_1 = \frac{25\ 000}{6(8)} = \frac{25\ 000}{48} = 521 \text{ psi (compression)}$$

As a result of the eccentricity, there will be a moment equal to Fe tending to bend the short column. The bending stress due to this moment will be

$$s_2 = \frac{Fec}{I}$$

where $F = 25\ 000$ lb

$\quad e = 1\frac{1}{2}$ in

$\quad c = 3$ in

$\quad I = \dfrac{bh^3}{12} = \dfrac{8(6^3)}{12} = 144$ in^4 (Note that the eccentric load tends to rotate the base about a centroidal axis perpendicular to the 6-in side.)

$\quad S_2 = \dfrac{25\ 000(1\frac{1}{2})(3)}{144} = 781$ psi

At side AB there will be compression due to the direct action of the load F and tension due to the bending effect. Since the stresses are not of the same type, we use the negative sign in Eq. (14-2).

$$S_{ab} = \frac{F}{A} - \frac{Fec}{I}$$

$$S_{ab} = 521 - 781 = -260 \text{ psi (tension)}$$

The negative sign indicates that the bending tensile stress will dominate and the resultant stress is tension.

At side CD there will be compression due to the direct action of the load F and compression due to the bending effect. Since both stresses are of the same type, we use the positive sign in Eq. (14-2).

$$S_{cd} = \frac{F}{A} + \frac{Fec}{I}$$

$$S_{cd} = 521 + 781 = 1302 \text{ psi}$$
$$\text{Say, } S_{cd} = 1300 \text{ psi (compression)}$$

Sample Problem 4   If for the conditions given in Sample Problem 3 the maximum compressive stress is specified as 1000 psi, what maximum load F can be applied?

Solution: We have seen in Sample Problem 3 that the stress at side AB will be tension. Therefore, we shall consider only side CD, where a compressive stress occurs.

From Eq. (14-2),

$$s = s_1 + s_2 = \frac{F}{A} + \frac{Fec}{I}$$

$$1000 = \frac{F}{48} + \frac{F(3)(1\frac{1}{2})}{144} = 0.0208F + 0.0313F$$

$$0.0521F = 1000$$

$$F = 19\ 200 \text{ lb (maximum)}$$

14-4 ECCENTRIC LOADING OF MACHINE MEMBERS

The preceding analysis for an eccentrically loaded short compression member can be extended to eccentric loading of machine members such as the press frame shown in Fig. 14-12a.

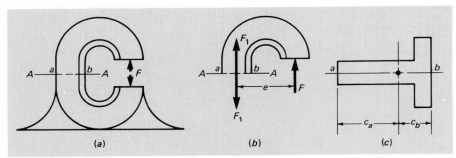

Fig. 14-12 (a) Eccentrically loaded C frame. (b) Introduction of equal and opposite forces at centroid of section A-A. (c) Section A-A.

Let us determine the stresses developed on plane AA at edges a and b as a result of the force F applied to the jaws of the C frame.

Figure 14-12b shows the portion of the press frame above the plane AA. At the centroid of the cross section, imagine that two equal and opposite forces $F_1 = F$ are applied. Then the downward force F_1 and the punching force F form a counterclockwise couple, or counterclockwise bending moment, equal to Fe. This moment will create a compressive stress at edge a and a tensile stress at edge b. The upward force F_1 creates tension evenly distributed across plane AA. As a result, tensile stresses are developed at edges a and b. The resultant stresses at a and b will be the sum of stresses due to tension and bending.

At edge a because of direct tension,

$$s_1 = \frac{F}{A}$$

At edge a because of bending (compression),

$$s_2 = \frac{Mc_a}{I} = \frac{Fec_a}{I}$$

where c_a = distance from neutral axis (centroid) to edge a, in
I = moment of inertia of the cross section, in^4

Therefore, the combined stress at edge a will be

$$s = s_1 - s_2$$

or
$$s = \frac{F}{A} - \frac{Fec_a}{I}$$

Whether the resultant stress is tension or compression will be governed by the values of s_1 and s_2.

At edge b because of bending (tension),

$$s_2 = \frac{Mc_b}{I} = \frac{Fec_b}{I}$$

where c_b = distance from neutral axis (centroid) to edge b, in
$\quad\quad I$ = moment of inertia of the cross section, in⁴

At edge b because of direct tension,

$$s_1 = \frac{F}{A}$$

Therefore, the combined stress at edge b will be

$$s = s_1 + s_2$$

or
$$s = \frac{F}{A} + \frac{Fec_b}{I}$$

This resultant stress will be tension since both s_1 and s_2 are tensile stresses.

Sample Problem 5   A punch press frame, as shown in Fig. 14-13a, applies a maximum load of $F = 15\ 000$ lb. The distance l from the line of action of the force to the inside surface of the frame is 12 in. The dimensions of the cross section AA are as given in Fig. 14-13b. Determine the stresses on the inside and outside surfaces of this section.

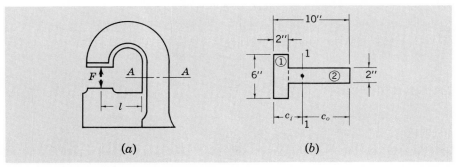

(a)   (b)

Fig. 14-13 Diagram for Sample Problem 5.

Solution: Since the force F is eccentric to the centroid of the cross section AA, this problem must be analyzed for combined stress. On section AA there will be a direct tensile force F pulling the section apart and a bending moment equal to F times the perpendicular distance from the line of action of the force to the centroid of the section AA. The direct tensile force will create a tensile stress on both the inside and outside surfaces. The bending moment will cause a tensile stress on the inside surface and a compressive stress on the outside surface. Therefore, the stress on the inside surface will be

$$s_i = \frac{F}{A} + \frac{Fec_i}{I}$$

and the stress on the outside surface will be

$$s_o = \frac{F}{A} - \frac{Fec_o}{I}$$

$$F = 15\ 000 \text{ lb}$$

$$A = A_1 + A_2 = (6 \times 2) + (8 \times 2) = 12 + 16 = 28 \text{ in}^2$$

In order to determine e, the location of the centroid from the inside surface must be found. Taking the moment of areas about this side

$$c_i = \frac{(12 \times 1) + (16 \times 6)}{28} = \frac{12 + 96}{28} = \frac{108}{28} = 3.86 \text{ in}$$

Therefore, $e = 12 + 3.86 = 15.86$ in.

To find I, we must find the moment of inertia of the entire figure about axis 1-1.

$$I_1 \text{ about axis 1-1} = \frac{6(2^3)}{12} + 12(2.86^2) = 4 + 98.2 = 102.2$$

$$I_2 \text{ about axis 1-1} = \frac{2(8^3)}{12} + 16(2.14^2) = 85.3 + 73.3 = 158.6$$

$$I = 260.8 \text{ in}^4 \qquad \text{say, } I = 261 \text{ in}^4$$

$$s_i = \frac{15\ 000}{28} + \frac{15\ 000(15.86)(3.86)}{261}$$

$$= 536 + 3520 = 4056 \text{ psi} \quad \text{say, } s_i = 4060 \text{ psi (tension)}$$

$$s_o = \frac{15\ 000}{28} - \frac{15\ 000(15.86)(6.14)}{261}$$

$$= 536 - 5600 = -5064 \text{ psi} \quad \text{say, } s_o = 5060 \text{ psi (compression)}$$

14-5 ECCENTRICALLY LOADED BOLTED JOINTS

Whenever possible, a bolted joint should be so constructed that the line of action of the resultant force on the joint passes through the centroid of the area of the group of bolts.

Load F, Fig. 14-14, has its line of action passing through the centroid of the area of the group of bolts A, B, C, and D. Each of the bolts will then resist a load of $F/4$, which could cause shear failure of the bolts. Thus, the resisting shear stress developed will be

$$s_1 = \frac{F}{4A}$$

where A = cross-sectional area of one bolt, in^2

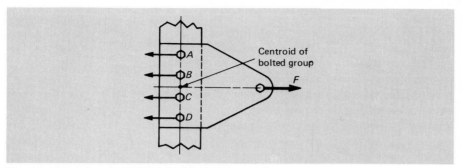

Fig. 14-14 Load on bolted joint without eccentricity.

In Fig. 14-15*a*, the line of action of force *F* *does not* pass through the centroid of the bolt group. The line of action is eccentric to the centroid by the distance *e*. This results in a direct pull on the bolts plus a counterclockwise moment about the centroid that is due to the eccentricity of the applied force. The direct shear stress developed is

$$s_1 = \frac{F}{4A}$$

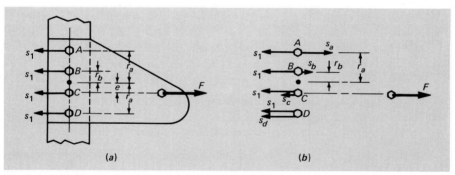

Fig. 14-15 (*a*) Eccentrically loaded bolted joint. (*b*) Direct and rotational shear stresses developed in the bolts.

The moment of the force *F* is

$$M = Fe$$

In order to maintain equilibrium, the sum of the resisting moments developed in the bolts must be equal and opposite to *Fe*. Hence, the resisting moments act in a clockwise direction.

$$Fe = M = M_a + M_b + M_c + M_d$$

where M_a = resisting moment of bolt *A*, in·lb
M_b = resisting moment of bolt *B*, in·lb
M_c = resisting moment of bolt *C*, in·lb
M_d = resisting moment of bolt *D*, in·lb

The resisting moment M_a on bolt A is

$$M_a = F_a r_a$$

where F_a = rotational resisting force of bolt A, lb
$\quad r_a$ = moment arm from line of action of F_a to the centroid, in

But the rotational resisting force will be a function of the resisting shear stress developed and the area of the bolt.

$$F_a = A_a s_a$$

where A_a = area of the bolt, in^2
$\quad s_a$ = resisting rotational shear stress in bolt A, psi

Therefore, $\qquad\qquad M_a = F_a r_a = A_a s_a r_a$

Since the bolts are of the same size and the distance of bolt D from the centroid is the same as that of bolt A, the resisting moment of bolt D will be equal to that of bolt A.

$$M_a = M_d$$

The rotational shear stress developed in bolt D will be equal and opposite to that in bolt A.

$$s_a = -s_d$$

Similarly, the resisting moment developed by bolt B equals

$$M_b = F_b r_b = A_b s_b r_b$$

where M_b = resisting moment of bolt B, in $\cdot$ lb
$\quad F_b$ = rotational resisting force of bolt B, lb
$\quad r_b$ = moment arm from line of action of F_b to centroid, in
$\quad A_b$ = area of bolt B, in^2
$\quad s_b$ = resisting rotational shear stress developed in bolt B, psi

The resisting moment of bolt C will be equal to that of bolt B, since the areas of the bolts and the distances from the centroid are the same. The stress in bolt C will be equal and opposite to that in bolt B.

$$s_b = -s_c$$

Therefore, $\qquad Fe = A_a s_a r_a + A_b s_b r_b + A_c s_c r_b + A_d s_d r_a$
$$= 2(A_a s_a r_a) + 2(A_b s_b r_b)$$

Figure 14-15b indicates the directions in which the stresses will act. The direct shear stress and rotational shear stress on each bolt can be added directly, since their lines of action are parallel. The resultant shear stresses in the bolts will be

$$\text{Bolt } A: s_s = s_1 - s_a$$
$$\text{Bolt } B: s_s = s_1 - s_b$$
$$\text{Bolt } C: s_s = s_1 + s_c$$
$$\text{Bolt } D: s_s = s_1 + s_d$$

Sample Problem 6   What is the greatest shearing stress in the bolted connection shown in Fig. 14-16? The bolts are $\frac{7}{8}$ in in diameter.

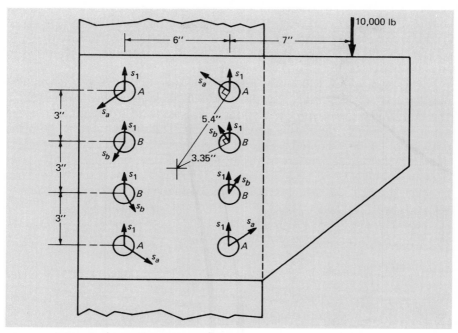

Fig. 14-16 Diagram for Sample Problem 6.

Solution: The load produces a direct shearing stress which is assumed to be the same for all bolts. The area for a $\frac{7}{8}$-in bolt $= 0.6$ in^2. Then the direct shear stress on each bolt is

$$s_1 = \frac{10\,000}{0.6(8)} = 2080 \text{ psi}$$

The load also tends to rotate the plate in a clockwise direction. The moment due to the load is

$$M = Fe = 10\,000(10) = 100\,000 \text{ in} \cdot \text{lb}$$

For equilibrium, the resisting moment of the bolts must be equal to 100 000 in·lb. The shearing stresses in the bolts will be proportional to the respective distances from the centroid of the group. Those distances are 5.4 in for bolts A and 3.35 in for bolts B. (The distances are calculated from other dimensions in the figure.)

Let s_a represent the rotational shearing stress due to the moment in the outermost bolts A; then the stress in bolts B will be $s_b = (3.35/5.4)s_a$. By taking moments about the centroid of the group, we have

$$Fe = 4(A_a s_a r_a) + 4(A_b s_b r_b)$$

$$100\,000 = 4[(0.6)s_a(5.4)] + 4\left[(0.6)\left(\frac{3.35}{5.4}\right)s_a(3.35)\right]$$

$$= 12.96s_a + 5.0s_a$$

$$17.96s_a = 100\,000$$

$$s_a = 5570 \text{ psi}$$

The actual stress will be the resultant of 2080 and 5570 psi. Before this resultant can be determined, the directions of the stresses must be known.

The 2080 psi is an upward stress for each bolt. The 5570 psi for each bolt A is perpendicular to the radius arm of the bolt. The direction varies with the position of the bolt, as can be seen from the figure. By inspection, it is evident that the resultant is greatest at the upper right-hand bolt A and the lower right-hand bolt A.

The stress of 5570 psi makes an angle θ with the horizontal that is equal to the angle that the radius makes with the vertical (see Fig. 14-17a).

$$\tan\theta = \frac{3}{4.5} = 0.667$$

$$\theta = 33.7°$$

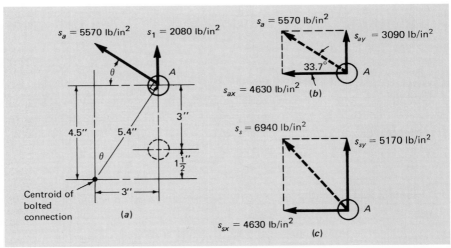

Fig. 14-17 (a) Stresses developed in bolt A. (b) Components of rotational shear stress in bolt A. (c) Resultant shear stress in bolt A.

On resolving $s_a = 5570$ psi into its vertical and horizontal components (Fig. 14-17b), we have

$$s_{ay} = 5570 \sin 33.7° = 3090 \text{ psi}$$

$$s_{ax} = 5570 \cos 33.7° = 4630 \text{ psi}$$

The vertical and horizontal components of the resultant are (Fig. 14-17c)

$$S_{sy} = 2080 + 3090 = 5170 \text{ psi}$$
$$S_{sx} = 4630 \text{ psi}$$

The resultant is

$$S_s = \sqrt{5170^2 + 4630^2} = 6940 \text{ psi (maximum)}$$

14-6 SHEAR STRESS DUE TO TENSION OR COMPRESSION

If an axial force is applied to a short compression member, the stress, which might be either tension or compression, is uniformly distributed over the normal cross section AA. There is no component of force parallel to the section; consequently, there is no shearing stress on the section.

If a section BB making some angle α with the normal section is chosen, as shown in Fig. 14-18, the axial force can be resolved into two components, one at right angles to section BB, called F_n, and the other parallel to section BB, called F_s. Then

$$F_n = F \cos \alpha$$
$$F_s = F \sin \alpha$$

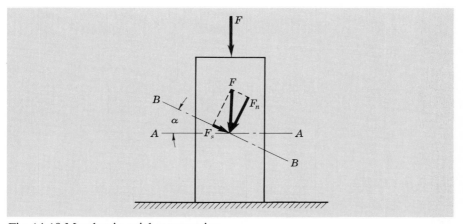

Fig. 14-18 Member in axial compression.

The area resisting those forces will be the area of the inclined section and equal to $A/\cos \alpha$, where A is the area of the normal section. Then the normal stress is

$$S_n = \frac{F_n}{A_n} = \frac{F \cos \alpha}{A/\cos \alpha} = \frac{F \cos^2 \alpha}{A} = \frac{F}{A} (\tfrac{1}{2} + \tfrac{1}{2} \cos 2\alpha)$$

$$S_n = \frac{F}{2A} + \frac{F}{2A} \cos 2\alpha \qquad (14\text{-}3)$$

This value is a maximum when the angle $\alpha = 0°$. That is, the maximum tension or compression will take place on the normal cross section. Also, the shearing stress is

$$s_s = \frac{F_s}{A_s} = \frac{F \sin \alpha}{A/\cos \alpha} = \frac{F}{A}(\sin \alpha \cos \alpha) = \frac{F}{A}\frac{\sin 2\alpha}{2}$$

$$s_s = \frac{F}{2A}\sin 2\alpha \tag{14-4}$$

The value of $\sin 2\alpha$ is largest when $\alpha = 45°$; therefore, the value of s_s is a maximum on a plane making an angle of $45°$ with the normal cross section.

14-7 MOHR'S CIRCLE—NORMAL AND SHEAR STRESSES DUE TO AXIAL LOAD

A graphical method for determining the normal and shearing stresses s_n and s_s created by an axial force on a plane other than the normal plane, Fig. 14-18, was developed in 1882 by Professor Otto Mohr, a German engineer.

As shown in Fig. 14-19, plot rectangular axes x and y with the x-axis representing the normal stress s_n and the y axis representing the shearing stress s_s. Draw a circle tangent to the y axis with its center O on the x axis and a radius equal to $F/2A$. The horizontal diameter of the circle AQ is

$$(s_n)_{\text{max}} = \frac{F}{A}$$

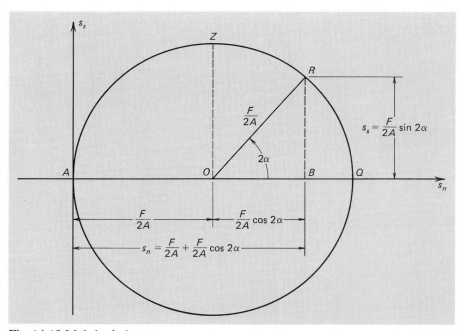

Fig. 14-19 Mohr's circle.

Turn the radius counterclockwise from the x axis and lay off an angle of 2α. The angle α is the angle between the normal plane and the plane being examined; see Fig. 14-18. Reference to Mohr's circle verifies that, as angle α increases from 0° to 45° (2α increases from 0° to 90°) the normal stress decreases from a maximum AQ to a minimum AO. The shearing stress increases from zero at point Q to a maximum at point Z.

Sample Problem 7   A short square steel compression member that is 3 in on a side is subjected to an axial load of 90 000 lb.

(a)   Find the maximum normal stress.
(b)   Find the normal and shearing stresses on planes making angles of 30° and 45° with the horizontal cross section.
(c)   Use Mohr's circle to solve (a) and (b).

Solution a:

$$F = 90\ 000\ \text{lb} \qquad A = 3(3) = 9\ \text{in}^2$$

$$(s_n)_{\text{max}} = \frac{F}{A} = \frac{90\ 000}{9} = 10\ 000\ \text{psi}$$

Solution b: On a 30° plane, from Eq. (14-3),

$$s_n = \frac{F}{2A} + \frac{F}{2A}\cos 2\alpha$$

$$= \frac{90\ 000}{2(9)} + \frac{90\ 000}{2(9)}\cos 60°$$

$$= 5000 + 5000(0.5)$$

$$s_n = 7500\ \text{psi}$$

From Eq. (14-4), $s_s = \dfrac{F}{2A}\sin 2\alpha$

$$= \frac{90\ 000}{2(9)}\sin 60° = 5000\ (0.866)$$

$$s_s = 4330\ \text{psi}$$

On a 45° plane, from Eq. (14-3),

$$s_n = \frac{F}{2A} + \frac{F}{2A}\cos 2\alpha$$

$$= \frac{90\ 000}{2(9)} + \frac{90\ 000}{2(9)}\cos 90°$$

$$= 5000 + 5000(0)$$

$$s_n = 5000\ \text{psi}$$

From Eq. (14-4),

$$s_s = \frac{F}{2A} \sin 2\alpha$$

$$= \frac{90\,000}{2(9)} \sin 90° = 5000(1)$$

$$s_s = 5000 \text{ psi}$$

Solution c: To construct Mohr's circle, calculate

$$\frac{F}{2A} = \frac{90\,000}{2(9)} = 5000 \text{ psi}$$

Draw a circle with a radius of 5000 psi, Fig. 14-20, and lay off angles of 2α. When $\alpha = 30°$, $2\alpha = 60°$. Read from the diagram:

$$s_n = AO + OB = 7500 \text{ psi}$$
$$s_s = BR = 4330 \text{ psi}$$

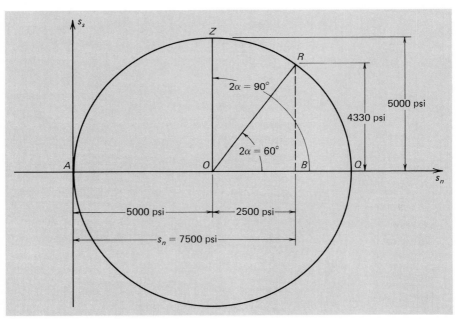

Fig. 14-20 Diagram for Sample Problem 7.

When $\alpha = 45°$, $2\alpha = 90°$. Read from the diagram:

$$s_n = AO = 5000 \text{ psi}$$
$$s_s = OZ = 5000 \text{ psi}$$

14-8 TENSION OR COMPRESSION DUE TO SHEAR

When a member such as the shaft shown in Fig. 14-21a is subjected to pure torsion, direct shear stresses s_1 are developed on a very small square surface area. From Sec. 12-2 we know that, to maintain equilibrium, other shear stresses, s_2 equal to s_1, are induced on the section at right angles to s_1, Fig. 14-21b. Analysis of the 45° plane AA (Fig. 14-22a) indicates that tensile stresses are acting to pull the section apart (Fig. 14-22b). On 45° plane BB, compressive stresses are acting to compress the section (Fig. 14-22c). No shear exists on these 45° planes. If some other planes, such as 15° or 20° planes, were analyzed, it would be found that combinations of shear and tension or compression would exist.

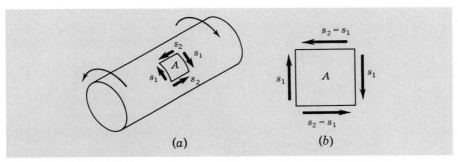

Fig. 14-21 (a) Shaft in pure torsion. (b) Shear stresses on small surface section A.

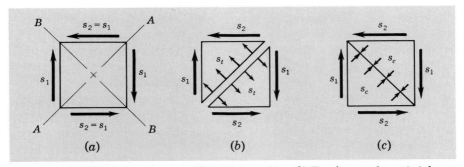

Fig. 14-22 (a) Shear stresses on small surface section. (b) Tension at plane A-A due to shear. (c) Compression at plane B-B due to shear.

14-9 COMBINED BENDING AND TORSION

A shaft which transmits or receives power by means of gear, belt, or chain drive develops not only torsional shear stress but also bending stress due to the forces acting on the shaft. Such a shaft is shown in Fig. 14-23.

Let us examine the stresses acting on an extremely small section A of the surface of the shaft. When this outer fiber section is at the bottom as a result of

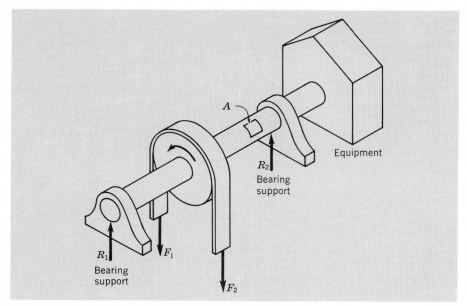

Fig. 14-23 Shaft subjected to combined bending and torsion.

the rotation of the shaft, there are both torsional and tensile effects on it. Torsion will create a shear stress s_s and bending tensile stress s_t (Fig. 14-24).

In Secs. 14-6 and 14-8 we have seen how shear can result in tensile and compressive stresses and how tensile and compressive loads can create shear stresses. In a similar manner, the combined effect of the shear and tensile stresses on section A (Fig. 14-24) will result in the following.

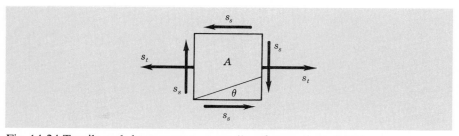

Fig. 14-24 Tensile and shear stresses on small surface section of shaft.

1. A maximum resultant tensile (or compressive) stress greater than the tensile (or compressive) stress due to bending alone. It is called the *principal stress.*

This maximum tensile (or compressive) stress is given by

$$(s_t)_{\text{max}} = \frac{s_t}{2} + \sqrt{s_s^2 + \left(\frac{s_t}{2}\right)^2} \qquad (14\text{-}5)$$

The angle θ of the plane on which the maximum resultant tensile (or compressive) stress will occur can be found from

$$\tan 2\theta = \frac{2s_s}{s_t} \tag{14-6}$$

For design purposes, Eq. (14-5) is recommended when *brittle* materials are involved.

2. A maximum value of shear stress greater than the shear stress due to torsion alone.

This *maximum shear stress* is given by

$$(s_s)_{\text{max}} = \sqrt{s_s^2 + \left(\frac{s_t}{2}\right)^2} \tag{14-7}$$

For design purposes, Eq. (14-7) is recommended when *ductile* materials are used. Also, the *ASME Boiler and Pressure Vessel Code* specifies that the design of members subjected to combined stresses is to be based on maximum shear stress.

For a solid or hollow shaft subjected to simultaneous torsion and bending, the shear stress due to the transmitted torque is

$$s_s = \frac{Tc}{J} = \frac{T}{S'}$$

and the tensile stress due to bending is

$$s_t = \frac{Mc}{I} = \frac{M}{S}$$

where $S' = 2S$ for solid or hollow circular cross sections. By substituting this information into Eq. (14-5), we have

$$(s_t)_{\text{max}} = \frac{M}{2S} + \sqrt{\frac{T^2}{(S')^2} + \frac{M^2}{4S^2}}$$

but

$$S' = 2S \quad \text{and} \quad (S')^2 = 4S^2$$

$$(s_t)_{\text{max}} = \frac{M}{2S} + \frac{1}{2S}\sqrt{T^2 + M^2}$$

$$M_e = (s_t)_{\text{max}}S = \frac{M}{2} + \frac{1}{2}\sqrt{T^2 + M^2} \tag{14-8}$$

Equation (14-8) is used to calculate M_e, the equivalent bending moment. Similarly, Eq. (14-7) can be transformed into the equation for equivalent torque.

$$T_e = (s_s)_{\text{max}}S' = \sqrt{T^2 + M^2} \tag{14-9}$$

where M_e = equivalent bending moment, in·lb; N·mm
$\quad\quad T_e$ = equivalent torque, in·lb; N·mm
$\quad\quad M$ = actual bending moment, in·lb; N·mm
$\quad\quad T$ = actual torque, in·lb; N·mm
$\quad\quad S$ = section modulus of a solid shaft $\pi d^3/32$

$\quad\quad\quad\quad$ or a hollow shaft $\dfrac{\pi}{32}\left(\dfrac{d_0{}^4 - d_i{}^4}{d_0}\right)$, in³; mm³

$\quad\quad S'$ = polar section modulus of a solid shaft $\pi d^3/16$

$\quad\quad\quad\quad$ or a hollow shaft $\dfrac{\pi}{16}\left(\dfrac{d_0{}^4 - d_i{}^4}{d_0}\right)$, in³; mm³

The equivalent bending moment M_e will always be greater than M, the actual bending moment, owing to the combined effect of bending with torsion. Similarly, T_e, the equivalent torque, is always greater than T, the actual torque.

*Sample Problem 8[1]   What diameter shaft is required to transmit a torque of 2.7 MN·mm with a maximum bending moment of 2.0 MN·mm:

(a)   If the allowable shear stress is 55 MPa?
(b)   If the allowable tensile stress is specified as 75 MPa?

Solution a: The maximum shear stress equation or equivalent torque expression is called for. From Eq. (14-9),

$$T_e = (s_s)_{max}S' = \sqrt{T^2 + M^2}$$

$$55\,\frac{\pi d^3}{16} = \sqrt{(2.7 \times 10^6)^2 + (2.0 \times 10^6)^2}$$

$$= \sqrt{(7.29 \times 10^{12}) + (4.0 \times 10^{12})} = \sqrt{11.29 \times 10^{12}}$$

$$= 3.36 \times 10^6 \text{ N·mm}$$

$$d^3 = \frac{(3.36 \times 10^6)(16)}{55\pi} = 0.311(10^6)$$

$$d = 67.8 \text{ mm}$$

Use a 70-mm-diameter shaft if the material is ductile.[2]

Solution b: The maximum resultant tensile stress equation or equivalent bending moment expression is called for. From Eq. (14-8),

$$M_e = (s_t)_{max}S = \frac{M}{2} + \frac{1}{2}\sqrt{T^2 + M^2}$$

$$75\,\frac{\pi d^3}{32} = \frac{2.0(10^6)}{2} + \frac{1}{2}(3.36)(10^6) = 2.68(10^6)$$

---

[1] See also Sample Computer Program 24, Chap. 17.

[2] Standard size taken from J. E. Shigley and C. R. Mischke, *Standard Handbook of Machine Design,* McGraw-Hill Book Company, New York, 1986.

$$d^3 = \frac{2.68(10^6)(32)}{75\pi} = 0.364(10^6)$$

$$d = 71.4 \text{ mm}$$

Use a 75-mm-diameter shaft if the material is brittle.[1]

Sample Problem 9 A $1\frac{1}{2}$-in Monel shaft (Fig. 14-23) transmits 30 hp at 250 rpm to a piece of equipment. Power is transmitted to the shaft by means of a belt drive. The shaft is supported on bearings 18 in apart, and the belt pulley is centrally located. The total belt pull on the shaft is 600 lb. Is the shaft satisfactory if a safety factor of 4 based on the ultimate is required?

Solution: From Table 1A, App. B, ultimate $s_s = 56\,000$ psi for Monel. Since Monel is a ductile material, the maximum shear stress expression (Eq. 14-7) will be used

$$(s_s)_{\text{max}} = \sqrt{s_s^2 + \left(\frac{s_t}{2}\right)^2}$$

$$s_s = \frac{T}{S'} \qquad \text{hp} = \frac{Tn}{63\,000} \qquad T = \frac{63\,000 \text{ hp}}{n}$$

$$T = \frac{63\,000(30)}{250} = 7560 \text{ in} \cdot \text{lb}$$

$$S' = \frac{\pi d^3}{16} = \frac{\pi(1.5^3)}{16} = 0.663 \text{ in}^3$$

$$s_s = \frac{7560}{0.663} = 11\,400 \text{ psi}$$

$$s_t = \frac{M}{S} \qquad M = \frac{FL}{4} = \frac{600(18)}{4} = 2700 \text{ in} \cdot \text{lb}$$

$$S = \frac{\pi d^3}{32} = 0.331 \text{ in}^3$$

$$s_t = \frac{2700}{0.331} = 8160 \text{ psi}$$

$$(s_s)_{\text{max}} = \sqrt{11\,400^2 + \left(\frac{8160}{2}\right)^2} = \sqrt{11\,400^2 + 4080^2}$$

$$(s_s)_{\text{max}} = \sqrt{(130 + 16.6)(10^6)}$$
$$= \sqrt{(146.6)(10^6)} = 12\,100 \text{ psi}$$

$$\text{Allowable stress} = \frac{56\,000}{4} = 14\,000 \text{ psi}$$

Therefore, the shaft is satisfactory.

[1] Standard size taken from J. E. Shigley and C. R. Mischke, *Standard Handbook of Machine Design*, McGraw-Hill Book Company, New York, 1986.

14-10 MOHR'S CIRCLE — COMBINED STRESS

The maximum resultant tensile (or compressive) stress, Eq. (14-5), and the maximum shear stress, Eq. (14-7), can be determined graphically by use of Mohr's circle. A point A on a shaft (Fig. 14-23) is stressed as shown in Fig. 14-24. The stresses are due to bending and torsion.

Referring to Fig. 14-25, locate the tensile stress due to bending on the x-axis, point B. Then locate point R by erecting a perpendicular to the x-axis, BR, at point B. The length of BR is the shear stress due to torsion. Point O is the center of Mohr's circle and the midpoint of AB. With OR as a radius, draw the circle. Analysis of Fig. 14-25 indicates that AQ is equal to the maximum resultant tensile stress and that the radius of the circle corresponds to the maximum shear stress.

$$(s_t)_{max} = AO + OR = \frac{s_t}{2} + \sqrt{\left(\frac{s_t}{2}\right)^2 + s_s^2}$$

$$(s_s)_{max} = OZ = \sqrt{\left(\frac{s_t}{2}\right)^2 + s_s^2}$$

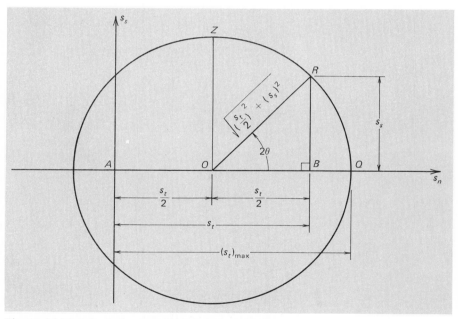

Fig. 14-25 Mohr's circle — combined stress.

The angle θ of the plane on which the maximum resultant tensile (or compressive) stress occurs can be determined from the diagram as follows:

$$\tan 2\theta = \frac{BR}{OB} = \frac{s_s}{s_t/2} = \frac{2s_s}{s_t}$$

Angle θ is one-half of the angle whose tangent is $2s_s/s_t$. Note that the above equations are those given as Eqs. (14-5), (14-6), and (14-7).

Sample Problem 10   For a shaft stressed as indicated in Fig. 14-24 with the tensile stress s_t due to bending equal to 12 000 psi and the shear stress s_s due to torsion equal to 5000 psi, determine the maximum resultant tensile stress and the maximum shear stress. Use Mohr's circle. Determine the angle of the plane on which the maximum resultant tensile stress occurs.

Solution: On the x-axis in Fig. 14-26 lay off $s_t = AB = 12\,000$ psi. Determine O, the center of Mohr's circle, by dividing s_t by 2.

$$AO = \frac{s_t}{2} = 6000 \text{ psi}$$

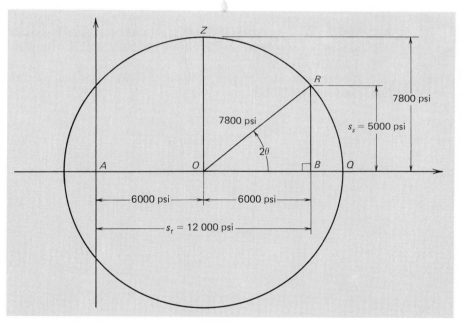

Fig. 14-26 Mohr's circle for Sample Problem 10.

Erect a perpendicular BR at point B whose length equals the shear stress due to torsion:

$$s_s = BR = 5000 \text{ psi}$$

Using OR as a radius, draw the circle. From the plot, OR is measured as 7800 psi. Since $AQ = (s_t)_{\text{max}}$

$$(s_t)_{\text{max}} = AO + OQ = 6000 + 7800 = 13\,800 \text{ psi}$$

and $(s_s)_{max} = OZ = OR = 7800 \text{ psi}$

$$\tan 2\theta = \frac{BR}{OB} = \frac{5000}{6000} = 0.8333$$

$$2\theta = 39.8°$$

Therefore, $\theta = \frac{39.8}{2} = 19.9°$

which is the angle of the plane of maximum resultant tensile stress.

PROBLEMS

*14-1. A beam has an axial tensile load of 45 kN and a concentrated load of 6.75 kN at each of the three quarter points of the beam. What would be the maximum and minimum stresses on a 100- by 150-mm (rough-sawn) beam simply supported on a 2.1-m span?

14-2. A simply supported rough-sawn timber beam 10 in wide and 12 ft long carries a uniform load of 700 lb/ft and an axial tensile load of 24 000 lb. What is the proper depth for the beam if the allowable stress is 1100 psi?

*14-3. Determine the maximum stress at the center cross section of the link shown in Fig. Prob. 14-3.

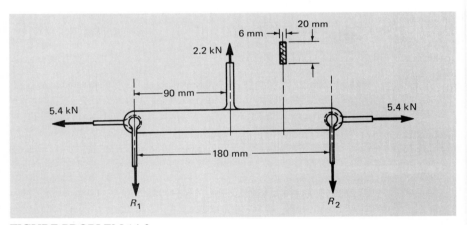

FIGURE PROBLEM 14-3

14-4. An S18 × 54.7 simply supported beam carries a concentrated load of 25 000 lb at the center of a 15-ft span and a total uniform load of 4500 lb, including the weight of the beam. What maximum axial tensile force can be applied to this beam if the allowable stress is 20 000 psi?

14-5. Find the maximum and minimum stresses in the wide-flanged member used as an eccentrically loaded short column as shown in Fig. Prob. 14-5.

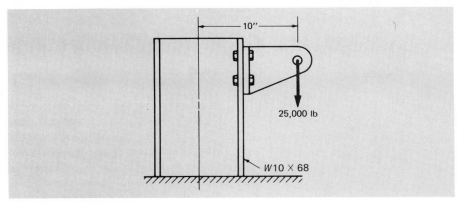

FIGURE PROBLEM 14-5

*14-6. The short column shown in Fig. Prob. 14-6 is subjected to an eccentric load of 790 kN. Determine the stresses at edges AB and CD.

14-7. A rectangular 3- by 6-ft concrete footing supports two columns whose loads act as shown in Fig. Prob. 14-7. Calculate the base pressure at each corner of the footing. Include the weight of the footing. (Use density of concrete as 150 lb/ft³.)

14-8. For the Class 20 cast-iron frame shown in Fig. Prob. 14-8, determine the maximum tensile and compressive stresses that would develop at section AA due to an applied load of $F = 12\ 000$ lb.

14-9. Find the maximum safe load F that can be applied to the Class 20 cast-iron machine frame of Fig. Prob. 14-8. Assume shock loading.

*14-10. What should be the diameter of the clamp at section AA for the arrangement shown in Fig. Prob. 14-10 if the allowable stress is 105 MPa?

*14-11. Determine the stresses acting at A and B for the machine link shown in Fig. Prob. 14-11.

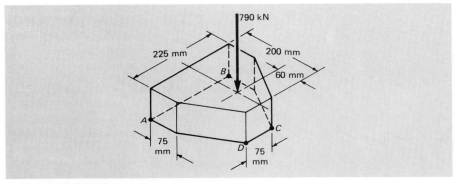

FIGURE PROBLEM 14-6

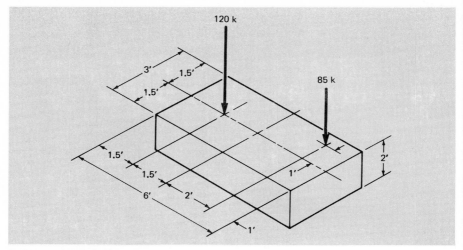

FIGURE PROBLEM 14-7

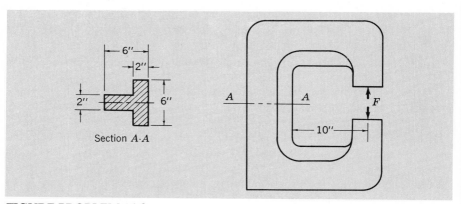

FIGURE PROBLEM 14-8

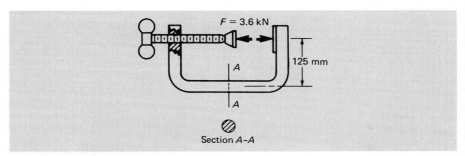

FIGURE PROBLEM 14-10

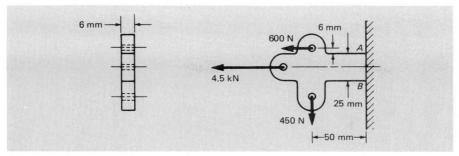

FIGURE PROBLEM 14-11

14-12. Determine the maximum force F that a structural steel channel loaded as shown in Fig. Prob. 14-12 can take if the allowable tensile and compressive stresses are 20 000 psi.

*14-13. The davits shown in Fig. Prob. 14-13 are to be designed to support the lifeboat, which has a mass of 816 kg and a capacity of 40 people. Assume an average mass of 77 kg per person. The davits are aluminum alloy 6061-T6 with a hollow circular cross section of 150 mm OD and 100 mm ID. Determine the factor of safety N_u for the davits.

14-14. The allowable tensile and compressive stresses at section AA of Fig. Prob. 14-14 are 11 000 psi. How large may the eccentricity e be?

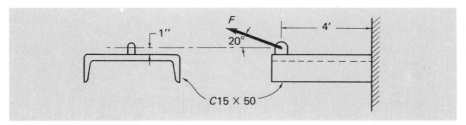

FIGURE PROBLEM 14-12

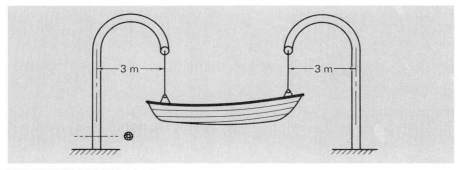

FIGURE PROBLEM 14-13

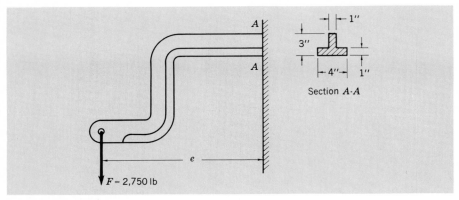

FIGURE PROBLEM 14-14

14-15. What shear stress will be developed in the bolts shown in Fig. Prob. 14-15? The bolts are $\frac{3}{4}$-in in diameter and are spaced $2\frac{1}{2}$ in from center to center.

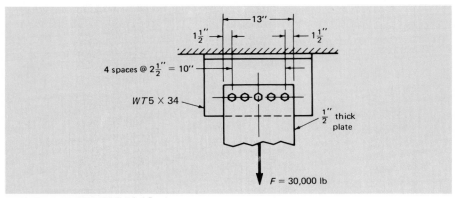

FIGURE PROBLEM 14-15

14-16. The bracket in Fig. Prob. 14-16 is connected by $\frac{7}{8}$-in-diameter bolts to a vertical member.
   a. Determine the shearing stress in each bolt.
   b. Omitting the center bolt, repeat part *a*.
14-17. In order to support a platform, it was found necessary to provide a bracket bolted to a column as shown in Fig. Prob. 14-17. If s_s is limited to 10 000 psi, what is a proper bolt size?[1]
14-18. For Fig. Prob. 14-18, if the bolts are 1 in in diameter and $\theta = 0°$, find the maximum shear stress and identify the bolt or bolts in which it occurs. $F = 10\ 000$ lb.

[1] See also Sample Computer Program 23, Chap. 17.

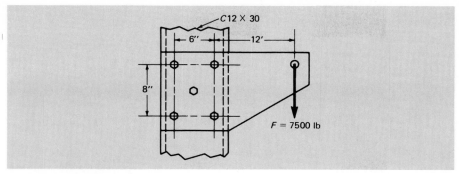

FIGURE PROBLEM 14-16

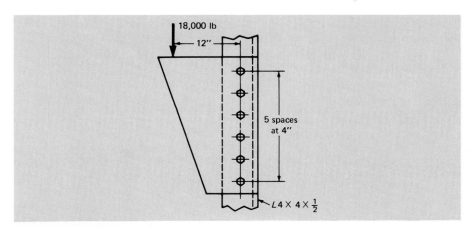

FIGURE PROBLEM 14-17

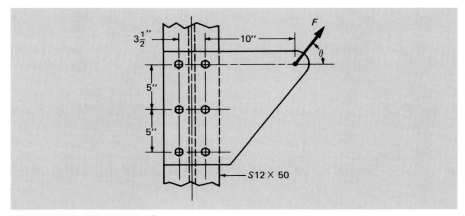

FIGURE PROBLEM 14-18

14-19. Same as Prob. 14-18, except $\theta = 30°$.

14-20. For Fig. Prob. 14-18, what maximum force F at $\theta = 90°$ can be applied to the 1-in-diameter bolts if the allowable shear stress is 15 000 psi?

14-21. The $\frac{1}{2}$-in bolts shown in Fig. Prob. 14-21 are spaced $2\frac{1}{2}$ in center to center.

    a. What maximum force F can be applied if the allowable shear stress is 15 000 psi?

    b. What maximum force F can be applied if bolt D is removed?

14-22. The allowable shear stress is 9000 psi for the $\frac{3}{8}$-in bolts shown in Fig. Prob. 14-22. What is the largest permissible distance l?

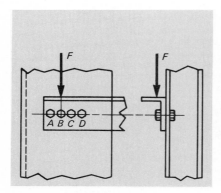

FIGURE PROBLEM 14-21

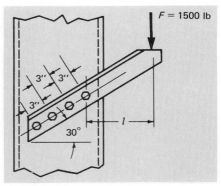

FIGURE PROBLEM 14-22

14-23. A steel shaft 3 in in diameter, making 150 rpm, is supported in bearings 6 ft apart. A load of 750 lb that acts downward at the middle of the shaft is due to belt pull. What horsepower can be transmitted by the shaft if the maximum shearing stress is 6000 psi?

*14-24. Find the diameter of a shaft subjected to a bending moment of 680 N·m and a twisting moment of 900 N·m if the bending stress does not exceed 70 MPa and the shear stress does not exceed 55 MPa.

14-25. Find the diameter of a steel shaft in bearings 6 ft apart that is to transmit 150 hp at 300 rpm. There is a load of 900 lb at the middle, and the maximum allowable stress is 10 000 psi in bending and 7000 psi in shear.

*14-26. A 100-mm shaft is transmitting 22 kW at 180 rpm and is subjected to a bending moment of 950 kN·mm. What will be the maximum tensile and shearing stresses developed?

14-27. If the bending moment is equal to the twisting moment, what should be the diameter, expressed in terms of the allowable stress, of a solid shaft that is to transmit 120 hp at 125 rpm?

    a. The shaft is made of a ductile material.

    b. The shaft is made of a brittle material.

*14-28. Determine the size of a hollow steel shaft whose inside diameter is equal to one-half the outside diameter if the bending moment developed is 1.4 MN · mm and the torque transmitted is 1.7 MN · mm. The allowable shear stress is 55 MPa.

*14-29. Same as Prob. 14-28 for a Class 40 cast-iron shaft with a factor of safety of 10 based on the ultimate stresses. Use SI units.

14-30. A short steel compression member that is $2\frac{1}{2}$ in square is loaded with an axial load of 75 000 lbs. Find:

    a. The maximum normal stress

    b. The normal and shearing stresses on planes making angles of 15° and 45° with the horizontal cross section. Use Mohr's circle.

14-31. Use the data in Sample Problem 9 to determine the maximum resultant tensile stress, the maximum shear stress, and the angle of the plane on which the maximum resultant tensile stress occurs. Use Mohr's circle.

15

Columns

15-1 INTRODUCTION

Short compression members which are subjected to axial loads can be treated by $s = F/A$ as was done in Chap. 7. When a short compression member is eccentrically loaded, it can be dealt with by $s = F/A \pm Fec/I$ as in Chap. 14. In both cases, the equations given produce reasonably reliable solutions for actual members. Minor discrepancies may appear because of the following.

1. Nonhomogeneous material
2. Unforeseen or accidental misalignment of loading
3. Slight variations in the straightness of the member
4. Presence of unknown initial stress in the member

These effects are usually negligible in short compression members as well as in tension members, torsion members, and beams. For relatively long compression members (columns) with axial loads, the above effects are of prime importance in determining and limiting the loads which such members may carry. The reader can demonstrate this by a simple experiment with a piece of ordinary gray cardboard (such as the backing of a pad of paper), say, $8\frac{1}{2}$ by 11 in, and several textbooks. When the full length of cardboard is used to support one end of a book (as shown in Fig. 15-1a), it will probably bend or buckle if a second book is added. After determining the book load for the 11-in length, cut the cardboard in half to produce two pieces $8\frac{1}{2}$ by $5\frac{1}{2}$ in. If the test is repeated by using one $8\frac{1}{2}$- by $5\frac{1}{2}$-in piece with the $5\frac{1}{2}$-in length supporting the books, approximately four books will be safely carried (see Fig. 15-1b). Thus, merely halving

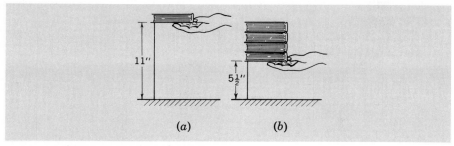

Fig. 15-1 Demonstration of column action.

the length of the compression member will permit a compressive load four times as large to be safely supported.

Column failures are characterized by sudden bending or local buckling. This *column action,* combined with the uncertain effects which contribute to failure, makes it difficult to calculate the actual stresses developed in the column material. Because of these special problems, column design formulas have been developed to find the maximum safe axial load and to relate this load to the shape and size of the column and the properties of the material.

15-2 SLENDERNESS RATIO

The experiment discussed in the preceding section demonstrates the importance of column length as a determinant of safe load. Length alone is not the only important factor. This can be shown by taking the unused $8\frac{1}{2}$- by $5\frac{1}{2}$-in piece of cardboard and rolling it into a cylindrical shape $5\frac{1}{2}$ in long and about 2 in in diameter (use tape to keep the cylinder from unrolling). If you repeat the experiment with this new shape, your supply of books might be exhausted before failure occurs. Apparently, the cross-sectional shape and size of the column play an important role in determining the safe load.

If you think back to the first two experiments, you may recall that the cardboard seemed to prefer to fail in a particular direction. That is, in both cases failure occurred as shown in Fig. 15-2. A horizontal cross section through the cardboard would look something like Fig. 15-3. In each of the first two experiments, the cardboard tended to buckle about axis YY. In neither case did buckling occur about axis XX. The reason for this is that the moment of inertia about axis YY is *less* than the moment of inertia about axis XX. To emphasize this point, let us calculate the moments of inertia and compare the results.

$$I_x = \frac{bh^3}{12} = \frac{0.03(8.5^3)}{12} = 1.54 \text{ in}^4$$

$$I_y = \frac{bh^3}{12} = \frac{8.5(0.03^3)}{12} = 0.0000191 \text{ in}^4$$

From these figures it is apparent that there is much less resistance to buckling about the YY axis.

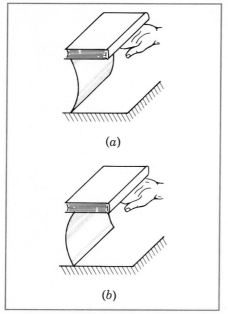

(a)

(b)

Fig. 15-2 Direction of buckling in demon-   Fig. 15-3 Cross section of cardboard.
stration.

Thus far, we have found that the important determinants of column strength for a given material are as follows.

1.   Length of column
2.   Cross-sectional shape
3.   Moment of inertia

However, from a known cross-sectional shape, the moment of inertia can be determined. This means that the two can be grouped together as one factor. The single factor used in column design that is related to both cross-sectional area and moment of inertia is called *radius of gyration,* the symbol for which is r. (The symbol k is sometimes used in place of r for radius of gyration.) Our list of column-strength determinants for a given material reduces to the following.

1.   Length of column l
2.   Radius of gyration r

If both terms are measured in the same units, their ratio is called *slenderness ratio.* Then

$$\text{Slenderness ratio} = \frac{l}{r} \qquad (15\text{-}1)$$

15-3 RADIUS OF GYRATION

To determine the radius of gyration of a plane figure such as a rectangle, let us consider the following problem. For the rectangle shown in Fig. 15-4a, calculate the moment of inertia about a centroidal axis and calculate the radius of gyration for that axis.

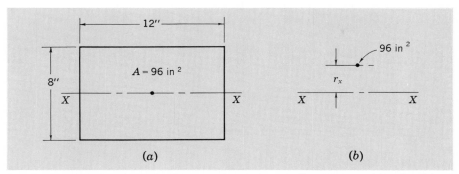

Fig. 15-4 (a) Rectangular area with centroidal axis XX. (b) Representation of radius of gyration about axis XX.

Moment of Inertia:

$$I_x = \frac{bh^3}{12} = \frac{12(8^3)}{12} = 512 \text{ in}^4$$

The radius of gyration is the distance from an axis to a point at which all the area of a plane figure can be imagined to be concentrated so that the moment of inertia is left unchanged. Moment of inertia can be thought of as the "second moment of an area" and was expressed in Chap. 10 as

$$I = \Sigma a \bar{y}^2$$

where $\bar{y}$ was the distance from the centroid of each segment of area to the axis. If all the area of the rectangle of Fig. 15-4a is concentrated at a point, as in Fig. 15-4b, then, from the above definition, the distance from that point to the axis is r, the radius of gyration. Therefore,

$$I = Ar^2$$

or
$$r = \sqrt{\frac{I}{A}} \qquad\qquad (15\text{-}2)$$

where I = moment of inertia, in⁴
    A = total area of figure, in²
    r = radius of gyration, in

Radius of Gyration:

$$r_x = \sqrt{\frac{I_x}{A}} = \sqrt{\frac{512}{96}} = \sqrt{5.33} = 2.31 \text{ in}$$

Let us now calculate I and r for the same figure about the YY centroidal axis (Fig. 15-5).

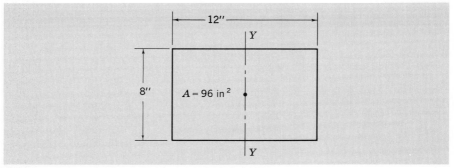

Fig. 15-5 Rectangular area with centroidal axis YY.

Moment of Inertia:

$$I_y = \frac{bh^3}{12} = \frac{8(12^3)}{12} = 1152 \text{ in}^4$$

Radius of Gyration:

$$r_y = \sqrt{\frac{I_y}{A}} = \sqrt{\frac{1152}{96}} = \sqrt{12} = 3.46 \text{ in}$$

From these calculations we can see that the axis which gives the smallest moment of inertia also gives the least radius of gyration. The same conclusion results from examination of Eq. (15-2).

In the experiments with cardboard columns, buckling occurred about the axis with the smallest moment of inertia. Thus, buckling of columns tends to occur about the axis with the least radius of gyration.

In determining the slenderness ratio of a column, the least radius of gyration is used.

$$\text{Slenderness ratio} = \frac{l}{r} \qquad (15\text{-}1)$$

where l = unsupported length of column, in
$\quad r$ = *least* radius of gyration, in

Values of r for common cross sections are given in Table 12, App. B.

15-4 CATEGORIES OF COLUMNS

Compression members are often subdivided into three categories according to slenderness ratio.

1. Short compression members
2. Intermediate columns
3. Long slender columns

These subdivisions can be visualized from an experimental curve obtained by plotting F/A (at failure) vs. l/r. This experiment can be done by compression-testing various lengths of a member of standard cross section. Figure 15-6 shows a curve of F/A (at failure) vs. l/r which might result from such an experiment. The broken lines indicate the probable variation which might be expected in experimental results. The experimental curve seems to have three distinct portions which correspond to the three categories mentioned earlier.

For low values of slenderness ratio, the experimental curve shows a horizontal straight-line portion. This means that F/A is constant in this range regardless of the slenderness ratio of the specimen. Specimens in this range are called *short compression members,* and their values of F/A upon failure are determined from the ultimate stress of the material.

$$\frac{F}{A} = s_u \qquad (15\text{-}3)$$

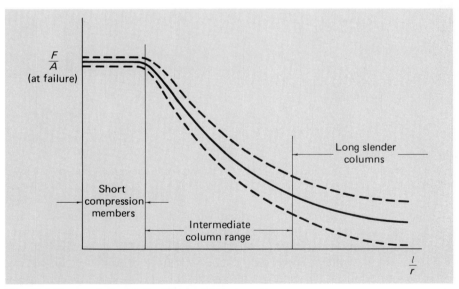

Fig. 15-6 Expected ranges for F/A (at failure) vs. l/r.

Since for short compression members the l/r ratio has no effect on strength, it is clear that *column action* does not become significant until a certain l/r value is exceeded. Briefly, then, short compression members require no special treatment as columns; instead, they are simply handled by $s = F/A$.

For members with large values of l/r, the experimental curve of Fig. 15-6 shows that very small F/A values can cause failure. Such members are categorized as *slender columns*. Experimental curves for slender columns closely fit an equation proposed by Euler,

$$\frac{F}{A} = \frac{\pi^2 E}{(l/r)^2} \qquad (15\text{-}4)$$

This equation takes account of the primary source of failure of slender columns: column action, or buckling.

In Fig. 15-7, Eqs. (15-3) and (15-4) are plotted as solid lines in the regions where they apply to columns and are extended as broken lines where they do not describe experimental results.

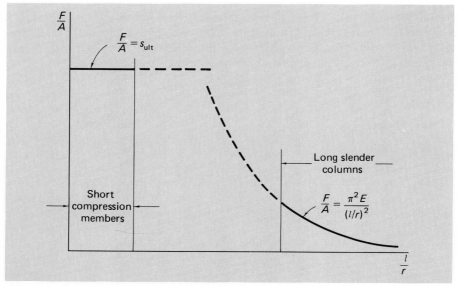

Fig. 15-7 Plot of F/A (at failure) vs. l/r for short compression members and long slender columns.

Intermediate columns are members whose l/r values lie in the range between short compression members and slender columns. The intermediate column range is characterized by a sharp decrease in the F/A value which causes failure as the l/r value is increased. The actual shape of the curve describing intermediate columns is difficult to determine, since minor differences in specimens cause large changes in F/A values. Because of this sensitivity to uncertainties, no one equation can be identified as the "correct" formula. Instead, several equally "correct" formulas are used for intermediate columns. The most important feature of such formulas is that they can be successfully used for design and selection of actual intermediate columns. In other words, they are justified because they work. The most common forms of intermediate-column equations are the following.

1. The straight-line form:

$$\frac{F}{A} = s - C\left(\frac{l}{r}\right) \tag{15-5}$$

2. The parabolic form:

$$\frac{F}{A} = s - C\left(\frac{l}{r}\right)^2 \tag{15-6}$$

Figure 15-8 shows the two forms of intermediate-column equations plotted as solid lines in their range of application and as broken lines outside that range. Note that both equations can be adapted to represent the experimental intermediate-column range. The constant term C in each equation can be given an appropriate value to ensure a good fit to experimental results.

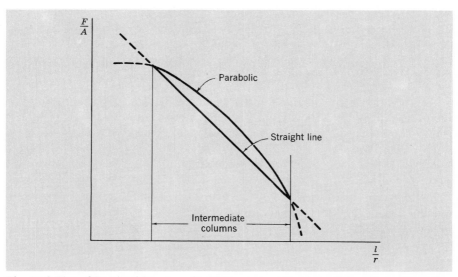

Fig. 15-8 Plot of F/A (at failure) vs. l/r for two types of intermediate column formulas.

15-5 END CONDITIONS

In addition to the slenderness ratio, the strength of a column is dependent upon the way in which the ends of the column are held. The various end conditions can be classified into the following four groups.

1. Both ends pinned or hinged (Fig. 15-9a)
2. Both ends fixed (Fig. 15-9b)
3. One end fixed, one end pinned (Fig. 15-9c)
4. One end fixed, one end free (Fig. 15-9d)

Other things being equal, column type 2 is strongest, type 3 is next in strength, type 1 next, and type 4 is the least strong. Each of these conditions may be approximated in actual practice, although certain end fixtures which appear to be securely fixed often permit some lateral or rotational movement. To

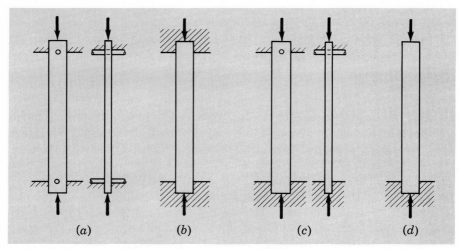

Fig. 15-9 End conditions: (*a*) both ends pinned; (*b*) both ends fixed; (*c*) one end fixed, one end pinned; (*d*) one end fixed, one end free.

account for the effect on the strength of a column due to the end conditions, the factor K is introduced. The factor is used in conjunction with the slenderness ratio l/r to produce an *effective* slenderness ratio. Thus,

$$\text{Effective slenderness ratio} = \frac{Kl}{r} \qquad (15\text{-}7)$$

Values of K recommended by the Column Research Council for use when end conditions are approximated in actual design are given in Table 15-1.

TABLE 15-1   RECOMMENDED VALUES OF K FOR VARIOUS COLUMN END CONDITIONS

End Conditions	Recommended K Value*
1. Both ends pinned or hinged (Fig. 15-9*a*)	1.0
2. Both ends fixed (Fig. 15-9*b*)	0.65
3. One end fixed, one end pinned (Fig. 15-9*c*)	0.8
4. One end fixed, one end free (Fig. 15-9*d*)	2.1

   * *Note:* Theoretical values of K for these conditions are 1.0, 0.5, 0.7, and 2.0, respectively.

15-6   COLUMN FORMULAS (METALS)

A great variety of column design formulas of the straight-line, parabolic, Euler, and other types have been proposed and used. A few of the more popular equations are presented in this section. The reader may wonder which is the best formula to use. The answer to this puzzle is that an *appropriate* formula must be used. In actual practice, a column design is often subjected to codes, laws, or company practices which dictate the design equations to be used, so that the designer has little or no choice. In situations which are not so restricted, the designer will select an equation, based on experience, which best fits the conditions of the design. The reader will usually not be asked to make such a choice; instead, the problems in this chapter will specify the formulas to be applied.

Several sets of column design equations for metal members are given in Table 15-2. All the equations are based on the *safe* load F. It is interesting to note that many of the sets of equations are not categorized specifically for short, intermediate, or slender columns. If you refer to Figs. 15-7 and 15-8, you will find that the parabolic curve (typical of the AISC equation for $Kl/r \le C_c$) flattens out and becomes nearly horizontal in the low l/r range. Thus, use of this equation for short compression members becomes valid. In a like manner, we can establish the reasoning behind the use of the various column equations that do not adhere strictly to the previously established categories.

As discussed in Sec. 15-5, K represents an end condition factor.

With reference to the AISC equations, C_c represents the maximum slenderness ratio at which a column can be designed by the parabolic equation. Above C_c, Euler's form is used. Take note that Table 15-2 indicates the value of C_c for various s_y.

The use of the AISC parabolic equation ($Kl/r \le C_c$) involves a factor of safety N. This is to be calculated from the given equation and has a range of 1.67, when $l/r = 0$, to 1.92, when $l/r = C_c$. Because A36 structural steel is most commonly used for columns and pinned ends are the most frequent design condition, Fig. 15-10 has been included so that N can be rapidly determined for various values of l/r.

Sample Problem 1 Determine the allowable axial compressive load F which the AZ61A-F magnesium alloy T section shown in Fig. 15-11 can carry if its ends are pinned and its length is 6 ft.

Solution: For this T section, the moments of inertia have been calculated in Sec. 10-9.

$$I_x = 57.9 \text{ in}^4 \qquad I_y = 38.7 \text{ in}^4 \qquad A = 20 \text{ in}^2$$

The least radius of gyration will be referred to the axis with the smallest moment of inertia. Therefore, from Eq. (15-2),

$$r_y = \sqrt{\frac{I_y}{A}} = \sqrt{\frac{38.7}{20}} = \sqrt{1.935} = 1.39 \text{ in}$$

$$l = 6(12) = 72 \text{ in}$$

$$\frac{l}{r} = \frac{72}{1.39} = 51.8$$

TABLE 15-2 DESIGN EQUATIONS FOR AXIALLY LOADED METAL COLUMNS

Designation	Short Compression Members	Intermediate Columns	Long Slender Columns
Machine design for any steel	$\max \dfrac{F}{A} = \dfrac{S_y}{N_y}$ $\dfrac{l}{r} < 40$	$\dfrac{F}{A} = \dfrac{S_y}{N_y}\left[1 - \dfrac{S_y\left(\dfrac{Kl}{r}\right)^2}{4\pi^2 E}\right]$ $40 \le \dfrac{l}{r} < \pi\sqrt{\dfrac{2E}{S_y K^2}}$	$\dfrac{F}{A} = \dfrac{\pi^2 E}{N_y\left(\dfrac{Kl}{r}\right)^2}$ $\dfrac{l}{r} \ge \pi\sqrt{\dfrac{2E}{S_y K^2}}$

Designation	AISC equations		
AISC for structural steels	$\dfrac{F}{A} = \dfrac{S_y}{N}\left[1 - \dfrac{\left(\dfrac{Kl}{r}\right)^2}{2C_c^2}\right]$		$\dfrac{F}{A} = \dfrac{149\,000\,000}{\left(\dfrac{Kl}{r}\right)^2}$

s_y, ksi*	C_c
36	126.1
42	116.7
46	111.6
50	107.0

$^\dagger N = \dfrac{5}{3} + \dfrac{3\left(\dfrac{Kl}{r}\right)}{8C_c} - \dfrac{\left(\dfrac{Kl}{r}\right)^3}{8C_c^3}$

$C_c = \sqrt{\dfrac{2\pi^2 E}{s_y}}$

[see C_c values at left]

$\dfrac{Kl}{r} \le C_c$ .... $\dfrac{Kl}{r} > C_c$ $\left[\max \dfrac{l}{r} = 200\right]$

Magnesium alloy AZ61A-F [for pinned ends]	$\dfrac{F}{A} = \dfrac{42\,800}{1 + \dfrac{42\,800\left(\dfrac{l}{r}\right)^2}{64.4 \times 10^6}}$

Cast iron [for pinned ends]	$\dfrac{F}{A} = 9000 - 40\left(\dfrac{l}{r}\right)$ $\dfrac{l}{r} \le 70$

* See Table 9-1 for recommended steels and their s_y.
† *Note:* Values of N for A36 structural steel and $K = 1$ can be determined from the graph in Fig. 15-10.

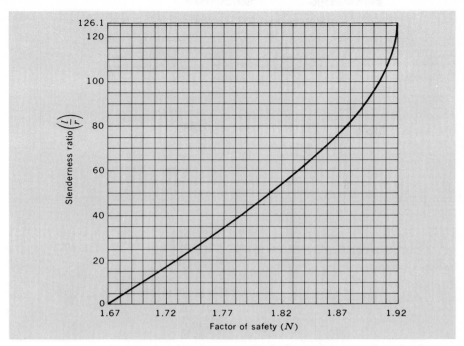

Fig. 15-10 Factor of safety N vs. slenderness ratio l/r for A36 structural steel columns, $K = 1$.

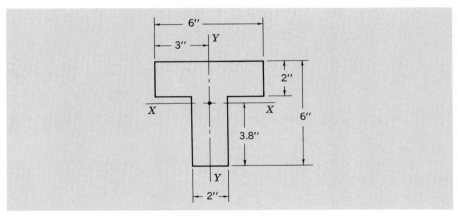

Fig. 15-11 Diagram for Sample Problem 1.

From Table 15-2,

$$\frac{F}{A} = \frac{42\ 800}{1 + \dfrac{42\ 800\left(\dfrac{l}{r}\right)^2}{64.4(10^6)}} = \frac{42\ 800}{1 + \dfrac{42\ 800(51.8^2)}{64.4(10^6)}}$$

$$= \frac{42\ 800}{1 + 1.78} = 15\ 400$$

$$F = 15\ 400(20) = 308\ 000\ \text{lb (maximum)}$$

Sample Problem 2 Figure 15-12 shows the cross section of a column (pinned ends) made by bolting two C15 × 40 sections to two 16- by $\frac{13}{16}$-in plates. Find the safe load it will carry when 20 ft long. Use the AISC formula. The material is A36 structural steel.

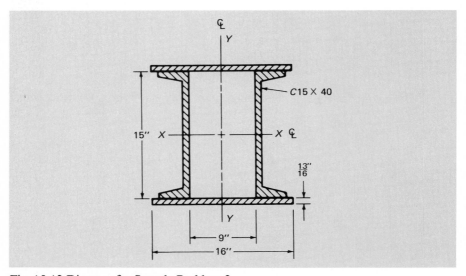

Fig. 15-12 Diagram for Sample Problem 2.

Solution: For the column,

$$I_x = 2(I_x)_{\text{channel}} + 2(I_x + Ad^2)_{\text{plate}}$$

But I_x for the plate will be very small and can be neglected.

$$I_x = 2(349) + 2(16)\left(\frac{13}{16}\right)\left(7.5 + \frac{13}{32}\right)^2 = 698 + 1625$$

$$= 2323\ \text{in}^4$$

$$I_y = 2\left[\frac{13(16^3)}{16(12)}\right] + 2[9.23 + 11.8(4.5 + 0.777)^2]$$

$$= 554 + 676 = 1230\ \text{in}^4$$

Least radius of gyration,

$$r_y = \sqrt{\frac{I_y}{A}}$$

$$A = 2(11.8) + 2(13) = 49.6 \text{ in}^2$$

$$r_y = \sqrt{\frac{1230}{49.6}} = \sqrt{24.8} = 4.98 \text{ in}$$

From Table 15-2 for A36 structural steel, $s_y = 36\ 000$ psi (Table 9-1) and $C_c = 126.1$. Since $K = 1.0$ for pinned ends and $r_y = 4.98$ in,

$$\frac{Kl}{r} = \frac{1.0(20)(12)}{4.98} = 48.2 \text{ (which is less than } C_c)$$

Using $\quad \dfrac{F}{A} = \dfrac{s_y}{N}\left[1 - \dfrac{\left(\dfrac{Kl}{r}\right)^2}{2C_c^2}\right]$

$$N = \frac{5}{3} + \frac{3(48.2)}{8(126.1)} - \frac{48.2^3}{8(126.1^3)} = 1.667 + 0.143 - 0.007 = 1.80$$

Or $\quad N = 1.80$ (from graph, Fig. 15-10)

Then $\quad \dfrac{F}{A} = \dfrac{36\ 000}{1.80}\left[1 - \dfrac{48.2^2}{2(126.1^2)}\right] = 20\ 000(1 - 0.073) = 18\ 500$ psi

$$F = 18\ 500(49.6) = 918\ 000 \text{ lb (maximum)}$$

Sample Problem 3 Select the most economical A36 structural steel W-shape column (pinned ends), 20 ft long, to carry an axial load of 300 000 lb. Use the AISC specifications.

Solution: In this type of problem, there are two unknowns, A and r, in the equation. Therefore, a trial-and-error solution is called for.

Assume that the column to be selected will have a $Kl/r < C_c$. From Table 15-2,

$$\frac{F}{A} = \frac{s_y}{N}\left[1 - \frac{\left(\dfrac{Kl}{r}\right)^2}{2C_c^2}\right]$$

This formula is limited to a range of Kl/r from 0 to 126.1 ($C_c = 126.1$ for $s_y = 36$ ksi). Assume Kl/r will fall in the middle of the range, say 63. Then

$$N = \frac{5}{3} + \frac{3\left(\dfrac{Kl}{r}\right)}{8C_c} - \frac{\left(\dfrac{Kl}{r}\right)^3}{8C_c^3} = 1.67 + 0.19 - 0.02 = 1.84$$

Or $\quad N = 1.84$ (from graph, Fig. 15-10)

$$\frac{F}{A} = \frac{36\,000}{1.84}\left[1 - \frac{63^2}{2(126.1^2)}\right] = 17\,120 \text{ psi}$$

$$A = \frac{F}{17\,120} = \frac{300\,000}{17\,120} = 17.5 \text{ in}^2$$

Now select a W-shape section whose area is approximately equal to the above value. There may be several sections with approximately that area in the tables. Since we shall eventually be interested in the most economical column, check one of lighter weight.

Try W12 $\times$ 65, $A = 19.1$ in^2, and $r_y = 3.02$ in.

$$\frac{Kl}{r} = \frac{1(20)(12)}{3.02} = 79.5$$

$$N = 1.67 + \frac{3(79.5)}{8(126.1)} - \frac{79.5^3}{8(126.1^3)} = 1.67 + 0.24 - 0.03 = 1.88$$

Or $\quad N = 1.88$ (from graph, Fig. 15-10)

$$\frac{F}{A} = \frac{36\,000}{1.88}\left[1 - \frac{79.5^2}{2(126.1^2)}\right] = \frac{36\,000}{1.88}(0.801) = 15\,340$$

$$F = 15\,340(19.1) = 293\,000 \text{ lb}$$

This column does not meet the 300 000-lb specification, although it comes quite close.

For our second trial, we want F to increase slightly without increasing A, if possible. Therefore, a section should be chosen with a larger r_y for approximately the same A. Thus the next most economical section would be W12 $\times$ 72, $A = 21.1$ in^2, and $r_y = 3.04$ in.

$$\frac{Kl}{r} = \frac{1(20)(12)}{3.04} = 79$$

$$N = 1.67 + \frac{3(79)}{8(126.1)} - \frac{79^3}{8(126.1^3)} = 1.87$$

Or $\quad N = 1.87$ (from graph, Fig. 15-10)

$$\frac{F}{A} = \frac{36\,000}{1.87}\left[1 - \frac{79^2}{2(126.1^2)}\right] = \frac{36\,000}{1.87}(0.804) = 15\,500$$

$$F = 15\,500(21.1) = 327\,000 \text{ lb}$$

The W12 $\times$ 72 is the most economical W-shape column section for the conditions specified. The Manual of Steel Construction (AISC) contains tables which permit rapid selection of economical column sections.

Sample Problem 4[1]

Design an 18-in-long linkage rod of circular cross section, assuming a varying axial compressive load of 4500 lb if the material is AISI 1045 steel. Use machine-design formulas.

[1] See also Sample Computer Program 25, Chap. 17.

Solution: From Table 1A, App. B, for AISI 1045 steel,

$$s_y = 60\ 000 \text{ psi} \qquad E = 30 \times 10^6 \text{ psi}$$

Assume pinned ends ($K = 1.0$). From Table 2, App. B,

$$N_y = 3$$

First determine the l/r ratio which separates intermediate and slender columns.

$$\frac{l}{r} = \pi\sqrt{\frac{2E}{s_y K^2}} = \pi\sqrt{\frac{2(30)(10^6)}{60\ 000(1^2)}} = \pi(31.6) = 99.3$$

At this stage, the actual value of l/r for the rod is not known. Let us assume that it will turn out to be an intermediate column ($40 \le l/r < 99.3$). From Table 15-2,

$$\frac{F}{A} = \frac{s_y}{N_y}\left[1 - \frac{s_y\left(\frac{Kl}{r}\right)^2}{4\pi^2 E}\right] \qquad \begin{array}{l} A = 0.785d^2 \\ r = 0.25d \end{array}$$

$$\frac{4500}{0.785d^2} = \frac{60\ 000}{3}\left[1 - \frac{60\ 000\left(\dfrac{1 \times 18}{0.25d}\right)^2}{4(3.14^2)(30)(10^6)}\right]$$

$$= 20\ 000\left(1 - \frac{0.263}{d^2}\right) = 20\ 000 - \frac{5260}{d^2}$$

$$\frac{4500}{0.785} = 20\ 000d^2 - 5260$$

$$20\ 000d^2 = 5730 + 5260 = 10\ 990$$

$$d^2 = 0.550$$

$$d = 0.742 \qquad \text{use } d = \tfrac{3}{4} \text{ in}$$

Check l/r: $\qquad r = 0.25d = 0.25(0.75) = 0.1875$ in

$$\frac{l}{r} = \frac{18}{0.1875} = 96$$

This confirms our use of the intermediate-column formula. If l/r had been larger than 99.3, the calculation would have had to be redone by using the slender-column equation.

15-7 COLUMN FORMULAS (TIMBER)

The National Forest Products Association (NFPA) recommends Eq. (15-8) for the design of simple solid columns made of stress-grade lumber. The formula may be used to design square, rectangular, and circular cross sections for pin-end or square-end conditions.

$$\frac{F}{A} = \frac{0.30E}{\left(\dfrac{l}{b}\right)^2} \qquad \frac{F}{A} \le s_{\text{allowable}} \qquad \frac{l}{b} \le 50 \qquad (15\text{-}8)$$

where F = axial compressive load on column, lb
  A = cross-sectional area of column, in²
  $s_{\text{allowable}}$ = allowable compressive stress parallel to grain, psi
  l = unsupported length of column, in
  b $\begin{cases} = \text{length of shortest side for rectangular cross section, in} \\ = 0.886 \text{ times the diameter for circular cross section, in} \end{cases}$
  E = modulus of elasticity, psi

Sample Problem 5   Use the NFPA formula to find the safe load that a nominal 6- by 6-in Douglas fir column is permitted to carry. The column is 9 ft long.

Solution:

$$A = 30.3 \text{ in}^2$$
$$b = 5.5 \text{ in} \qquad l = 9(12) = 108 \text{ in}$$
$$E = 1.2 \times 10^6 \text{ psi}$$
$$\frac{F}{A} = \frac{0.30E}{\left(\dfrac{l}{b}\right)^2} \qquad F = \frac{A(0.30)E}{\left(\dfrac{l}{b}\right)^2}$$
$$F = \frac{30.3(0.30)(1.2)(10^6)}{\left(\dfrac{108}{5.5}\right)^2} = \frac{10.9(10^6)}{386}$$
$$F = 28\ 300 \text{ lb (maximum safe load)}$$

Sample Problem 6   Find the diameter of a Sitka spruce compression member 28 in long that is to carry a load of 1000 lb. Use the NFPA formula.

Solution: For spruce, $s_c = 875$ psi, $E = 1.3 \times 10^6$ psi. For a circular cross section, $b = 0.886d$.

$$\frac{F}{A} = \frac{0.3E}{(l/b)^2}$$
$$A = 0.785d^2$$
$$\frac{1000}{0.785d^2} = \frac{0.3(1.3)(10^6)}{(28/0.886d)^2} = \frac{0.39(10^6)d^2}{1000}$$
$$d^4 = \frac{1000(1000)}{0.785(0.39)(10^6)} = 3.27$$
$$d = 1.34 \text{ in (minimum)}$$

Check the stress with maximum allowable stress:

$$\frac{F}{A} = \frac{1000}{0.785(1.34^2)} = 710 \text{ psi}$$

This does not exceed 875 psi; therefore, the calculated d is acceptable.

PROBLEMS

15-1. An S15 × 42.9 section is used as a 12-ft (pinned ends) column. Find the safe load it will carry, using AISC formulas. The material is A36 structural steel.

15-2. Find the safe load that a C15 × 40 section will carry when used as a 6-ft column (fixed ends) using AISC formulas. The material is A36 structural steel.

15-3. Find the safe load that a L6 × 6 × 1 section will carry when used as a column (pinned ends) 15 ft long. Use AISC formulas. The material is A36 structural steel.

*15-4. A 12-mm-diameter 302 stainless-steel pin, 200 mm long with pinned ends, is to support a varying load of 9.0 kN in compression. Is the member satisfactory?

15-5. Determine the maximum safe length of a hollow cast-iron column whose outside diameter is 8 in and inside diameter is 6 in, with an axial load of 139 000 lb.

15-6. A hollow round cast-iron column 20 ft long is to carry a load of 250 tons. If the external diameter is 15 in, what is the inside diameter?

15-7. Find the safe load that a column (pinned ends) 24 ft long, consisting of a 12- by $\frac{1}{2}$-in web plate, four L6 × 4 × $\frac{3}{4}$ sections, and two 13- by $\frac{5}{8}$-in cover plates, will support. Short legs of the angles are to be bolted to the web plate. Use the AISC code. All material is A36 structural steel.

15-8. Find the dimensions of a square AZ61A-F pin-ended magnesium alloy strut 10 ft long to carry a compressive load of 20 000 lb.

15-9. A hollow rectangular strut, pin-ended and 7 ft long, is made of AZ61-A-F magnesium alloy. The outside dimensions of the strut are 3 by 4 in, and the wall thickness is $\frac{1}{2}$ in. Determine the safe axial compressive load it can support.

15-10. A piece of seamless AISI 1045 steel tubing is used as a brace, which requires it to support an axial compressive load of 12 000 lb. The tube outside diameter is 2.875 in, and the wall thickness is 0.203 in. Find the maximum permissible length.

15-11. The upper-chord compression member of a truss must safely carry an axial load of 22 kips. The member will be 11 ft long. Select two equal-leg A36 structural steel angles which, when welded back to back, will meet AISC specifications. Assume pinned-end conditions.

15-12. What should be the diameter of a solid AISI 1095 steel rod, pin-ended and 8 ft long, which is required to support a compressive load of 20 000 lb?

15-13. Select the most economical S section to carry an axial load of 50 000 lb for a column length of 12 ft with pinned ends. Use AISC formulas. The material is A36 structural steel.

15-14. Select the most economical W section, according to AISC, to carry a central compressive load of 225 000 lb if the member is 15 ft long with pinned ends. The material is A36 structural steel.

15-15. A link with pinned ends, 15.75 in long, is subject to a compressive load of 4700 lb. The link is made of AISI 4130 steel; the cross section is rectangular; and the width is twice the depth. Find the dimensions of the cross section, assuming varying load.

15-16. Choose the proper W section for a 15-ft column, with pinned ends, to carry a central load of 230 000 lb under AISC specifications. The material is A36 structural steel.

15-17. A Sitka spruce column that is 10 ft long and 6 by 6 in dressed in cross section supports a load of 9000 lb. Is this arrangement safe according to the NFPA specifications?

15-18. What safe load will an eastern white pine post that is 12 ft long and 4 by 6 in dressed in cross section support according to the NFPA formulas?

15-19. Find the dressed size of a square column of Douglas fir, 20 ft long, to carry a load of 100 000 lb. Use the NFPA formula.

15-20. Find the dressed size of a square ponderosa pine timber to be used as a column 25 ft long to carry a load of 75 000 lb. Use the NFPA formula.

15-21. A floor beam which carries a uniform load of 1000 lb/ft on a 12-ft simple span is supported by an 18-ft-long column at the left end and by an 11-ft-long column at the right end. What size Douglas fir columns should be used for the supports according to the NFPA formula?

16

Indeterminate Beams

6-1 TYPES OF STATICALLY INDETERMINATE BEAMS

The beams discussed in Chap. 11 were statically determinate; that is, the unknown reactions and moments were found by using the conditions for static equilibrium: $\Sigma F_x = 0$, $\Sigma F_y = 0$, and $\Sigma M = 0$. Certain types of beams require more equations for their solution than are available from the static equilibrium formulas because there are too many unknowns. Such beams are called *statically indeterminate*. Other methods are used to find the unknown reactions and moments on these beams. Some typical indeterminate beams are shown in Fig. 16-1.

6-2 BEAM WITH ONE END FIXED, ONE END SUPPORTED

Figure 16-2*a* shows a beam with the right end fixed in a wall and the left end on a support; the beam carries a uniform load. The elastic curve for this beam is shown in Fig. 16-2*b*. If the support were removed, the beam would be a cantilever with a maximum downward deflection at the free end. In the actual beam (Fig. 16-2), there is no deflection at the left end. Therefore, the supporting force R_l exactly counteracts the deflection which would occur without the support. This fact can be used to help solve this type of indeterminate beam.

Imagine that the left support is removed and the beam becomes a cantilever, as shown in Fig. 16-3*a*. The elastic curve of the cantilever (Fig. 16-3*b*) shows the deflection y_L at the free end. From Eq. (12-19),

$$y_L = \frac{WL^3}{8EI} \qquad \text{for the free end of a cantilever with uniform load}$$

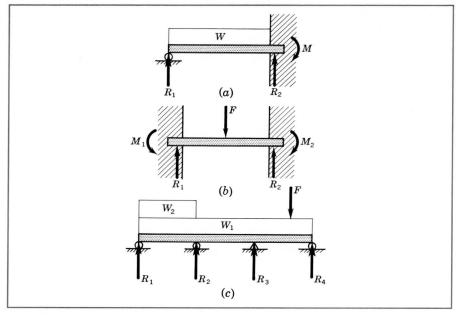

Fig. 16-1 Some typical statically indeterminate beams: (*a*) One end fixed, one end supported; R_1, R_2, and M are unknown. (*b*) Both ends fixed; R_1, R_2, M_1, and M_2 are unknown. (*c*) Continuous beam; R_1, R_2, R_3, and R_4 are unknown.

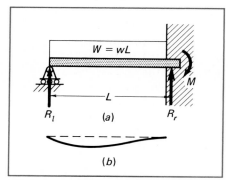

Fig. 16-2 (*a*) Beam with one end fixed, one end supported; R_l, R_r, and M are unknown. (*b*) Elastic curve for the beam.

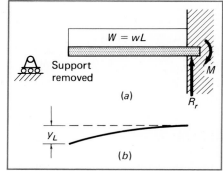

Fig. 16-3 (*a*) Beam of Fig. 16-2 with left support removed. (*b*) Elastic curve for this condition.

Now we may ask what force at the free end is required to eliminate y_L (because in the actual beam, Fig. 16-2, $y_L = 0$). Imagine the same beam without the uniform load (as in Fig. 16-4*a*) but with an upward force R_l at the free end which produces a deflection y_L' (Fig. 16-4*b*). From Eq. (12-16),

$$y_L' = \frac{R_l L^3}{3EI} \qquad \text{for the free end of a cantilever with load at end}$$

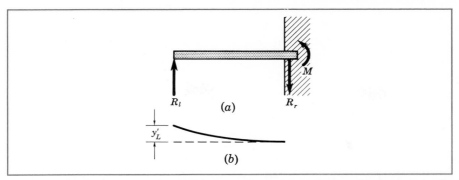

Fig. 16-4 (*a*) Beam of Fig. 16-2 with only one support force acting. (*b*) Elastic curve for this condition.

Since y_L is to be eliminated by the effect of R_l,

$$y_L = y_L'$$

or

$$\frac{WL^3}{8EI} = \frac{R_l L^3}{3EI}$$

from which

$$R_l = \frac{3}{8} W$$

With this new piece of information, the solution of the actual beam can proceed. Figure 16-5*a* shows the beam with only two unknowns, R_r and M. The remainder of the solution follows the usual steps of finding R_r from $\Sigma F_y = 0$ and finding M from $\Sigma M = 0$. The shear diagram (Fig. 16-5*c*) indicates the location of maximum bending moments as before, while the moment diagram (Fig. 16-5*d*) shows how the moments vary along this beam.

The principle of compensated deflection at the end support can be used for this type of beam with any loading. In one of the following sample problems, a beam of this kind with a concentrated load is solved. The method is easily extended to beams in which the support is caused to settle by the loading, as in the following sample problem.

Sample Problem 1 A machine base rests on two W8 × 31 beams, each of which is mounted with one end embedded in concrete and the other end resting on a vibration damping pad (see Fig. 16-6). After the machine is put into place, the supported end settles $\frac{1}{2}$ in. Find the reaction on one of the beams that is due to the vibration pad, and find the maximum stress in the beam.

Solution: The uniform load on one beam is

$$W_1 = \frac{10(2000)}{2} = 10\ 000 \text{ lb (load is shared by two beams)}$$

$$W_2 = 20(31) = 620 \text{ lb (weight of beam)}$$

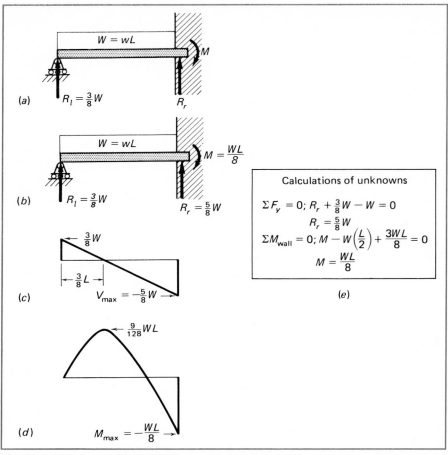

Fig. 16-5 (*a*) Beam of Fig. 16-2. (*b*) Beam with computed values of R_l, R_r, and M. (*c*) Shear force diagram. (*d*) Bending moment diagram. (*e*) Calculations for R_r and M.

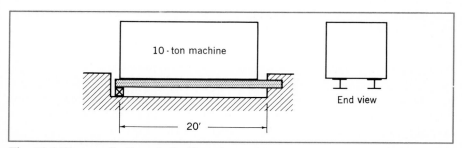

Fig. 16-6 Diagram for Sample Problem 1.

Therefore, $W = 10\,000 + 620 = 10\,620$ lb

The deflection which would occur with the support removed is given by

$$y_L = \frac{WL^3}{8EI} = \frac{10\,620(240^3)}{8(30)(10^6)(110)} = 5.56 \text{ in}$$

The upward deflection of an unloaded beam due to a concentrated reaction force R_l at the left end is

$$y_L' = \frac{R_l L^3}{3EI} = \frac{R_l}{3}\left[\frac{(240^3)}{30(10^6)(110)}\right] = 0.0014 R_l$$

The left end of the actual beam, when fully loaded, settles $\frac{1}{2}$ in; therefore,

$$y_L = y_L' + \frac{1}{2}$$

from which $5.56 = 0.0014 R_l + 0.5$

$$R_l = 3620 \text{ lb} \qquad \text{(reaction of vibration pad)}$$

The problem can now be solved in the usual way.

$\Sigma F_y = 0$ $\qquad\qquad\qquad\qquad R_l + R_r - 10\,620 = 0$

$\qquad\qquad\qquad\qquad\qquad\qquad\qquad R_r = 7000 \text{ lb}$

$\Sigma M_{\text{wall}} = 0$ $\qquad\quad M - 10\,620(10) + 3620(20) = 0$

$\qquad\qquad\qquad\qquad\qquad\qquad M = 33\,800 \text{ ft} \cdot \text{lb (at wall)}$

The shear diagram (Fig. 16-7b) indicates a maximum moment at x feet from the left support. By similar triangles,

$$\frac{x}{3620} = \frac{20 - x}{7000}$$

$$7000x = 72\,400 - 3620x$$

$$x = \frac{72\,400}{10\,620} = 6.8 \text{ ft}$$

The bending moment at this section in the beam is

$$M_{6.8} = 3620(6.8) - 10\,620\left(\frac{6.8}{20}\right)(3.4)$$

$$= 24\,620 - 12\,280 = 12\,340 \text{ ft} \cdot \text{lb}$$

The maximum bending moment occurs at the wall; $M_{\text{max}} = 33\,800 \text{ ft} \cdot \text{lb}$ (see Fig. 16-7c).

For a W8 × 31, the section modulus $S = 27.5$ in^3. Thus,

$$s = \frac{M}{S} = \frac{33\,800(12)}{27.5} = 14\,700 \text{ psi (maximum stress)}$$

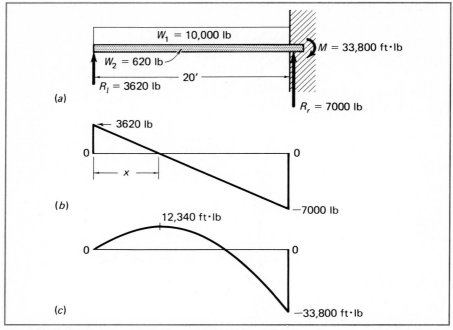

Fig. 16-7 (*a*) Beam diagram for Sample Problem 1. (*b*) Shear force diagram. (*c*) Bending moment diagram.

Sample Problem 2

A beam has one end fixed and the other end resting on a support. The beam carries a concentrated load of 10 000 lb at the center of its 12-ft span. Select the most economical standard A36 structural steel S beam (Fig. 16-8).

Solution: If the support is removed, the deflection at the free end is given by Eq. (12-17).

$$y_L = \frac{F}{6EI} (2L^3 - 3aL^2 + a^3)$$

$$= \frac{10\ 000}{6EI} [2(144^3) - 3(72)(144^2) + (72^3)]$$

$$= \frac{31.1(10^8)}{EI}$$

The upward force R_l at the support would produce a deflection on an unloaded beam equal to

$$y_L' = \frac{R_l(144^3)}{3EI} = \frac{995\ 000R_l}{EI}$$

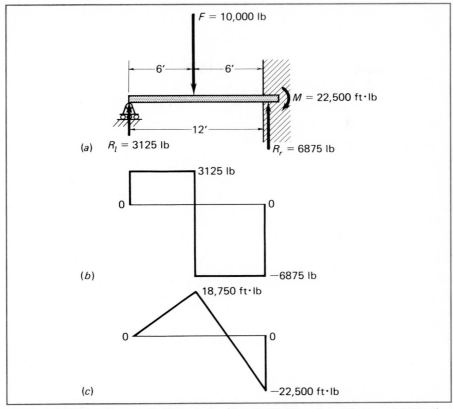

Fig. 16-8 (*a*) Beam diagram for Sample Problem 2. (*b*) Shear force diagram. (*c*) Bending moment diagram.

Equating these deflections,

$$\frac{31.1(10^8)}{EI} = \frac{995\,000R_l}{EI}$$

$$R_l = 3125 \text{ lb}$$

$$\Sigma F_y = 0 \qquad R_l + R_r - 10\,000 = 0$$

$$R_r = 6875 \text{ lb}$$

$$\Sigma M_{\text{wall}} = 0$$

$$M - 10\,000(6) + 3125(12) = 0$$

$$M = 60\,000 - 37\,500$$

$$= 22\,500 \text{ ft} \cdot \text{lb (at wall)}$$

$$M_6 = 3125(6) = 18\,750 \text{ ft} \cdot \text{lb (at 6-ft section)}$$

$$M_{\text{max}} = 22\,500 \text{ ft} \cdot \text{lb (at wall)}$$

$$s = \frac{M}{S}$$

$$S = \frac{M}{s} = \frac{22\ 500(12)}{24\ 000} = 11.25 \text{ in}^3$$

Select the S8 $\times$ 18.4 ($S = 14.4$ in³).

16-3   BEAM WITH BOTH ENDS FIXED

Beams with both ends fixed can be treated by imagining them to be separated into three portions forming a simply supported beam in the central portion and cantilever beams at both ends. The point of imagined separation is the point at which the bending moment diagram crosses zero. This point, called the *point of inflection of the elastic curve,* locates a section in the beam where the moment is zero. Since the end of a simply supported beam and the free end of a cantilever beam have zero bending moments, the imagined separation is valid if it is done at a zero moment section.

Beam with Concentrated Load at Center of Span: Consider the beam shown in Fig. 16-9. There are two unknown forces R_l and R_r and two unknown moments M_l and M_r. The points of inflection are located on the elastic curve. These locations are derived in more advanced books. The problem is to express the reaction forces and moments in terms of F and L.

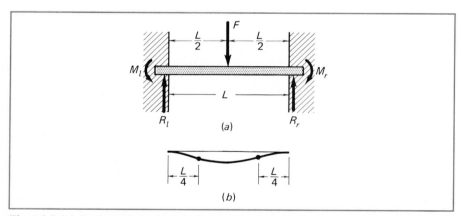

Fig. 16-9 (*a*) Beam with both ends fixed and concentrated load at center of span. (*b*) Elastic curve showing location of inflection points.

Imagine this beam to be separated at the points of inflection into three portions as shown in Fig. 16-10. Although the beam appears to be cut into three portions, the shear forces acting at the separation sections are not zero and must be considered in order to maintain equivalence to the original beam. V_a represents the shear force in the beam at a distance $L/4$ from the left wall. When we

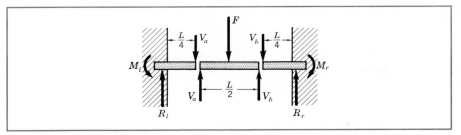

Fig. 16-10 Beam of Fig. 16-9 separated at the inflection points.

separate the beam at this section, V_a is treated as an external effect of the adjacent portion. V_b is treated similarly. We have converted a statically indeterminate beam into three statically determinate beams.

Let us examine the central portion (Fig. 16-11). The portion is a simply supported beam $L/2$ long with a concentrated load F at the center of the span. From the symmetry, we note that

$$V_a = \frac{F}{2}$$

and

$$V_b = \frac{F}{2}$$

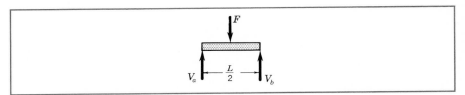

Fig. 16-11 Central portion of the beam of Fig. 16-10.

Furthermore, it was shown in Chap. 11 that the maximum bending moment for this type of beam occurs at midspan and equals the load times one-fourth of the span. Since the span in Fig. 16-11 is $L/2$, the maximum moment in this portion is

$$M = \frac{F(L/2)}{4} = \frac{FL}{8}$$

Now consider one of the cantilever portions. Either end portion can be analyzed and the results can be applied to both ends because of the symmetrical loading on the original beam. The right end is shown in Fig. 16-12. Let us apply the conditions for static equilibrium to this portion of the original beam, noting that $V_b = F/2$ from the preceding step.

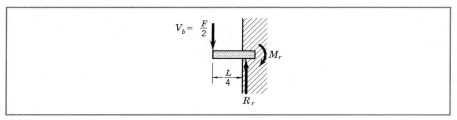

Fig. 16-12 End portion of the beam of Fig. 16-10.

$$\Sigma F_y = 0 \qquad\qquad R_r - \frac{F}{2} = 0$$

$$R_r = \frac{F}{2}$$

$$\Sigma M_{\text{wall}} = 0 \qquad\qquad M_r - \frac{F}{2}\left(\frac{L}{4}\right) = 0$$

$$M_r = \frac{FL}{8}$$

The reaction and moment at the left end of the beam are equal to R_r and M_r, respectively. Therefore,

$$R_r = R_l = \frac{F}{2}$$

$$M_r = M_l = \frac{FL}{8}$$

and, as previously calculated, the moment at the center of the span is

$$M = \frac{FL}{8}$$

The shear force and bending moment diagrams for the beam with both ends fixed and carrying a concentrated load at midspan are shown in Fig. 16-13. These diagrams should be compared with Fig. 11-20 for a simply supported beam.

The maximum deflection for the fixed-ends beam is obtained by adding the deflection of the central portion (Fig. 16-11) and the deflection at the free end of the cantilever portion (Fig. 16-12). By applying the deflection formulas, Eqs. (12-11) and (12-16), to the portions of the beam, we obtain

$$y_{\text{max}} = \frac{F_1 L_1{}^3}{48EI} + \frac{F_2 L_2{}^3}{3EI}$$

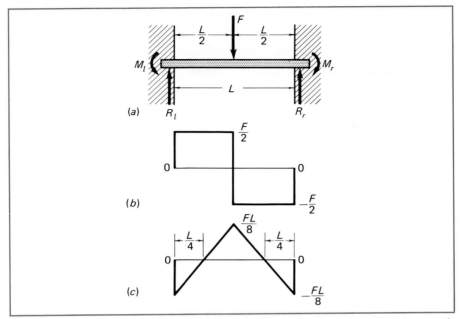

Fig. 16-13 (*a*) Beam with both ends fixed and concentrated load at center of span. (*b*) Shear force diagram. (*c*) Bending moment diagram.

but $\qquad F_1 = F \qquad L_1 = \dfrac{L}{2} \qquad F_2 = \dfrac{F}{2} \qquad L_2 = \dfrac{L}{4}$

$$y_{max} = \frac{F\left(\dfrac{L^3}{8}\right)}{48EI} + \frac{\dfrac{F}{2}\left(\dfrac{L^3}{64}\right)}{3EI}$$

$$y_{max} = \frac{FL^3}{192EI} \tag{16-1}$$

Beam with Uniform Load: A similar procedure can be used to solve fixed-ends beams with uniform loading. The results of such a procedure are summarized in Fig. 16-14.

Sample Problem 3[1] An overhead crane rides on a single rail which is rigidly fixed in concrete at the ends to provide a span of 20 ft. An allowable bending stress of 9000 psi is specified for the rail material. The rated capacity of the crane is 20 tons, but an overload of 25 percent is to be provided for. The depth of the rail is limited to 14 in by other structural interference. Assuming a static design situation with the crane at midspan, determine the following.

[1] See also Sample Computer Program 26, Chap. 17.

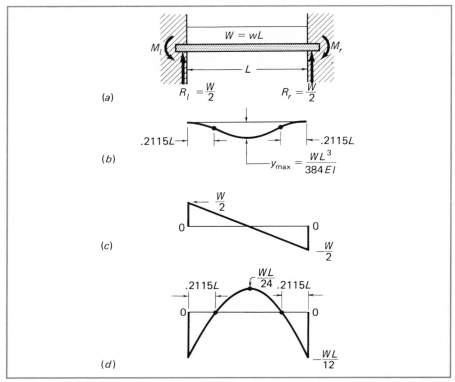

Fig. 16-14 (*a*) Beam with both ends fixed and uniformly distributed load. (*b*) Elastic curve showing location of inflection points. (*c*) Shear force diagram. (*d*) Bending moment diagram.

(a) The minimum centroidal moment of inertia of a 14-in-deep rail if the cross section is symmetrical about the *XX* axis

(b) The maximum deflection of the rail, using the moment of inertia from part *a*, if the material is an alloy steel

Solution a: This physical system is represented by Fig. 16-15; the total load of $20 + 20(0.25) = 25$ tons is concentrated at midspan. The weight of the rail is

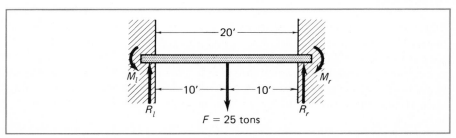

Fig. 16-15 Beam diagram for Sample Problem 3.

assumed negligible for simplification. From Fig. 16-13,

$$R_l = R_r = \frac{F}{2} = \frac{25(2000)}{2} = 25\ 000\ \text{lb}$$

$$M_l = M_r = M_{\text{midspan}} = \frac{FL}{8} = \frac{50\ 000(20)}{8} = 125\ 000\ \text{ft}\cdot\text{lb}$$

$$s = \frac{Mc}{I}$$

$$I = \frac{Mc}{s}$$

$$c = 7\ \text{in}$$

$$s = 9000\ \text{psi}$$

$$I = \frac{125\ 000(12)(7)}{9000} = 1167\ \text{in}^4\ (\text{minimum})$$

Solution b: From Eq. (16-1),

$$y_{\text{max}} = \frac{FL^3}{192EI}$$

where $F = 50\ 000$ lb
  $L = 20(12) = 240$ in
  $E = 30 \times 10^6$ psi
  $I = 1167$ in^4

$$y_{\text{max}} = \frac{50\ 000(240^3)}{192(30)(10^6)(1167)} = 0.103\ \text{in}$$

PROBLEMS

16-1.   A balcony is supported by S6 × 17.25 A36 structural steel beams spaced 10 ft on centers. It projects 6 ft from the wall and carries 200 psf on the floor area. Deflection at its end is prevented by AISI 1020 steel rods fastened from above. Find the size of rods necessary and the maximum stress in the beams. Rod spacing is 10 ft.

16-2.   A 10- by 14-in dressed mountain hemlock beam is fixed at one end and supported at the other. It carries a load of 400 lb per linear foot. The support settles $\frac{1}{2}$ in under the action of the load. Find the reaction of the support and the maximum fiber stress if the span is 17 ft.
  Hint: If the support were removed, the deflection of the end would be

$$y_1 = +\frac{1}{8}\frac{WL^3}{EI}$$

The support brings the end up:

$$y_2 = \frac{1}{3}\frac{RL^3}{EI}$$

Then

$$y_1 = y_2 + \tfrac{1}{2}\,\text{in}$$

Now assume that the support does not settle. Find maximum fiber stress and compare with the first case.

16-3. What are the bending and shearing stresses in a beam 9 ft long that is fixed at one end and supported at the same level at the other? The rough-sawn cross section of the beam is 6 by 12 in. The beam carries a uniform load, including its own weight, of 820 lb per linear foot. Sketch the shear and moment diagrams. $E = 1\,200\,000$ psi.

16-4. Compute the reactions and bending and shearing stresses if the end support in Prob. 16-3 settles 0.4 in.

16-5. If the end support in Prob. 16-3 is pushed 0.4 in above its previous level, what will be the stresses in the beam?

16-6. A run of schedule 40 seamless steel pipe is rigidly fixed into two concrete walls with a clear span of 15 ft between the walls. The outside diameter is 5.563 in, and the wall thickness of the pipe is 0.258 in. A load of 1 ton is carried at midspan. Determine the following.
   a. The maximum bending stress in the pipe
   b. The maximum deflection of the pipe

16-7. A platform is supported by two beams as shown in Fig. Prob. 16-7. Each beam consists of two L4 × 4 × $\frac{3}{8}$ sections welded back to back to form a T section. The platform is designed to carry 220 psf. What maximum bending stress is developed in the beams, and how much will they deflect?

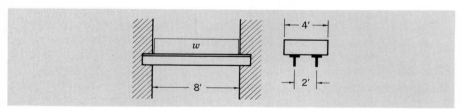

FIGURE PROBLEM 16-7

17

Computer Programs

17-1 INTRODUCTION TO SAMPLE COMPUTER PROGRAMS

This chapter includes 26 sample computer programs which demonstrate problem-solving applications for a selection of topics associated with statics and strength of materials. These programs are written in BASIC for use on an IBM-PC. To use these programs on another personal computer, it may be necessary to adjust the syntax and commands to conform to the requirements of the particular form of BASIC used by that machine.

All of these programs can be rewritten to become more sophisticated and more compact. However, they are not intended to be overly efficient or elegant. They have been made intentionally explicit for use by persons who are relatively inexperienced with computer programming and computer applications problem solving. Thus, there is no intention to teach computer programming; rather, the intention is to provide selected applications suitable to this field of study.

To run a sample computer program, call up BASIC and carefully type in the program as given. Then type RUN (or press function key F2) followed by pressing a carriage return or ENTER key.

To produce hard copy on a printer connected to the computer, press Ctrl and PrtSc simultaneously before running the program. Then the program can be RUN (or F2) and will communicate through the monitor display while recording the same information on the printer. The same technique can be used to print a copy of the program itself by typing LIST (or pressing function key F1) followed by a carriage return or ENTER.

Most of these programs can be expanded to increase the range of problems which can be treated. More advanced readers may wish to modify the programs of some of these samples or write new programs for topics not covered by the programs in this chapter.

17-2 SAMPLE COMPUTER PROGRAMS

Each sample computer program in this chapter has been given a file name which includes a number followed by .BAS, which the computer includes for BASIC programs. For example, there are five programs dealing with force systems which range from FORCES1.BAS to FORCES5.BAS. If an additional program is written for force systems, a similar name with a different number can be used to continue the identification with that topic.

Most sample computer programs have been cross-referenced to a specific sample problem or end-of-chapter problem in one of the earlier chapters.

The following sample computer programs are included in this chapter:

Chapter Reference	Sample Computer Program			Cross-Reference or Description
	No.	Name	Page	
General	1	ANGLES1.BAS	403	Convert degrees to or from radians
Chapter 2	2	FORCES1.BAS	403	Sec. 2-9; Concurrent forces
	3	FORCES2.BAS	404	Concurrent forces; variable angle
Chapter 4	4	FORCES3.BAS	405	Sec. 4-2; Nonconcur coplanar forces
	5	TRUSSES1.BAS	407	Sample Prob 5; Pratt Truss
Chapter 5	6	FORCES4.BAS	408	Samp Prob 2; Concur noncopln forces
	7	FORCES5.BAS	409	Samp Prob 4; Concur noncopln forces
Chapter 6	8	FRICTN1.BAS	409	Samp Prob 7; Least force for motion
Chapter 7	9	RODYOKE1.BAS	410	Samp Prob 5; Stresses in rod & yoke
Chapter 8	10	TENTEST1.BAS	411	Sec. 8-3; Tension test calculations
	11	TIEROD1.BAS	413	Samp Prob 5; Design of tie rod
Chapter 9	12	JOINTS1.BAS	414	Samp Probs 1 & 3; Bolted lap joint
	13	JOINTS2.BAS	415	Samp Prob 2; Bolted butt joint
Chapter 10	14	SHAPES1.BAS	416	Samp Prob 3; Centroid location
	15	SHAPES2.BAS	418	Moment of inertia for composite shape
Chapter 11	16	BEAMS1.BAS	420	Samp Probs 1 & 2; Beam reactions
	17	BEAMS2.BAS	421	Overhanging beam reactions
	18	BEAMS3.BAS	422	Samp Prob 8; Shear & moments
Chapter 12	19	BEAMS4.BAS	424	Samp Prob 3; Beam flexure stresses
	20	BEAMS5.BAS	427	Samp Prob 8; Beam design
Chapter 13	21	SHAFTS1.BAS	428	Samp Prob 4; Power transmission
	22	SHAFTS2.BAS	429	Samp Probs 6 & 7; Flange coupling
Chapter 14	23	ECCENTR1.BAS	431	Prob 14-17; Eccentric load on joint
	24	COMBINE1.BAS	432	Samp Prob 7; Shaft bending & torsion
Chapter 15	25	COLUMN1.BAS	433	Samp Prob 4; Design rod as column
Chapter 16	26	BEAMS6.BAS	435	Samp Prob 3; Beam fixed both ends

Sample Computer Program 1 — ANGLES1.BAS
This program will:

(a) Convert a given angle in degrees to radians
(b) Convert a given angle in radians to degrees

```
LIST
10 CLS: PRINT"ANGLES1:  CONVERT DEGREES TO RADIANS AND VICE VERSA": PRINT
20 PRINT"FOR DEGREES TO RADIANS, TYPE '1'; FOR RADIANS TO DEGREES, TYPE '2'
30 PRINT"TO EXIT PROGRAM, TYPE '3'
40 INPUT K
50 PI = 3.141593
60 IF K = 2 THEN 130 ELSE 70
70 IF K <> 1 THEN END ELSE 80
80 PRINT:PRINT"TYPE VALUE OF ANGLE (DEGREES)"
90 INPUT AD
100 AR = AD * PI/180  : REM EQUATION TO CONVERT DEGREES TO RADIANS
110 CLS :PRINT"THE ANGLE ";AD;" DEGREES = ";AR;" RADIANS"
120 END
130 PRINT:PRINT"TYPE VALUE OF ANGLE (RADIANS)"
140 INPUT AR
150 AD = AR * 180/PI  : REM EQUATION TO CONVERT RADIANS TO DEGREES
160 CLS :PRINT"THE ANGLE ";AR;" RADIANS = ";AD;" DEGREES"
170 END
Ok

RUN
ANGLES1:  CONVERT DEGREES TO RADIANS AND VICE VERSA

FOR DEGREES TO RADIANS, TYPE '1'; FOR RADIANS TO DEGREES, TYPE '2'
TO EXIT PROGRAM, TYPE '3'
? 1

TYPE VALUE OF ANGLE (DEGREES)
? 45
THE ANGLE  45  DEGREES =  .7853983  RADIANS
Ok

RUN
ANGLES1:  CONVERT DEGREES TO RADIANS AND VICE VERSA

FOR DEGREES TO RADIANS, TYPE '1'; FOR RADIANS TO DEGREES, TYPE '2'
TO EXIT PROGRAM, TYPE '3'
? 2

TYPE VALUE OF ANGLE (RADIANS)
? 1.15
THE ANGLE  1.15  RADIANS =  65.89014  DEGREES
Ok
```

Sample Computer Program 2 — FORCES1.BAS
This program will determine the magnitude and direction of the resultant of two concurrent coplanar forces. (See also Sec. 2-9.)

```
LIST
10 CLS: PRINT"FORCES1:  RESULTANT OF 2 CONCURRENT COPLANAR FORCES": PRINT
20 PRINT"TYPE MAGNITUDE (LB) & ANGLE (DEG) OF ONE FORCE.  SEPARATE WITH COMMAS.
30 INPUT F1,A1
40 PRINT:PRINT"TYPE MAGNITUDE & ANGLE OF SECOND FORCE.  SEPARATE WITH COMMAS."
50 INPUT F2,A2
60 PI = 3.141593
70 CD = ABS(180 - A2 + A1) : REM ANGLE OPPOSITE RESULTANT IN FORCE TRIANGLE
80 C = CD * PI/180 : REM CONVERT THAT ANGLE TO RADIANS FOR CALCULATIONS
90 R = SQR((F1^2 + F2^2) - (2 * F1 * F2 * COS (C))) : REM LAW OF COSINES
100 S = (F1/R) * SIN (C) : REM LAW OF SINES FOR SINE OF ANGLE OPPOSITE F1
110 AS = ATN (S/SQR(1 - S * S)) : REM COMPUTER FORMULA FOR ARC SINE
120 B = PI -AS -C  : BD = B * 180/PI : REM ANGLE OPPOSITE F2 IN FORCE TRIANGLE
130 IF (A2 - A1) <= 180 THEN 140 ELSE 150
140 AR = BD + A1  :  GOTO 160
```

```
150 AR = ABS(360 - (BD - A1))
160 IF AR >= 360 THEN 170 ELSE 180
170 AR = (AR - 360)
180 PRINT
190 PRINT"RESULTANT = ";R;" LB AT ";AR;" DEGREES FROM HORIZONTAL AXIS"
200 END
Ok

RUN
FORCES1:  RESULTANT OF 2 CONCURRENT COPLANAR FORCES

TYPE MAGNITUDE (LB) & ANGLE (DEG) OF ONE FORCE.  SEPARATE WITH COMMAS.
? 100, 30

TYPE MAGNITUDE & ANGLE OF SECOND FORCE.  SEPARATE WITH COMMAS.
? 250, 95

RESULTANT =  305.9917  LB AT  77.77127  DEGREES FROM HORIZONTAL AXIS
Ok
```

Sample Computer Program 3 — FORCES2.BAS

This program will:

(a) Determine the magnitude and direction of the resultant for two concurrent coplanar forces, one of which acts at a varying angle, at several incremental positions of the variable angle

(b) Determine the magnitude and direction of the maximum resultant as well as the angle of the movable force which produces the maximum resultant

```
LIST
10 CLS: PRINT"FORCES2:  RESULTANT OF 2 CONCURRENT COPLANAR FORCES"
20 PRINT"             1 WITH VARIABLE ANGLE" : PRINT
30 DIM R(360), AR(360), AS(360), A(360), B(360), BD(360), C(360), CD(360)
40 DIM I(360), S(360)
50 PRINT"GIVE VALUE OF FORCE F1 (LB) & ITS ANGLE (DEG) WITH THE HORIZONTAL"
60 PRINT:PRINT"SEPARATE THE VALUES WITH COMMAS."
70 INPUT F1, AO
80 PRINT:PRINT"GIVE VALUE OF FORCE F2"
90 INPUT F2
100 PRINT:PRINT"GIVE VALUES OF BEGINNING & ENDING ANGLES OF FORCE F2"
110 PRINT"SEPARATE THE VALUES WITH COMMAS."
120 INPUT AB, AE
130 LET PI = 3.141593 : LET RM = 0 : LET A = 0
140 IF (AE - AB) < 100 THEN 150 ELSE 160 : REM ESTABLISH NUMBER OF INCREMENTS
150 LET N = 5   :  GOTO 190
160 IF (AE - AB) < 200 THEN 170 ELSE 180
170 LET N = 10  : GOTO 190
180 IF (AE - AB) < 300 THEN N = 15 ELSE N = 20
190 CLS :PRINT"    RESULTANT" TAB(17) "ANGLE OF R WITH X-AXIS";
200 PRINT TAB(43) "ANGLE OF F2 WITH X-AXIS"
210 PRINT"      (LB)" TAB(25) "(DEG)" TAB(51) "(DEG)"
220 FOR I = AB TO AE STEP N
230 LET CD(I) = ABS(180-A-AB+AO):REM ANGLE OPPOSITE RESULTANT IN FORCE TRIANGLE
240 LET C(I) = CD(I) * PI/180 : REM CONVERT ANGLE TO RADIANS FOR CALCULATIONS
250 LET R(I) = SQR((F1^2 + F2^2) - (2*F1*F2* COS(C(I)))) : REM LAW OF COSINES
260 LET S(I) = (F1/R(I)) * SIN(C(I)) : REM LAW OF SINES FOR ANGLE OPPOSITE F1
270 LET AS(I) = ATN(S(I)/SQR(1-S(I)*S(I))) : REM COMPUTER EQUATION FOR ARC SINE
280 LET B(I) = PI-AS(I)-C(I) : LET BD(I) = B(I) *180/PI :REM ANGLE OPPOSITE F2
290 IF ((AB + A) - AO) <= 180 THEN 300 ELSE 310
300 LET AR(I) = BD(I) + AO  : GOTO 320
310 LET AR(I) = ABS(360 - (BD(I) - AO))
320 IF AR(I) > 360 THEN 330 ELSE 340
330 LET AR(I) = ABS(AR(I) - 360)
340 IF R(I) > RM THEN 450 : REM CHECK FOR MAXIMUM RESULTANT VALUE
350 PRINT"   ";R(I) TAB(24) AR(I) TAB(51) (AB + A)
360 IF I >= AE THEN 420  : REM TABULATION COMPLETE
370 IF I > (AE - N) THEN 380 ELSE 390
380 LET I = AE  :  LET A = AE - AB  :  GOTO 230
390 LET A = A + N
400 NEXT I
```

```
410 END
420 PRINT:PRINT"THE MAXIMUM RESULTANT = ";RM;" LB AT ";AM;" DEG
430 PRINT"          WITH FORCE F2 AT ";AF;" DEG
440 END
450 LET RM = R(I)
460 LET AM = AR(I)
470 LET AF = (AB + A)
480 GOTO 350
Ok
```

```
RUN
FORCES2:   RESULTANT OF 2 CONCURRENT COPLANAR FORCES
              1 WITH VARIABLE ANGLE

GIVE VALUE OF FORCE F1 (LB) & ITS ANGLE (DEG) WITH THE HORIZONTAL

SEPARATE THE VALUES WITH COMMAS.
? 100, 15

GIVE VALUE OF FORCE F2
? 300

GIVE VALUES OF BEGINNING & ENDING ANGLES OF FORCE F2
SEPARATE THE VALUES WITH COMMAS.
? 30, 195
     RESULTANT      ANGLE OF R WITH X-AXIS    ANGLE OF F2 WITH X-AXIS
       (LB)               (DEG)                     (DEG)
      397.4362           26.26614                     30
      392.9103           33.82527                     40
      386.1983           41.45892                     50
      377.3943           49.20093                     60
      366.626            57.08944                     70
      354.0581           65.1685                      80
      339.8958           73.48999                     90
      324.3908           82.11562                    100
      307.8485           91.11928                    110
      290.6387          100.5888                     120
      273.2086          110.6263                     130
      256.0965          121.3455                     140
      239.945           132.8607                     150
      225.5014          145.2646                     160
      213.592           158.588                      170
      205.0474          172.7486                     180
      200.57            187.5095                     190
      200               195                          195

THE MAXIMUM RESULTANT =   397.4362   LB AT   26.26614   DEG
          WITH FORCE F2 AT   30   DEG
Ok
```

Sample Computer Program 4 — FORCES3.BAS

This program will determine the magnitude, direction, moment arm, and direction of rotation about a given axis of the resultant of three nonconcurrent coplanar forces. (See also Sec. 4-2.)

```
LIST
10 CLS: PRINT"FORCES3:   RESULTANT OF 3 NONCONCURRENT COPLANAR FORCES" : PRINT
20 PRINT"FOR FORCE F1, GIVE MAGNITUDE (LB), ANGLE (DEGREES), & MOMENT ARM (IN)"
30 PRINT:PRINT"SEPARATE THE VALUES WITH COMMAS."
40 INPUT F1,A1,L1
50 PRINT: PRINT"GIVE THE MAGNITUDE, ANGLE & MOMENT ARM FOR FORCE F2."
60 PRINT"SEPARATE THE VALUES WITH COMMAS."
70 INPUT F2,A2,L2
80 PRINT: PRINT"GIVE THE SAME INFORMATION FOR FORCE F3.   SEPARATE WITH COMMAS."
90 INPUT F3,A3,L3
100 PRINT: PRINT"GIVE THE MOMENT DIRECTION OF F1 AS EITHER CW (FOR CLOCKWISE)"
110 PRINT"OR CCW (FOR COUNTERCLOCKWISE). IF NO MOMENT ARM, USE CW."
120 INPUT F1$
130 PRINT: PRINT"GIVE THE MOMENT DIRECTION (CW OR CCW) OF F2."
140 INPUT F2$
```

```
150 PRINT: PRINT"GIVE THE MOMENT DIRECTION OF F3."
160 INPUT F3$
170 M1 = F1 * L1 : REM CALCULATE MOMENTS OF THE THREE FORCES
180 M2 = F2 * L2
190 M3 = F3 * L3
200 IF F1$ = "CW" THEN 220 ELSE 210 : REM INCLUDE ROTATION DIRECTION OF MOMENTS
210 M1 = - M1
220 IF F2$ = "CW" THEN 240 ELSE 230
230 M2 = - M2
240 IF F3$ = "CW" THEN 260 ELSE 250
250 M3 = - M3
260 M = M1 + M2 + M3 : REM CALCULATE THE NET MOMENT OF ALL THREE MOMENTS
270 PI = 3.141593
280 B1 = (A1 * PI)/180 : REM CONVERT ANGLES TO RADIANS FOR CALCULATIONS
290 B2 = (A2 * PI)/180
300 B3 = (A3 * PI)/180
310 X1 = F1 * COS (B1) : REM CALCULATE HORIZONTAL COMPONENT OF F1
320 Y1 = F1 * SIN (B1) : REM CALCULATE VERTICAL COMPONENT OF F1
330 X2 = F2 * COS (B2)
340 Y2 = F2 * SIN (B2)
350 X3 = F3 * COS (B3)
360 Y3 = F3 * SIN (B3)
370 FX = X1 + X2 + X3 : REM SUM THE HORIZONTAL COMPONENTS OF THE THREE FORCES
380 FY = Y1 + Y2 + Y3 : REM SUM THE VERTICAL COMPONENTS OF THE THREE FORCES
390 R = SQR(FX^2 + FY^2) : REM PYTHAGOREAN THEOREM TO FIND RESULTANT
400 B = ATN(ABS(FY/FX)) : REM FIND ANGLE OF RESULTANT FROM ARC TAN; RADIAN UNIT
410 BD = ((B * 180)/PI) : REM CONVERT ANGLE TO DEGREES
420 IF FY < .001 AND FX > .001   THEN BD = BD + 270
430 IF FY < .001   AND FX < .001   THEN BD = BD + 180
440 IF FY > .001   AND FX < .001   THEN BD = BD + 90
450 IF FY > .001   AND FX > .001   THEN BD = BD
460 IF FY < .001 AND FY > -.001   AND FX > .001   THEN BD = 0
470 IF FY > .001   AND FX < .001 AND FX > -.001   THEN BD = 90
480 IF FY < .001   AND FY > -.001   AND FX < .001   THEN BD = 180
490 IF FY < .001 AND FX < .001 AND FX > -.001   THEN BD = 270
500 L = ABS(M)/R : REM FIND MOMENT ARM OF RESULTANT
510 CLS :PRINT:PRINT"RESULTANT FORCE IS ";R;" LB AT ";L;" IN. FROM THE AXIS"
520 PRINT
530 PRINT"ACTING AT AN ANGLE OF ";USING "###.# ";BD;
540 PRINT"DEGREES CAUSING A MOMENT OF ";M;" IN-LB"
550 PRINT:PRINT"IF THE MOMENT IS +, IT IS CLOCKWISE; IF -, COUNTERCLOCKWISE"
560 END
Ok




RUN
FORCES3:   RESULTANT OF 3 NONCONCURRENT COPLANAR FORCES

FOR FORCE F1, GIVE MAGNITUDE (LB), ANGLE (DEGREES), & MOMENT ARM (IN)

SEPARATE THE VALUES WITH COMMAS.
? 100,0,6

GIVE THE MAGNITUDE, ANGLE & MOMENT ARM FOR FORCE F2.
SEPARATE THE VALUES WITH COMMAS.
? 40,90,4

GIVE THE SAME INFORMATION FOR FORCE F3.   SEPARATE WITH COMMAS.
? 250,30,5

GIVE THE MOMENT DIRECTION OF F1 AS EITHER CW (FOR CLOCKWISE)
OR CCW (FOR COUNTERCLOCKWISE). IF NO MOMENT ARM, USE CW.
? CCW

GIVE THE MOMENT DIRECTION (CW OR CCW) OF F2.
? CCW

GIVE THE MOMENT DIRECTION OF F3.
? CW

RESULTANT FORCE IS   356.9331  LB AT   1.372806  IN. FROM THE AXIS

ACTING AT AN ANGLE OF   27.5 DEGREES CAUSING A MOMENT OF   490  IN-LB

IF THE MOMENT IS +, IT IS CLOCKWISE; IF -, COUNTERCLOCKWISE
Ok
```

Sample Computer Program 5 — TRUSSES1.BAS

For a four-panel Pratt truss, this program will:

(a)   Determine the support reactions
(b)   Determine the forces in each member of the truss
      See also Sample Problem 5, Chap. 4.

```
LIST
10 CLS : PRINT"TRUSSES1:  METHOD OF JOINTS FOR 4 PANEL PRATT TRUSS" : PRINT
20 PRINT"MAKE A SKETCH OF THE PRATT TRUSS WITH LOADS AND LENGTHS."
30 PRINT"LABEL THE JOINTS STARTING FROM THE LEFT REACTION AS 'A' WITH THE"
40 PRINT"TOP JOINTS AS 'B', 'C', 'D' (LEFT TO RIGHT), THE LEFT REACTION "
50 PRINT"AS 'E' AND THE LOWER JOINTS AS 'F', 'G', 'H' (RIGHT TO LEFT)."
60 PRINT:PRINT"THE REACTIONS WILL BE 'RA' AND 'RE'."
70 PRINT:PRINT"THE LOADING IS ASSUMED TO BE VERTICAL ACTING AT JOINTS AND"
80 PRINT"EACH LOAD WILL BE LABELLED FOR ITS JOINT, AS 'PA','PB','PC',ETC."
90 PRINT:PRINT:PRINT"GIVE SPAN LENGTH & VERTICAL HEIGHT OF TRUSS (FT)."
100 PRINT:PRINT"SEPARATE THE VALUES WITH COMMAS."
110 INPUT L, H
120 PRINT:PRINT"GIVE VALUES OF LOADS 'PA','PB','PC','PD' & 'PE' (KIP)."
130 PRINT"SEPARATE THE FIVE VALUES WITH COMMAS."
140 INPUT PA, PB, PC, PD, PE
150 PRINT:PRINT"GIVE VALUES OF LOADS 'PF','PG' & 'PH' (KIP)."
160 PRINT"SEPARATE THE THREE VALUES WITH COMMAS."
170 INPUT PF, PG, PH
180 LET LH = SQR(H^2 + (L/4)^2) : REM CALCULATE SLANT MEMBER LENGTH
190 LET M = (PA*L)+((PB+PH)*.75*L)+((PC+PG)*.5*L)+((PD+PF)*.25*L)
200 LET RA = M/L : REM SOLVE FOR RA IN EQUATION FOR MOMENTS ABOUT RE
210 LET FY = PA + PB + PC + PD + PE + PF + PG + PH
220 LET RE = FY - RA : REM SOLVE FOR RE IN EQUATION FOR VERTICAL FORCES
230 LET AB = (PA - RA) * (LH/H) : REM SOLUTION AT JOINT A BY SIMILAR TRIANGLES
240 LET AH = - AB * ((L/4)/LH)
250 LET BH = PH : LET HG = AH : REM JOINT H BY INSPECTION
260 LET BG = - AB - ((PB + BH) * (LH/H)) : REM SOLUTION AT JOINT B
270 LET BC = (AB - BG) * ((L/4)/LH)
280 LET CD = BC : LET CG = - PC : REM JOINT C BY INSPECTION
290 LET DG = ((PG - CG) * (LH/H)) - BG : REM SOLUTION AT JOINT G
300 LET GF = HG + ((BG - DG) * ((L/4)/LH))
310 LET FE = GF : LET DF = PF : REM JOINT F BY INSPECTION
320 LET DE = (PE - RE) * (LH/H) : REM SOLUTION AT JOINT E
330 PRINT:PRINT:PRINT : REM PRINT ANSWERS
340 PRINT"REACTIONS (KIPS) ARE " TAB(25) "RA = ";RA; TAB(45) "RE = ";RE
350 PRINT:PRINT"FORCES (KIPS) IN MEMBERS ARE:"
360 PRINT TAB(25) "BC = ";BC; TAB(45) "CD = ";CD
370 PRINT"AB = ";AB; TAB(20) "BG = ";BG; TAB(40) "DG = ";DG; TAB(60) "DE = ";DE
380 PRINT TAB(10) "BH = ";BH; TAB(30) "CG = ";CG; TAB(50) "DF = ";DF
390 PRINT"AH = ";AH; TAB(20) "HG = ";HG; TAB(40) "GF = ";GF; TAB(60) "FE = ";FE
400 PRINT:PRINT"NEGATIVE RESULTS ARE IN COMPRESSION, POSITIVE IN TENSION."
410 END
Ok
```

```
RUN
TRUSSES1:  METHOD OF JOINTS FOR 4 PANEL PRATT TRUSS

MAKE A SKETCH OF THE PRATT TRUSS WITH LOADS AND LENGTHS.
LABEL THE JOINTS STARTING FROM THE LEFT REACTION AS 'A' WITH THE
TOP JOINTS AS 'B', 'C', 'D' (LEFT TO RIGHT), THE LEFT REACTION
AS 'E' AND THE LOWER JOINTS AS 'F', 'G', 'H' (RIGHT TO LEFT).

THE REACTIONS WILL BE 'RA' AND 'RE'.

THE LOADING IS ASSUMED TO BE VERTICAL ACTING AT JOINTS AND
EACH LOAD WILL BE LABELLED FOR ITS JOINT, AS 'PA','PB','PC',ETC.

GIVE SPAN LENGTH & VERTICAL HEIGHT OF TRUSS (FT).

SEPARATE THE VALUES WITH COMMAS.
? 64, 30

GIVE VALUES OF LOADS 'PA','PB','PC','PD' & 'PE' (KIP).
SEPARATE THE FIVE VALUES WITH COMMAS.
? 0,5,10,5,0
```

```
GIVE VALUES OF LOADS 'PF','PG' & 'PH' (KIP).
SEPARATE THE THREE VALUES WITH COMMAS.
? 10,20,10


REACTIONS (KIPS) ARE     RA =  30              RE =  30

FORCES (KIPS) IN MEMBERS ARE:
                        BC = -24         CD = -24
AB = -34         BG =  17         DG =  17         DE = -34
        BH =  10         CG = -10         DF =  10
AH =  16         HG =  16         GF =  16         FE =  16

NEGATIVE RESULTS ARE IN COMPRESSION, POSITIVE IN TENSION.
Ok
```

Sample Computer Program 6 — FORCES4.BAS

This program will determine the magnitude and direction angles of the resultant of a system of concurrent-noncoplanar forces represented by their x, y, and
z components. (See also Sample Problem 2, Chap. 5.)

```
LIST
10 CLS: PRINT"FORCES4:  SYSTEM OF CONCURRENT NONCOPLANAR FORCES"
20 PRINT"            SAMPLE PROBLEM 2, CHAPTER 5" : PRINT
30 PRINT"GIVE THE SUM OF THE X-DIRECTION FORCE COMPONENTS (LB)."
40 INPUT FX
50 PRINT:PRINT"GIVE THE SUM OF THE Y-DIRECTION FORCE COMPONENTS (LB)."
60 INPUT FY
70 PRINT:PRINT"GIVE THE SUM OF THE Z-DIRECTION FORCE COMPONENTS (LB)."
80 INPUT FZ
90 PI = 3.141593
100 R = SQR (((FX)^2) + ((FY)^2) + ((FZ)^2)) : REM R FROM PYTHAGOREAN THEOREM
110 X = FX/R : REM CALCULATE DIRECTION COSINE FOR R FROM X-AXIS
120 Y = FY/R : REM CALCULATE DIRECTION COSINE FOR R FROM Y-AXIS
130 Z = FZ/R : REM CALCULATE DIRECTION COSINE FOR R FROM Z-AXIS
140 K = 90
150 M = 180/PI
160 AX = K - (ATN(X/SQR(1-X*X))) * M : REM COMPUTER FORMULA TO CALCULATE ANGLE
170 AY = K - (ATN(Y/SQR(1-Y*Y))) * M
180 AZ = K - (ATN(Z/SQR(1-Z*Z))) * M
190 CLS :PRINT"THE RESULTANT FORCE IS ";R;" LB, WITH DIRECTION ANGLES OF"
200 PRINT:PRINT"      ";AX;" DEGREES FROM THE X-AXIS"
210 PRINT:PRINT"      ";AY;" DEGREES FROM THE Y-AXIS"
220 PRINT:PRINT"      ";AZ;" DEGREES FROM THE Z-AXIS"
230 END
Ok

RUN
FORCES4:  SYSTEM OF CONCURRENT NONCOPLANAR FORCES
            SAMPLE PROBLEM 2, CHAPTER 5

GIVE THE SUM OF THE X-DIRECTION FORCE COMPONENTS (LB).
? 45

GIVE THE SUM OF THE Y-DIRECTION FORCE COMPONENTS (LB).
? 90

GIVE THE SUM OF THE Z-DIRECTION FORCE COMPONENTS (LB).
? 67.5
THE RESULTANT FORCE IS  121.1662  LB, WITH DIRECTION ANGLES OF

      68.1986  DEGREES FROM THE X-AXIS

      42.03112  DEGREES FROM THE Y-AXIS

      56.1455  DEGREES FROM THE Z-AXIS
Ok
```

Sample Computer Program 7 — FORCES5.BAS

This program will determine the concurrent-noncoplanar forces in the members of the system shown in Sample Problem 4, Chap. 5, as well as the lengths of the unknown members.

```
LIST
10 CLS : PRINT"FORCES5:  CONCURRENT NONCOPLANAR FORCE SYSTEM"
20 PRINT"            SAMPLE PROBLEM 4, CHAPTER 5" : PRINT
30 PRINT"WHAT IS THE LOAD (LB) AT THE END OF THE BOOM?"
40 INPUT W
50 PRINT:PRINT"NOTE THAT TRIANGLES OBC AND OCD ARE EACH 3:4:5 RIGHT TRIANGLES."
60 PRINT:PRINT"TYPE THE LENGTH (FT) OF MEMBER BC."
70 INPUT L1 : REM NOTE THAT L1 = LENGTH OF BC = LENGTH OF CD
80 PRINT:PRINT"TYPE THE LENGTH (FT) OF MEMBER OC."
90 INPUT L2 : REM NOTE THAT L2 = LENGTH OF OC
100 LET L3 = L1*(5/3) :REM L3=LENGTH OF OB=LENGTH OF OD.  5/3 FROM TRIANGLE OBC
110 CLS:PRINT"THE LENGTH OF MEMBER OB = ";L3;" FT AND NOTE THAT OB = OD"
120 PRINT:PRINT"NOTE THAT TRIANGLE OAC IS A 45 DEG RIGHT TRIANGLE"
130 LET L4 = L2 * (SQR(2)/1) : REM L4 = LENGTH OF OA
140 PRINT:PRINT"THE LENGTH OF MEMBER OA = ";L4;" FT"
150 PRINT:PRINT"THE FORCES ARE PROJECTED ON THE VERTICAL PLANE OAC,"
160 PRINT"        AND THE SUM OF THE VERTICAL FORCES IS SET TO ZERO."
170 LET OA = W/(1/SQR(2)) : REM NOTE THAT SIN 45 DEG = 1/(SQ ROOT OF 2)
180 PRINT:PRINT"THE FORCES ARE ALSO PROJECTED ON THE HORIZONTAL PLANE OBD,"
190 PRINT"        AND THE SUM OF THE HORIZONTAL FORCES IS SET TO ZERO."
200 REM OA * (COS 45) - 2*(OB*COS<BOC) = 0  AND  2 * OB * (L2/L3) = OA * (L2/L4
210 LET OB = (OA * (L3/L4))/2 : REM NOTE THAT FORCE IN OD = FORCE IN OB
220 PRINT:PRINT"THE FORCE IN MEMBER OA = ";OA;" LB (COMPRESSION)"
230 PRINT:PRINT"THE FORCE IN MEMBERS OB AND OD = ";OB;" LB (TENSION)"
240 END
Ok

RUN
FORCES5:  CONCURRENT NONCOPLANAR FORCE SYSTEM
            SAMPLE PROBLEM 4, CHAPTER 5

WHAT IS THE LOAD (LB) AT THE END OF THE BOOM?
? 200

NOTE THAT TRIANGLES OBC AND OCD ARE EACH 3:4:5 RIGHT TRIANGLES.

TYPE THE LENGTH (FT) OF MEMBER BC.
? 9

TYPE THE LENGTH (FT) OF MEMBER OC.
? 12
THE LENGTH OF MEMBER OB =  15  FT AND NOTE THAT OB = OD

NOTE THAT TRIANGLE OAC IS A 45 DEG RIGHT TRIANGLE

THE LENGTH OF MEMBER OA =  16.97056  FT

THE FORCES ARE PROJECTED ON THE VERTICAL PLANE OAC,
        AND THE SUM OF THE VERTICAL FORCES IS SET TO ZERO.

THE FORCES ARE ALSO PROJECTED ON THE HORIZONTAL PLANE OBD,
        AND THE SUM OF THE HORIZONTAL FORCES IS SET TO ZERO.

THE FORCE IN MEMBER OA =  282.8427  LB (COMPRESSION)

THE FORCE IN MEMBERS OB AND OD =  125  LB (TENSION)
Ok
```

Sample Computer Program 8 — FRICTN1.BAS

For any load and coefficient of friction in the system shown in Sample Problem 7, Chap. 6, this program will:

(a) Determine the least horizontal force to cause motion against friction
(b) Determine the least force in any direction to cause motion against friction

```
LIST
10 CLS: PRINT"FRICTN1:  LEAST FORCE TO CAUSE MOTION AGAINST FRICTION"
20 PRINT"            SAMPLE PROBLEM 7, CHAPTER 6" : PRINT
30 PRINT"GIVE THE VALUE OF THE LOAD W (LB)."
40 INPUT W
50 PRINT:PRINT"GIVE THE VALUE OF THE COEFFICIENT OF FRICTION."
60 INPUT CF
70 PI = 3.141593
80 FMAX = W * CF : REM CALCULATE HORIZONTAL FORCE TO OVERCOME FRICTION
90 CLS :PRINT"LEAST HORIZONTAL FORCE P = ";FMAX;" LB"
100 PRINT:PRINT: AMAX = ATN(CF) : REM ARC TAN OF COEFFICIENT OF FRICTION
110 PRINT:PRINT: P = W * SIN(AMAX) : REM CALCULATE LEAST FORCE
120 PRINT:PRINT"THE LEAST FORCE IN ANY DIRECTION TO CAUSE MOTION IS "
130 PRINT:PRINT"    P = ";P;" LB AT ";((AMAX)*180/PI);" DEGREES FROM HORIZONTAL"
140 END
Ok

RUN
FRICTN1:   LEAST FORCE TO CAUSE MOTION AGAINST FRICTION
            SAMPLE PROBLEM 7, CHAPTER 6

GIVE THE VALUE OF THE LOAD W (LB).
? 50

GIVE THE VALUE OF THE COEFFICIENT OF FRICTION.
? .3
LEAST HORIZONTAL FORCE P =  15  LB




THE LEAST FORCE IN ANY DIRECTION TO CAUSE MOTION IS

    P =  14.3674  LB AT  16.69924  DEGREES FROM HORIZONTAL
Ok
```

Sample Computer Program 9 — RODYOKE1.BAS

For the system shown in Sample Problem 5, Chap. 7, for any tensile load on the rod and yoke with any diameters of the pin and the rod, this program will:

(a) Determine the shear area and the average shear stress in the pin

(b) Determine the tension area and the average tensile stress in the rod

```
LIST
10 CLS: PRINT"RODYOKE1:  ROD AND YOKE SIMPLE STRESS ANALYSIS"
20 PRINT"            SAMPLE PROBLEM 5, CHAPTER 7" : PRINT
30 PRINT"UNITS WILL BE: FORCE LB, LENGTH IN., STRESS PSI"
40 PRINT:PRINT"GIVE THE TENSILE LOAD ON THE ROD & YOKE"
50 INPUT F
60 PRINT:PRINT"GIVE THE DIAMETER OF THE PIN"
70 INPUT DP
80 PRINT:PRINT"GIVE THE DIAMETER OF THE ROD AT SECTION A-A"
90 INPUT DR
100 PI = 3.141593
110 AP = 2 * (PI * (DP^2))/4 : REM CALCULATE DOUBLE SHEAR AREA IN PIN
120 AR = (PI * (DR^2))/4 : REM CALCULATE AREA IN ROD AT SECTION A-A
130 SP = F/AP : REM CALCULATE AVERAGE SHEAR STRESS IN PIN
140 SR = F/AR : REM CALCULATE AVERAGE TENSILE STRESS IN ROD
150 CLS :PRINT"AVERAGE SHEAR STRESS IN PIN = ";SP;" PSI"
160 PRINT"      ON THE DOUBLE SHEAR AREA OF ";AP;" SQ IN"
170 PRINT:PRINT"AVERAGE TENSILE STRESS IN ROD = ";SR;" PSI"
180 PRINT"      ON ROD AT SECTION A-A WITH AREA OF ";AR;" SQ IN"
190 END
Ok

RUN
RODYOKE1:   ROD AND YOKE SIMPLE STRESS ANALYSIS
            SAMPLE PROBLEM 5, CHAPTER 7

UNITS WILL BE: FORCE LB, LENGTH IN., STRESS PSI
```

```
GIVE THE TENSILE LOAD ON THE ROD & YOKE
? 11000

GIVE THE DIAMETER OF THE PIN
? .5

GIVE THE DIAMETER OF THE ROD AT SECTION A-A
? .875
AVERAGE SHEAR STRESS IN PIN =  28011.27  PSI
     ON THE DOUBLE SHEAR AREA OF  .3926991  SQ IN

AVERAGE TENSILE STRESS IN ROD =  18293.07  PSI
     ON ROD AT SECTION A-A WITH AREA OF  .6013205  SQ IN
Ok
```

Sample Computer Program 10 — TENTEST1.BAS

For a typical tension test on a metallic specimen, this program will:

(a) Determine stress and strain for each set of tension test data provided
(b) Determine the yield stress
(c) Determine the ultimate stress
(d) Determine the apparent rupture stress
(e) Determine the true rupture stress
(f) Determine the modulus of elasticity
(g) Determine the percent elongation
(h) Determine the percent reduction in area

A series of pairs of load elongation values must be added to the program as DATA, after which the program, when RUN, will request additional information from the tension test. (See also Sec. 8-3.)

```
LIST
10 CLS : PRINT"TENTEST1:  TENSILE TEST CALCULATIONS" : PRINT
20 DEFINT S
30 DIM L3(100), F(100), S(100), E(100)
40 PRINT"HAS TEST DATA BEEN ADDED TO PROGRAM? TYPE 'Y' FOR YES OR 'N' FOR NO."
50 INPUT Z$
60 IF Z$ = "Y" OR Z$ = "y" THEN 180 ELSE 70
70 PRINT"ADD DATA BY INTERRUPTING THE PROGRAM AND STARTING WITH"
80 PRINT" LINE '2000 DATA' FOLLOWED BY PAIRS OF LOAD/ELONGATION VALUES"
90 PRINT:PRINT"CONTINUE WITH LINES '2010 DATA', '2020 DATA', ETC.
100 PRINT" UNTIL ALL PAIRS OF LOAD/ELONGATION VALUES ARE ENTERED."
110 PRINT
120 PRINT"THE UNITS OF LOAD MUST BE 'KIP' & ELONGATION MUST BE 'IN'"
130 PRINT"WHEN DATA IS COMPLETE, PRESS 'RUN' & 'RETURN' TO START PROGRAM"
140 PRINT:PRINT
150 PRINT"WHEN YOU FINISH READING THE MESSAGE ABOVE, TYPE 'OK'"
160 INPUT C$ : IF C$ = "OK" OR C$ = "ok" THEN END ELSE 150
170 DEFSNG L, E
180 PRINT:PRINT"GIVE INITIAL & FINAL VALUES OF DIAMETER"
190 INPUT D1,D2
200 PRINT"GIVE ORIGINAL & FINAL VALUES OF GAGE LENGTH"
210 INPUT L1,L2
220 PRINT"HOW MANY PAIRS OF LOAD/ELONGATION DATA VALUES ARE THERE?"
230 INPUT N1
240 PRINT"GIVE THE YIELD LOAD, THE ULTIMATE LOAD, & THE RUPTURE LOAD"
250 INPUT YL, UL, RL
260 PI = 3.141593
270 A1 = (PI * (D1^2))/4 : REM CALCULATE ORIGINAL AREA
280 A2 = (PI * (D2^2))/4 : REM CALCULATE FINAL AREA
290 FOR I = 1 TO N1
300 READ F(I),L3(I) : REM READ IN THE LOAD/ELONGATION PAIRS OF DATA
310 IF I = 1 THEN 380 ELSE 320: REM PRINT COLUMN HEADINGS FOR TABULATION
320 S(I) = F(I)/A1 : REM CALCULATION OF STRESS FOR EACH LOAD
330 E(I) = L3(I)/L1 : REM CALCULATION OF STRAIN FOR EACH ELONGATION
```

```
340 PRINT"   ";F(I) TAB(18) L3(I) TAB(48) S(I) TAB(60) E(I)
350 IF I >= (N1*.75) AND I <= ((N1*.75)+1) THEN 440 ELSE 360:REM MODULUS POINT
360 IF I = N1 THEN 480 ELSE 370 : REM CHECK WHETHER TABULATION IS COMPLETE
370 NEXT I
380 PRINT:PRINT
390 PRINT"   LOAD "," ELONGATION ",," STRESS "," STRAIN"
400 PRINT"   KIP "," IN.",," KSI "," IN/IN"
410 S1 = F(1)/A1 : REM CALCULATE INITIAL STRESS AT FIRST LOAD VALUE
420 E1 = L3(1)/L1 : REM CALCULATE INITIAL STRAIN AT FIRST ELONGATION VALUE
430 GOTO 320
440 SP = S(I) : REM STRESS AT POINT FOR MODULUS OF ELASTICITY CALCULATION
450 EP = E(I) : REM STRAIN AT POINT FOR MODULUS OF ELASTICITY CALCULATION
460 ME = (SP - S1)/(EP - E1) : REM MODULUS OF ELASTICITY FROM 2 DATA POINTS
470 GOTO 360
480 PRINT:PRINT
490 PE = ((L2 - L1)/L1) * 100 : REM PERCENTAGE ELONGATION CALCULATION
500 PA = ((A1 - A2)/A1) * 100 : REM PERCENTAGE REDUCTION IN AREA CALCULATION
510 YS = YL/A1 : US = UL/A1 : RS = RL/A1 : TR = RL/A2
520 PRINT"THE YIELD STRESS = ";YS;" KSI"
530 PRINT:PRINT"THE ULTIMATE STRESS = ";US;" KSI"
540 PRINT:PRINT"THE APPARENT RUPTURE STRESS = ";RS;" KSI"
550 PRINT:PRINT"THE TRUE RUPTURE STRESS = ";TR;" KSI"
560 PRINT:PRINT"THE MODULUS OF ELASTICITY = ";ME;" KSI"
570 PRINT:PRINT"THE PERCENTAGE ELONGATION = ";PE;" %"
580 PRINT:PRINT"THE PERCENTAGE REDUCTION IN AREA = ";PA;" %"
590 PRINT:PRINT:PRINT
600 PRINT"TO FIND THE PROPORTIONAL LIMIT OR ELASTIC LIMIT,"
610 PRINT"    PLOT THE STRESS-STRAIN CURVE AND SELECT THE POINT VISUALLY."
620 END
Ok




RUN
TENTEST1:  TENSILE TEST CALCULATIONS

HAS TEST DATA BEEN ADDED TO PROGRAM? TYPE 'Y' FOR YES OR 'N' FOR NO.
? N
ADD DATA BY INTERRUPTING THE PROGRAM AND STARTING WITH
 LINE '2000 DATA' FOLLOWED BY PAIRS OF LOAD/ELONGATION VALUES

CONTINUE WITH LINES '2010 DATA', '2020 DATA', ETC.
 UNTIL ALL PAIRS OF LOAD/ELONGATION VALUES ARE ENTERED.

THE UNITS OF LOAD MUST BE 'KIP' & ELONGATION MUST BE 'IN'

WHEN DATA IS COMPLETE, PRESS 'RUN' & 'RETURN' TO START PROGRAM


WHEN YOU FINISH READING THE MESSAGE ABOVE, TYPE 'OK
? OK
Ok

2000 DATA .5,0, 1.5,.0003, 2.5,.0007, 3.5,.001, 4.5,.0013
2010 DATA 5.5,.0016, 6.5,.002, 7.5,.0023, 8.5,.0026, 9.5,.003
2020 DATA 10, .0032, 10.5,.0034, 11,.0036, 11.5,.0039, 12,.0043
2030 DATA 12.5,.0046, 13,.0051, 13.5,.006, 14,.0073, 14.5,.0088
2040 DATA 15,.0118




RUN
TENTEST1:  TENSILE TEST CALCULATIONS

HAS TEST DATA BEEN ADDED TO PROGRAM? TYPE 'Y' FOR YES OR 'N' FOR NO.
? Y

GIVE INITIAL & FINAL VALUES OF DIAMETER
? .503, .312
GIVE ORIGINAL & FINAL VALUES OF GAGE LENGTH
? 2.0, 2.328
HOW MANY PAIRS OF LOAD/ELONGATION DATA VALUES ARE THERE?
? 21
GIVE THE YIELD LOAD, THE ULTIMATE LOAD, & THE RUPTURE LOAD
? 14.5, 15.83, 12.5
```

LOAD	ELONGATION		STRESS	STRAIN
KIP	IN.		KSI	IN/IN
.5	0		3	0
1.5	.0003		8	.00015
2.5	.0007		13	.00035
3.5	.001		18	.0005
4.5	.0013		23	.00065
5.5	.0016		28	.0008
6.5	.002		33	.001
7.5	.0023		38	.00115
8.5	.0026		43	.0013
9.5	.003		48	.0015
10	.0032		50	.0016
10.5	.0034		53	.0017
11	.0036		55	.0018
11.5	.0039		58	.00195
12	.0043		60	.00215
12.5	.0046		63	.0023
13	.0051		65	.00255
13.5	.006		68	.003
14	.0073		70	.00365
14.5	.0088		73	.0044
15	.0118		75	.0059

```
THE YIELD STRESS =  72.96962  KSI

THE ULTIMATE STRESS =  79.66269  KSI

THE APPARENT RUPTURE STRESS =  62.90484  KSI

THE TRUE RUPTURE STRESS =  163.4974  KSI

THE MODULUS OF ELASTICITY =  26086.96  KSI

THE PERCENTAGE ELONGATION =  16.4  %

THE PERCENTAGE REDUCTION IN AREA =  61.52548  %


TO FIND THE PROPORTIONAL LIMIT OR ELASTIC LIMIT,
     PLOT THE STRESS-STRAIN CURVE AND SELECT THE POINT VISUALLY.
Ok
```

Sample Computer Program 11 — TIEROD1.BAS

For the system described in Sample Problem 5, Chap. 8, this program will:

(a) Determine the minimum safe tie rod diameter for any given axial load and allowable tensile stress

(b) Determine the minimum safe tie rod diameter for any given temperature change and axial load, as well as the stresses due to the load and to the temperature change

```
LIST
10 CLS : PRINT"TIEROD1:  TIE ROD WITH AXIAL LOAD AND TEMPERATURE CHANGE"
20 PRINT"            SAMPLE PROBLEM 5, CHAPTER 8" : PRINT
30 PRINT"WHAT IS THE AXIAL TENSILE LOAD (KN) ON THE TIE ROD?"
40 INPUT F
50 PRINT:PRINT"WHAT IS THE ALLOWABLE TENSILE STRESS (N/MM^2)"
60 INPUT SA
70 LET PI = 3.141593
80 LET AR = (F * (10^3))/SA : REM CALCULATE MINIMUM AREA REQUIRED
90 LET DR = SQR ((4 * AR)/PI) : REM CALCULATE MINIMUM DIAMETER
100 PRINT:PRINT"MINIMUM SAFE TIE ROD DIAMETER = ";DR;" MM"
110 PRINT"        AT A TENSILE STRESS OF ";SA;" N/MM^2"
120 PRINT:PRINT:PRINT
130 PRINT"GIVE ROD MATERIAL MODULUS OF ELASTICITY (N/MM^2); DON'T USE COMMAS"
140 INPUT E
150 PRINT:PRINT"GIVE ROD MATERIAL COEFFICIENT OF LINEAR EXPANSION (PER DEG C)"
```

```
160 INPUT EX
170 PRINT:PRINT"GIVE THE TEMPERATURE CHANGE (DEG C); USE MINUS FOR DROP"
180 INPUT T  :  LET TD = - T
190 LET SX = E * EX * TD : REM CALCULATE STRESS DUE TO TEMPERATURE
200 LET ST = SA - SX : REM MAXIMUM SAFE STRESS DUE TO AXIAL LOAD
210 LET AR = (F * (10^3))/ST : REM CALCULATE MINIMUM AREA REQUIRED
220 LET DR = SQR ((4 * AR)/PI) : REM CALCULATE MINIMUM DIAMETER
230 CLS :PRINT:PRINT"MINIMUM SAFE TIE ROD DIAMETER = ";DR;" MM  WHERE THE"
240 PRINT:PRINT"       TEMPERATURE STRESS IS ";SX;" N/MM^2 AND THE"
250 PRINT:PRINT"       STRESS DUE TO AXIAL LOAD IS ";ST;" N/MM^2"
260 END
Ok


RUN
TIEROD1:  TIE ROD WITH AXIAL LOAD AND TEMPERATURE CHANGE
          SAMPLE PROBLEM 5, CHAPTER 8

WHAT IS THE AXIAL TENSILE LOAD (KN) ON THE TIE ROD?
? 18

WHAT IS THE ALLOWABLE TENSILE STRESS (N/MM^2)
? 160

MINIMUM SAFE TIE ROD DIAMETER =  11.96827  MM
      AT A TENSILE STRESS OF  160  N/MM^2



GIVE ROD MATERIAL MODULUS OF ELASTICITY (N/MM^2); DON'T USE COMMAS
? 200000

GIVE ROD MATERIAL COEFFICIENT OF LINEAR EXPANSION (PER DEG C)
? .0000117

GIVE THE TEMPERATURE CHANGE (DEG C); USE MINUS FOR DROP
? -38

MINIMUM SAFE TIE ROD DIAMETER =  17.95633  MM  WHERE THE

     TEMPERATURE STRESS IS  88.92  N/MM^2 AND THE

     STRESS DUE TO AXIAL LOAD IS  71.08  N/MM^2
Ok
```

Sample Computer Program 12 — JOINTS1.BAS

For a single-bolted lap joint of either the friction or bearing type when the dimensions and allowable stresses of the bolts and plates are input by the user, this program will:

(a)   Determine the maximum safe load on the joint
(b)   Determine the efficiency of the joint

See also Sample Problems 1 and 3, Chap. 9.

```
LIST
10 CLS: PRINT"JOINTS1:  SINGLE BOLTED LAP JOINT -- SAFE LOAD & EFFICIENCY"
20 PRINT"            SAMPLE PROBLEMS 1 & 3, CHAPTER 9" : PRINT
30 PRINT"IS THE JOINT FRICTION-TYPE? TYPE 'Y' FOR YES OR 'N' FOR NO."
40 INPUT JT$
50 PRINT:PRINT"GIVE PLATE WIDTH OR PITCH OF REPEATING SECTION (IN); DECIMALS."
60 INPUT L
70 PRINT"GIVE THE BOLT DIAMETER & PLATE THICKNESS (IN); USE DECIMALS."
80 INPUT DB, PT
90 PRINT"GIVE THE NUMBER OF SHEAR AREAS."
100 INPUT NS
110 PRINT"GIVE THE ALLOWABLE SHEAR STRESS ON THE BOLTS (PSI); NO COMMAS."
120 INPUT SS
130 AS = 3.141593*(DB^2)/4 : FS = NS*AS*SS : REM SAFE LOAD -- SHEAR
140 PRINT"GIVE THE ALLOWABLE TENSILE STRESS ON THE PLATE; NO COMMAS."
150 INPUT ST
```

```
160 DH = DB+(1/16) : N1 = NS : AT = (L-(N1*DH))*PT : REM AREA OF PLATE AT HOLES
170 FT = AT*ST : FP = L*PT*ST : REM SAFE LOAD-TENSION; SAFE LOAD WITHOUT HOLES
180 IF JT$ = "N" OR JT$ = "n" THEN 260 ELSE 190
190 IF FT > FS THEN 220 ELSE 200
200 PRINT:PRINT"MAXIMUM SAFE LOAD ON THE JOINT = ";FT;"LB BASED ON TENSION"
210 EF = (FT/FP) * 100 : GOTO 240: REM JOINT EFFICIENCY
220 PRINT:PRINT"MAXIMUM SAFE LOAD ON THE JOINT = ";FS;" LB BASED ON SHEAR "
230 EF = (FS/FP) * 100 : REM JOINT EFFICIENCY
240 PRINT:PRINT"THE EFFICIENCY OF THE JOINT = ";EF;" %"
250 END
260 NC = NS : AC = DB*PT : REM ONE BEARING AREA-PROJECTED AREA OF BOLT IN PLATE
270 PRINT:PRINT"GIVE THE ALLOWABLE BEARING STRESS ON THE PLATE/BOLT; NO COMMAS.
280 INPUT SC
290 LET FC = NC * AC * SC : REM SAFE LOAD ON JOINT BASED ON BEARING ONLY
300 IF FC > FS THEN 190 EDIT 310
310 IF FC > FT THEN 190 ELSE 320
320 PRINT:PRINT"MAXIMUM SAFE LOAD ON THE JOINT = ";FC;" LB BASED ON BEARING"
330 LET EF = (FC/FP) * 100 : GOTO 240: REM JOINT EFFICIENCY
Ok

RUN
JOINTS1:   SINGLE BOLTED LAP JOINT -- SAFE LOAD & EFFICIENCY
           SAMPLE PROBLEMS 1 & 3, CHAPTER 9

IS THE JOINT FRICTION-TYPE? TYPE 'Y' FOR YES OR 'N' FOR NO.
? Y

GIVE PLATE WIDTH OR PITCH OF REPEATING SECTION (IN); DECIMALS.
? 12
GIVE THE BOLT DIAMETER & PLATE THICKNESS (IN); USE DECIMALS.
? .875, .5
GIVE THE NUMBER OF SHEAR AREAS.
? 4
GIVE THE ALLOWABLE SHEAR STRESS ON THE BOLTS (PSI); NO COMMAS.
? 17500
GIVE THE ALLOWABLE TENSILE STRESS ON THE PLATE; NO COMMAS.
? 29000

MAXIMUM SAFE LOAD ON THE JOINT =  42092.44   LB BASED ON SHEAR

THE EFFICIENCY OF THE JOINT =  24.19105   %
Ok
```

Sample Computer Program 13 — JOINTS2.BAS

For a double-bolted butt joint of either the friction or the bearing type when the dimensions of the bolts and plates are input by the user and the AISC code applies, this program will:

(a) Determine the maximum safe load on the joint

(b) Determine the efficiency of the joint

See also Sample Problem 2, Chap. 9.

```
LIST
10 CLS: PRINT"JOINTS2: DOUBLE BOLTED BUTT JOINT -- AISC SAFE LOAD & EFFICIENCY"
20 PRINT"              SAMPLE PROBLEM 2, CHAPTER 9" : PRINT
30 PRINT:PRINT"IS THE JOINT FRICTION-TYPE?  TYPE 'Y' FOR YES OR 'N' FOR NO."
40 INPUT JT$ : PRINT
50 PRINT"GIVE WIDTH OF MAIN PLATE OR PITCH OF REPEATING SECTION (IN); DECIMALS"
60 INPUT L
70 PRINT:PRINT"GIVE BOLT DIAMETER & MAIN PLATE THICKNESS (IN); USE DECIMALS."
80 INPUT DB, PT
90 PRINT:PRINT"GIVE THE TOTAL NUMBER OF SHEAR AREAS ON HALF THE JOINT."
100 INPUT NS
110 PRINT:PRINT"GIVE THE ALLOWABLE SHEAR STRESS ON THE BOLTS (PSI); NO COMMAS."
120 INPUT SS
130 AS = 3.141593*(DB^2)/4 : FS = NS*AS*SS : REM SAFE LOAD - SHEAR
140 PRINT:PRINT"GIVE THE ALLOWABLE TENSILE STRESS ON THE PLATE; NO COMMAS."
150 INPUT ST
160 FP = L * PT * ST : REM SAFE LOAD ON PLATE IN TENSION WITHOUT HOLES
```

```
170 DH = DB + (1/16) : REM BOLT HOLE DIAMETER FOR AISC CODE
180 PRINT:PRINT"GIVE THE NUMBER OF HOLES IN OUTER ROW (ROW 1) OF MAIN PLATE."
190 INPUT N1
200 A1 = (L-(N1*DH))*PT : F1 = A1*ST : REM SAFE LOAD BASED ON ROW 1 TENSION
210 PRINT:PRINT"GIVE THE NUMBER OF HOLES IN INNER ROW (ROW 2) OF MAIN PLATE."
220 INPUT N2
230 NT = N1+N2 : A2 = (L-(N2*DH))*PT : REM TENSION AREA AT ROW 2
240 F2 = (NT/(NT - N1)) * A2 * ST : REM SAFE LOAD ON JOINT BASED ON ROW 2
250 IF JT$ = "Y" OR JT$ = "Y" THEN 260 ELSE 380
260 PRINT: IF F2 < F1 THEN 280 ELSE 270
270 FT = F1 : GOTO 290
280 FT = F2
290 IF FT > FS THEN 360 ELSE 300
300 IF F2 < F1 THEN 340 ELSE 310
310 PRINT"MAXIMUM SAFE LOAD ON JOINT = ";F1;" LB IN TENSION BASED ON ROW 1"
320 EF = (F1/FP) * 100 : REM JOINT EFFICIENCY
330 PRINT:PRINT"THE EFFICIENCY OF THE JOINT = ";EF;" %" : END
340 PRINT"MAXIMUM SAFE LOAD ON JOINT = ";F2;" LB IN TENSION BASED ON ROW 2"
350 EF = (F2/FP) * 100 : GOTO 330 : REM JOINT EFFICIENCY
360 PRINT"MAXIMUM SAFE LOAD ON JOINT = ";FS;" LB IN SHEAR"
370 EF = (FS/FP) * 100 : GOTO 330 : REM JOINT EFFICIENCY
380 NC = NT : AC = DB*PT : REM BEARING AREA IS PROJECTED AREA OF BOLT ON PLATE
390 PRINT:PRINT"GIVE THE ALLOWABLE BEARING STRESS ON THE PLATE/BOLT; NO COMMAS.
400 INPUT SC
410 FC = NC*AC*SC : REM SAFE LOAD ON JOINT BASED ON BEARING ONLY
420 IF FC > FS THEN 260 ELSE 430
430 IF FC > FT THEN 260 ELSE 440
440 PRINT:PRINT"MAXIMUM SAFE LOAD ON THE JOINT = ";FC;" LB BASED ON BEARING"
450 EF = (FC/FP) * 100 : GOTO 330 : REM JOINT EFFICIENCY
Ok


RUN
JOINTS2: DOUBLE BOLTED BUTT JOINT -- AISC SAFE LOAD & EFFICIENCY
            SAMPLE PROBLEM 2, CHAPTER 9


IS THE JOINT FRICTION-TYPE?  TYPE 'Y' FOR YES OR 'N' FOR NO.
? N

GIVE WIDTH OF MAIN PLATE OR PITCH OF REPEATING SECTION (IN); DECIMALS
? 10

GIVE BOLT DIAMETER & MAIN PLATE THICKNESS (IN); USE DECIMALS.
? 1, .75

GIVE THE TOTAL NUMBER OF SHEAR AREAS ON HALF THE JOINT.
? 8

GIVE THE ALLOWABLE SHEAR STRESS ON THE BOLTS (PSI); NO COMMAS.
? 40000

GIVE THE ALLOWABLE TENSILE STRESS ON THE PLATE; NO COMMAS.
? 33500

GIVE THE NUMBER OF HOLES IN OUTER ROW (ROW 1) OF MAIN PLATE.
? 2

GIVE THE NUMBER OF HOLES IN INNER ROW (ROW 2) OF MAIN PLATE.
? 2

GIVE THE ALLOWABLE BEARING STRESS ON THE PLATE/BOLT; NO COMMAS.
? 105000

MAXIMUM SAFE LOAD ON JOINT =  197859.4  LB IN TENSION BASED ON ROW 1

THE EFFICIENCY OF THE JOINT =  78.75  %
Ok
```

Sample Computer Program 14 — SHAPES 1.BAS

For a composite shape consisting of rectangles, triangles, and semicircles, this program will:

(a) Determine the total area

(b) Determine the location of the centroid

See also Sample Problem 3, Chap. 10.

```
LIST
10 CLS: PRINT"SHAPES1:  COMPOSITE SHAPE CENTROID LOCATION FOR SAMPLE PROBLEM 3";
20 PRINT", CHAPTER 10" : PRINT
30 DEFINT A : REM PROGRAM FOR SHAPES OF RECTANGLES, TRIANGLES & SEMI-CIRCLES
40 PRINT "SELECT X & Y REFERENCE AXES"
50 PRINT "GIVE UNITS FOR LENGTH (IN, FT, M, MM)?  TYPE UNIT."
60 INPUT L$
70 PRINT: PRINT "TYPE THE NUMBER OF RECTANGLES. IF NONE, TYPE 0."
80 INPUT NR
90 PRINT TAB(20) "HOW MANY TRIANGLES?"
100 INPUT NT
110 PRINT TAB(20) "HOW MANY SEMI-CIRCLES?"
120 INPUT NS
130 PRINT:PRINT"ENTER DATA FOR EACH SEPARATE SHAPE IN THE SEQUENCE SPECIFIED."
140 PRINT:PRINT"GIVE ALL DATA FOR ANY ONE SHAPE ON ONE LINE SEPARATED BY COMMAS"
150 PRINT:PRINT "FOR EACH AREA WHICH IS A HOLE,"
160 PRINT "ENTER ONE DIMENSION AS A NEGATIVE NUMBER."
170 IF NR = 0 THEN 250 ELSE 180
180 PRINT:PRINT"FOR EACH RECTANGLE, GIVE HORIZONTAL & VERTICAL LENGTHS &"
190 PRINT"THE DISTANCES OF ITS CENTROID FROM THE VERTICAL REFERENCE AXIS"
200 PRINT:PRINT "AND FROM THE HORIZONTAL REFERENCE AXIS."
210 FOR I = 1 TO NR
220 INPUT BR(I), HR(I), XR(I), YR(I)
230 AR(I) = (BR(I))*(HR(I)) : IF I = NR THEN 250 ELSE 240
240 NEXT I
250 IF NT = 0 THEN 330 ELSE 260
260 PRINT
270 PRINT:PRINT"FOR EACH TRIANGLE, GIVE ITS HORIZONTAL BASE & VERTICAL ALTITUDE"
280 PRINT"AND THE DISTANCES OF ITS CENTROID FROM THE Y-AXIS & FROM THE X-AXIS."
290 FOR J = 1 TO NT
300 INPUT BT(J), HT(J), XT(J), YT(J)
310 AT(J) = (BT(J))*(HT(J))/2 : IF J = NT THEN 330 ELSE 320
320 NEXT J
330 IF NS = 0 THEN 400 ELSE 340
340 PRINT:PRINT"FOR EACH SEMI-CIRCLE, GIVE ITS RADIUS AND THE DISTANCES"
350 PRINT"OF ITS CENTROID FROM THE Y-AXIS & FROM THE X-AXIS."
360 FOR K = 1 TO NS
370 INPUT RS(K), XS(K), YS(K)
380 AS(K) = (3.141593)*(RS(K))*ABS (RS(K))/2 : IF K = NS THEN 400 ELSE 390
390 NEXT K
400 SX = 0: SY = 0: AA = 0
410 IF NR = 0 THEN 470 ELSE 420 : FOR I = 1 TO NR
420 AR(I) = (BR(I))*(HR(I)) : AA = AA + AR(I)
430 UR(I) = (AR(I))*(XR(I)) : VR(I) = (AR(I))*(YR(I))
440 SX = SX + UR(I) : SY = SY + VR(I)
450 IF I = NR THEN 470 ELSE 460
460 NEXT I
470 IF NT = 0 THEN 530 ELSE 480 : FOR J = 1 TO NT
480 AT(J) = (BT(J))*(HT(J))/2 : AA = AA + AT(J)
490 UT(J) = (AT(J))*(XT(J)) : VT(J) = (AT(J))*(YT(J))
500 SX = SX + UT(J) : SY = SY + VT(J)
510 IF J = NT THEN 530 ELSE 520
520 NEXT J
530 IF NS = 0 THEN 590 ELSE 540 : FOR K = 1 TO NS
540 AS(K) = (3.141593)*(RS(K))*ABS (RS(K))/2 : AA = AA + AS(K)
550 US(K) = (AS(K))*(XS(K)) : VS(K) = (AS(K))*(YS(K))
560 SX = SX + US(K) : SY = SY + VS(K)
570 IF K = NS THEN 590 ELSE 580
580 NEXT K
590 CLS:PRINT TAB(20) "TOTAL AREA      = ";AA;" SQ. ";L$
600 PRINT:PRINT TAB(20) "SUM OF AX     = ";SX
610 PRINT:PRINT TAB(20) "SUM OF AY     = ";SY
620 XO = SX/AA : YO = SY/AA
630 PRINT:PRINT:PRINT TAB(20) "X-BAR = ";USING "#######.##";XO;
640 PRINT" ";L$
650 PRINT: PRINT TAB(20) "Y-BAR = ";USING "#######.##";YO;
660 PRINT" ";L$
670 END
Ok
```

```
RUN
SHAPES1:   COMPOSITE SHAPE CENTROID LOCATION FOR SAMPLE PROBLEM 3, CHAPTER 10

SELECT X & Y REFERENCE AXES
GIVE UNITS FOR LENGTH (IN, FT, M, MM)?  TYPE UNIT.
? IN

TYPE THE NUMBER OF RECTANGLES. IF NONE, TYPE 0.
? 1
                        HOW MANY TRIANGLES?
? 1
                        HOW MANY SEMI-CIRCLES?
? 1

ENTER DATA FOR EACH SEPARATE SHAPE IN THE SEQUENCE SPECIFIED.

GIVE ALL DATA FOR ANY ONE SHAPE ON ONE LINE SEPARATED BY COMMAS

FOR EACH AREA WHICH IS A HOLE,
ENTER ONE DIMENSION AS A NEGATIVE NUMBER.

FOR EACH RECTANGLE, GIVE HORIZONTAL & VERTICAL LENGTHS &
THE DISTANCES OF ITS CENTROID FROM THE VERTICAL REFERENCE AXIS

AND FROM THE HORIZONTAL REFERENCE AXIS.
? 4, 6, 5, 3


FOR EACH TRIANGLE, GIVE ITS HORIZONTAL BASE & VERTICAL ALTITUDE
AND THE DISTANCES OF ITS CENTROID FROM THE Y-AXIS & FROM THE X-AXIS.
? 3, 6, 8, 4

FOR EACH SEMI-CIRCLE, GIVE ITS RADIUS AND THE DISTANCES
OF ITS CENTROID FROM THE Y-AXIS & FROM THE X-AXIS.
? 3, 1.725, 3              TOTAL AREA      =  47  SQ. IN

                          SUM OF AX       =  216.15

                          SUM OF AY       =  150


                          X-BAR =         4.60 IN

                          Y-BAR =         3.19 IN
Ok
```

Sample Computer Program 15 — SHAPES2.BAS

For a composite shape consisting of rectangles, triangles, and semicircles, this program will:

(a) Determine the composite centroidal moment of inertia about the x-axis
(b) Determine the composite centroidal moment of inertia about the y-axis
(c) Determine the individual x- and y-axis moments of inertia and their transfer moments of inertia for each different shape

     See also Sample Computer Program 14.

```
LIST
10 CLS:PRINT"SHAPES2:  MOMENT OF INTERIA FOR COMPOSITE RECTANGLES, TRIANGLES &";
20 PRINT" SEMI-CIRCLES" : PRINT
30 LET IX = 0: LET IY = 0: LET TU = 0: LET TV = 0: LET TX = 0: LET TY = 0
40 PRINT "WHAT UNITS FOR LENGTH?" : INPUT L$
50 PRINT:PRINT"HOW MANY RECTANGLES IN THE COMPOSITE SHAPE?" : INPUT NR
60 PRINT:PRINT"HOW MANY TRIANGLES ARE IN THE COMPOSITE SHAPE?" : INPUT NT
70 PRINT:PRINT "HOW MANY SEMI-CIRCLES ARE IN THE COMPOSITE SHAPE?" : INPUT NS
80 IF NR = 0 THEN 240 ELSE 90
90 PRINT:PRINT"FOR EACH RECTANGLE, GIVE THE HORIZONTAL & VERTICAL LENGTHS AND"
100 PRINT"THE X & Y DISTANCES BETWEEN THE CENTROIDAL AXES OF RECTANGLE AND THE"
```

```
110 PRINT"COMPOSITE SHAPE CENTROIDAL AXES.  SEPARATE DATA ITEMS WITH COMMAS."
120 PRINT"FOR HOLES (MISSING AREAS) GIVE ONE LENGTH AS NEGATIVE VALUE."
130 FOR I = 1 TO NR
140 INPUT BR(I), HR(I), WR(I), ZR(I)
150 IF I = NR THEN 170 ELSE 160
160 NEXT I
170 FOR I = 1 TO NR
180 AR(I) = (BR(I)) * (HR(I)) : UR(I) = (BR(I)) * ((HR(I))^3)/12
190 VR(I) = (HR(I)) * ((BR(I))^3)/12 : XR(I) = (AR(I)) * ((ZR(I))^2)
200 YR(I) = (AR(I)) * ((WR(I))^2) : IX =IX+ UR(I)+ XR(I) : IY =IY+ VR(I)+ YR(I)
210 TU = TU + UR(I) : TV = TV + VR(I) : TX = TX + XR(I) : TY = TY + YR(I)
220 IF I = NR THEN 240 ELSE 230
230 NEXT I
240 IF NT = 0 THEN 360 ELSE 250
250 PRINT:PRINT"FOR EACH TRIANGLE, GIVE THE HORIZONTAL BASE & VERTICAL ALTITUDE
260 PRINT"AND THE X & Y DISTANCES BETWEEN THE CENTROIDAL AXES OF THE TRIANGLE &
270 PRINT"THE COMPOSITE SHAPE CENTROIDAL AXES.  SEPARATE ITEMS WITH COMMAS."
280 FOR J = 1 TO NT
290 INPUT BT(J), HT(J), WT(J), ZT(J)
300 AT(J) = (BT(J)) * (HT(J))/2 : UT(J) = (BT(J)) * ((HT(J))^3)/36
310 VT(J) = (HT(J)) * ((BT(J))^3)/36 : XT(J) = (AT(J)) * ((ZT(J))^2)
320 YT(J) = (AT(J)) * ((WT(J))^2): IX =IX +UT(J) +XT(J): IY = IY +VT(J) +YT(J)
330 TU = TU + UT(J) : TV = TV + VT(J) : TX = TX + XT(J) : TY = TY + YT(J)
340 IF J = NT THEN 360 ELSE 350
350 NEXT J
360 IF NS = 0 THEN 480 ELSE 370
370 PRINT:PRINT"FOR EACH SEMI-CIRCLE, GIVE THE RADIUS AND THE X & Y DISTANCES"
380 PRINT"BETWEEN THE SEMI-CIRCLE CENTROIDAL AXES AND THE COMPOSITE SHAPE"
390 PRINT"CENTROIDAL AXES.  SEPARATE DATA ITEMS WITH COMMAS."
400 FOR K = 1 TO NS
410 INPUT RS(K), WS(K), ZS(K)
420 AS(K) = 3.141593*(RS(K))* ABS(RS(K))/2: US(K) = .11*((RS(K))^3)* ABS(RS(K))
430 VS(K) = .393 * ((RS(K))^3) * ABS(RS(K)) : XS(K) = (AS(K)) * ((ZS(K))^2)
440 YS(K) = (AS(K)) * ((WS(K))^2): IX = IX +US(K) +XS(K): IY = IY +VS(K) +YS(K)
450 TU = TU + US(K) : TV = TV + VS(K) : TX = TX + XS(K) : TY = TY + YS(K)
460 IF K = NS THEN 480 ELSE 470
470 NEXT K
480 CLS :PRINT:PRINT
490 PRINT"TOTAL OF INDIVIDUAL X-AXIS MOMENTS OF INERTIA = ";TU;" ";L$;"^4"
500 PRINT
510 PRINT"TOTAL OF INDIVIDUAL Y-AXIS MOMENTS OF INERTIA = ";TV;" ";L$;"^4"
520 PRINT
530 PRINT"TOTAL OF INDIVIDUAL X-AXIS TRANSFER MOMENTS   = ";TX;" ";L$;"^4"
540 PRINT
550 PRINT"TOTAL OF INDIVIDUAL Y-AXIS TRANSFER MOMENTS   = ";TY;" ";L$;"^4"
560 PRINT:PRINT:PRINT TAB(20) "COMPOSITE CENTROIDAL IX = ";IX ;L$;"^4"
570 PRINT: PRINT TAB(20) "COMPOSITE CENTROIDAL IY = ";IY ;L$;"^4"
580 END
Ok


RUN
SHAPES2:  MOMENT OF INTERIA FOR COMPOSITE RECTANGLES, TRIANGLES & SEMI-CIRCLES

WHAT UNITS FOR LENGTH?
? IN

HOW MANY RECTANGLES IN THE COMPOSITE SHAPE?
? 2

HOW MANY TRIANGLES ARE IN THE COMPOSITE SHAPE?
? 2

HOW MANY SEMI-CIRCLES ARE IN THE COMPOSITE SHAPE?
? 2

FOR EACH RECTANGLE, GIVE THE HORIZONTAL & VERTICAL LENGTHS AND
THE X & Y DISTANCES BETWEEN THE CENTROIDAL AXES OF RECTANGLE AND THE
COMPOSITE SHAPE CENTROIDAL AXES.  SEPARATE DATA ITEMS WITH COMMAS.
FOR HOLES (MISSING AREAS) GIVE ONE LENGTH AS NEGATIVE VALUE.
? 4, 6, .4, .19
? -1.5, .75, .4, .19

FOR EACH TRIANGLE, GIVE THE HORIZONTAL BASE & VERTICAL ALTITUDE
AND THE X & Y DISTANCES BETWEEN THE CENTROIDAL AXES OF THE TRIANGLE &
THE COMPOSITE SHAPE CENTROIDAL AXES.  SEPARATE ITEMS WITH COMMAS.
? 3, 6, 3.4, .81
? -1, 3, 3.4, .81
```

FOR EACH SEMI-CIRCLE, GIVE THE RADIUS AND THE X & Y DISTANCES
BETWEEN THE SEMI-CIRCLE CENTROIDAL AXES AND THE COMPOSITE SHAPE
CENTROIDAL AXES.   SEPARATE DATA ITEMS WITH COMMAS.
? 3, 2.875, .19
? -1.25, 2.875, .19

TOTAL OF INDIVIDUAL X-AXIS MOMENTS OF INERTIA = 97.83871 IN^4

TOTAL OF INDIVIDUAL Y-AXIS MOMENTS OF INERTIA = 67.07926 IN^4

TOTAL OF INDIVIDUAL X-AXIS TRANSFER MOMENTS   = 6.168287 IN^4

TOTAL OF INDIVIDUAL Y-AXIS TRANSFER MOMENTS   = 186.9257 IN^4

                   COMPOSITE CENTROIDAL IX = 104.007 IN^4

                   COMPOSITE CENTROIDAL IY = 254.0049 IN^4
Ok

Sample Computer Program 16 — BEAMS1.BAS

This program will determine the support reactions of any simply supported
beam with one concentrated load and one uniformly distributed load over all or
part of the span. (See also Sample Problems 1 and 2, Chap. 11.)

```
LIST
10 CLS: PRINT"BEAMS1:  REACTIONS FOR SIMPLY SUPPORTED BEAM -- 1 CONCENTRATED &";
20 PRINT" 1 UNIFORM LOAD" : PRINT
30 PRINT"GIVE UNITS FOR SPAN (FT OR M), FOR CONCENTRATED LOAD (LB, KIP, N, KN)"
40 PRINT:PRINT"AND FOR UNIFORMLY DISTRIBUTED LOAD (LB/FT,KIP/FT,N/M,KN/M)."
50 PRINT:PRINT"ENTER UNITS IN SEQUENCE SPECIFIED WITH ITEMS SEPARATED BY COMMAS.
60 INPUT L$, F$, W$
70 PRINT:PRINT"GIVE THE SPAN LENGTH (DISTANCE BETWEEN SUPPORTS). OMIT UNITS."
80 INPUT L
90 PRINT:PRINT"FOR CONCENTRATED LOAD, GIVE ITS VALUE & DISTANCE FROM LEFT END"
100 INPUT P, X
110 PRINT:PRINT"FOR UNIFORM LOAD, GIVE ITS VALUE AND THE DISTANCES OF ITS"
120 PRINT:PRINT"LEFT END & OF ITS RIGHT END FROM THE LEFT END OF THE BEAM."
130 INPUT W, XL, XR
140 RL = ((P * (L - X)) + (W * (XR - XL)) * ((.5 * (XR - XL)) + L - XR))/L
150 RR = P + (W * (XR - XL)) - RL
160 CLS:PRINT"FOR A SIMPLY SUPPORTED BEAM WITH SPAN = ";L;" ";L$
170 PRINT:PRINT"CONCENTRATED LOAD = ";P;" ";F$;" AT ";X;" ";L$;" FROM LEFT END"
180 PRINT:PRINT"UNIFORM LOAD = ";W;" ";W$;" & ";(XR-XL);" ";L$;" LONG";
190 PRINT" STARTING ";XL;" ";L$;" FROM LEFT END"
200 PRINT:PRINT TAB(20) "LEFT REACTION  = ";RL;" ";F$
210 PRINT:PRINT TAB(20) "RIGHT REACTION = ";RR;" ";F$
220 END
Ok

RUN
BEAMS1:  REACTIONS FOR SIMPLY SUPPORTED BEAM -- 1 CONCENTRATED & 1 UNIFORM LOAD

GIVE UNITS FOR SPAN (FT OR M), FOR CONCENTRATED LOAD (LB, KIP, N, KN)

AND FOR UNIFORMLY DISTRIBUTED LOAD (LB/FT,KIP/FT,N/M,KN/M).

ENTER UNITS IN SEQUENCE SPECIFIED WITH ITEMS SEPARATED BY COMMAS.
? FT, LB, LB/FT

GIVE THE SPAN LENGTH (DISTANCE BETWEEN SUPPORTS). OMIT UNITS.
? 25

FOR CONCENTRATED LOAD, GIVE ITS VALUE & DISTANCE FROM LEFT END
? 15000, 8

FOR UNIFORM LOAD, GIVE ITS VALUE AND THE DISTANCES OF ITS

LEFT END & OF ITS RIGHT END FROM THE LEFT END OF THE BEAM.
? 450, 3, 18.5
FOR A SIMPLY SUPPORTED BEAM WITH SPAN =  25   FT
```

```
CONCENTRATED LOAD =   15000   LB AT   8   FT FROM LEFT END

UNIFORM LOAD =   450   LB/FT &   15.5   FT LONG STARTING   3   FT FROM LEFT END

                        LEFT REACTION  =   14175.75   LB

                        RIGHT REACTION =   7799.25   LB
Ok
```

Sample Computer Program 17 — BEAMS2.BAS

This program will determine the support reactions of any overhanging beam for incremental positions of a single concentrated load from one end of the beam to the other when the weight of the beam is neglected.

```
LIST
10 CLS: PRINT"BEAMS2:  EFFECT OF LOCATION OF CONCENTRATED LOAD ON OVERHANGING";
20 PRINT" BEAM" : PRINT
30 DEFINT R : DIM N(75), RL(75), RR(75), X(75)
40 PRINT"GIVE UNITS FOR BEAM LENGTH (FT OR M) & FOR LOAD (LB, KIP, N, KN)."
50 INPUT L$, F$
60 PRINT:PRINT"GIVE SPAN LENGTH, LENGTH OF LEFT OVERHANG, TOTAL LENGTH OF BEAM"
70 INPUT L, LO, TL
80 PRINT:PRINT"GIVE VALUE OF CONCENTRATED LOAD (TO BE MOVED ACROSS THE BEAM)"
90 INPUT P
100 CLS :PRINT:PRINT
110 IF L$ = "M" OR L$ = "m" THEN N = .25 ELSE N = 1
120 PRINT TAB(6) "LOCATION OF P" TAB(30) "LEFT REACTION" TAB(50) "RIGHT REACTION
130 PRINT TAB(5) L$;" FROM LEFT" TAB(36) F$ TAB(57) F$
140 IF LO = 0 THEN 210 ELSE 150
150 FOR X = 0 TO LO STEP N
160 GOSUB 340
170 IF X >= LO THEN 210 ELSE 180
180 IF X > (LO - N) THEN 190 ELSE 200
190 X = LO : GOTO 160
200 NEXT X
210 FOR X = LO TO (L + LO) STEP N
220 GOSUB 340
230 IF X >= (L + LO) THEN 270 ELSE 240
240 IF X > (L + LO - N) THEN 250 ELSE 260
250 X = L + LO : GOTO 220
260 NEXT X
270 IF (L + LO) = TL THEN END ELSE 280
280 FOR X = (L + LO) TO TL STEP N
290 GOSUB 340
300 IF X >= TL THEN END ELSE 310
310 IF X >= (TL - N) THEN 320 ELSE 330
320 X = TL  : GOTO 290
330 NEXT X
340 RL(X) = P * (L + LO - X)/L
350 RR(X) = P - RL(X)
360 PRINT TAB(6) USING"####.##";X;
370 PRINT TAB(35) RL(X) TAB(56) RR(X)
380 RETURN
Ok
```

```
RUN
BEAMS2:  EFFECT OF LOCATION OF CONCENTRATED LOAD ON OVERHANGING BEAM

GIVE UNITS FOR BEAM LENGTH (FT OR M) & FOR LOAD (LB, KIP, N, KN).
? FT, KIP

GIVE SPAN LENGTH, LENGTH OF LEFT OVERHANG, TOTAL LENGTH OF BEAM
? 12, 3.4, 17.2

GIVE VALUE OF CONCENTRATED LOAD (TO BE MOVED ACROSS THE BEAM)
? 14400
```

LOCATION OF P FT FROM LEFT	LEFT REACTION KIP	RIGHT REACTION KIP
0.00	18480	-4080
1.00	17280	-2880
2.00	16080	-1680
3.00	14880	-480
3.40	14400	0
3.40	14400	0
4.40	13200	1200
5.40	12000	2400
6.40	10800	3600
7.40	9600	4800
8.40	8400	6000
9.40	7200	7200
10.40	6000	8400
11.40	4800	9600
12.40	3600	10800
13.40	2400	12000
14.40	1200	13200
15.40	0	14400
15.40	0	14400
16.40	-1200	15600
17.20	-2160	16560

Ok

Sample Computer Program 18 — BEAMS3.BAS

For any simply supported beam with one concentrated load at any location and one full-length uniformly distributed load, this program will:

(a) Determine the support reactions
(b) Determine the shear forces and bending moments at incremental sections along the beam
(c) Determine the maximum shear force and its location
(d) Determine the maximum bending moment and its location

See also Sample Problem 8, Chap. 11.

```
LIST
10 CLS: PRINT"BEAMS3:  SHEAR FORCES & BENDING MOMENTS FOR SIMPLY SUPPORTED BEAM"
20 PRINT
30 REM: WITH 1 CONCENTRATED LOAD AND ONE FULL LENGTH UNIFORM LOAD
40 DIM X(75), V(75), M(75)
50 PRINT"GIVE THE UNITS FOR SPAN LENGTH (FT OR M)"
60 INPUT L$
70 PRINT:PRINT"GIVE THE UNITS FOR THE CONCENTRATED LOAD (LB, KIP, N, KN)"
80 INPUT F$
90 PRINT:PRINT"GIVE THE UNITS FOR THE UNIFORM LOAD (LB/FT, KIP/FT, N/M, KN/M)"
100 INPUT W$
110 PRINT:PRINT"GIVE THE UNITS FOR BENDING MOMENT (FT-LB, FT-KIP, N-M, KN-M)
120 INPUT M$
130 PRINT:PRINT"GIVE VALUES OF SPAN LENGTH (";L$;") & UNIFORM LOAD (";W$;")."
140 INPUT L, W
150 PRINT:PRINT"GIVE VALUE OF THE CONCENTRATED LOAD (";F$;") & ITS DISTANCE"
160 PRINT:PRINT"FROM THE LEFT END OF THE BEAM (";L$;")."
170 INPUT P, XP
180 RL = (W * L/2) + P * (L - XP)/L : REM CALCULATE LEFT REACTION
190 RR = (W * L) + P - RL : REM CALCULATE RIGHT REACTION
200 CLS:PRINT"THE REACTIONS ARE  RL = ";RL;"";F$;" AND RR = ";RR;"";F$
210 MMAX = 0 : VMAX = 0 : XV = 0 : XM = 0
220 PRINT:PRINT:PRINT TAB(10) "SECTION -- ";L$ TAB(30) "SHEAR FORCE -- ";F$;
230 PRINT TAB(50) "MOMENT -- ";M$
240 IF L$ = "FT" OR L$ = "ft" THEN N = 1
250 IF L$ = "M" OR L$ = "m" THEN N = .25
260 FOR X = 0 TO XP STEP N
270 V(X) = RL - (W * X) : REM SHEAR FORCE EQUATION FOR X = 0 TO XP
280 M(X) = (RL * X) - (W * X * X/2) : REM MOMENT EQUATION FOR X = 0 TO XP
290 IF ABS(V(X)) > VMAX THEN 300 ELSE 310
300 VMAX = V(X) : XV = X
```

```
310 IF ABS(M(X)) > MMAX THEN 320 ELSE 330
320 MMAX = M(X) : XM = X
330 PRINT TAB(15) X TAB(38) V(X) TAB(56) M(X)
340 IF X => XP THEN 360 ELSE 350
350 NEXT X
360 FOR X = XP TO L STEP N
370 V(X) = RL - (W * X) - P : REM SHEAR FORCE EQUATION FOR X = XP TO L
380 M(X) = (RL*X) - (W*X*X/2) - P*(X-XP) : REM MOMENT EQUATION FOR X = XP TO L
390 IF ABS(V(X)) > VMAX THEN 400 ELSE 410
400 VMAX = V(X) : XV = X
410 IF ABS(M(X)) > MMAX THEN 420 ELSE 430
420 MMAX = M(X) : XM = X
430 PRINT TAB(15) X TAB(38) V(X) TAB(56) M(X)
440 IF X > = L THEN 460 ELSE 450
450 NEXT X
460 PRINT:PRINT"THE MAXIMUM SHEAR FORCE = ";VMAX;" ";F$;" AT ";XV;" ";L$
470 PRINT:PRINT"THE MAXIMUM BENDING MOMENT = ";MMAX;" ";M$;" AT ";XM;" ";L$
480 END
Ok


RUN
BEAMS3:  SHEAR FORCES & BENDING MOMENTS FOR SIMPLY SUPPORTED BEAM

GIVE THE UNITS FOR SPAN LENGTH (FT OR M)
? FT

GIVE THE UNITS FOR THE CONCENTRATED LOAD (LB, KIP, N, KN)
? LB

GIVE THE UNITS FOR THE UNIFORM LOAD (LB/FT, KIP/FT, N/M, KN/M)
? LB/FT

GIVE THE UNITS FOR BENDING MOMENT (FT-LB, FT-KIP, N-M, KN-M)
? FT-LB

GIVE VALUES OF SPAN LENGTH (FT) & UNIFORM LOAD (LB/FT).
? 25, 450

GIVE VALUE OF THE CONCENTRATED LOAD (LB) & ITS DISTANCE

FROM THE LEFT END OF THE BEAM (FT).
? 9500, 10
THE REACTIONS ARE  RL =  11325 LB AND RR =  9425 LB
```

SECTION -- FT	SHEAR FORCE -- LB	MOMENT -- FT-LB
0	11325	0
1	10875	11100
2	10425	21750
3	9975	31950
4	9525	41700
5	9075	51000
6	8625	59850
7	8175	68250
8	7725	76200
9	7275	83700
10	6825	90750
10	-2675	90750
11	-3125	87850
12	-3575	84500
13	-4025	80700
14	-4475	76450
15	-4925	71750
16	-5375	66600
17	-5825	61000
18	-6275	54950
19	-6725	48450
20	-7175	41500
21	-7625	34100
22	-8075	26250
23	-8525	17950
24	-8975	9200
25	-9425	0

```
THE MAXIMUM SHEAR FORCE =  11325  LB AT  0  FT

THE MAXIMUM BENDING MOMENT =  90750  FT-LB AT  10  FT
Ok
```

```
RUN
BEAMS3:   SHEAR FORCES & BENDING MOMENTS FOR SIMPLY SUPPORTED BEAM

GIVE THE UNITS FOR SPAN LENGTH (FT OR M)
? M

GIVE THE UNITS FOR THE CONCENTRATED LOAD (LB, KIP, N, KN)
? KN

GIVE THE UNITS FOR THE UNIFORM LOAD (LB/FT, KIP/FT, N/M, KN/M)
? KN/M

GIVE THE UNITS FOR BENDING MOMENT (FT-LB, FT-KIP, N-M, KN-M)
? KN-M

GIVE VALUES OF SPAN LENGTH (M) & UNIFORM LOAD (KN/M).
? 6, 28

GIVE VALUE OF THE CONCENTRATED LOAD (KN) & ITS DISTANCE

FROM THE LEFT END OF THE BEAM (M).
? 250, 2.7
THE REACTIONS ARE   RL =   221.5 KN AND RR =   196.5 KN


          SECTION -- M        SHEAR FORCE -- KN    MOMENT -- KN-M
                0                   221.5              0
               .25                  214.5             54.5
               .5                   207.5            107.25
               .75                  200.5            158.25
              1                     193.5            207.5
              1.25                  186.5            255
              1.5                   179.5            300.75
              1.75                  172.5            344.75
              2                     165.5            387
              2.25                  158.5            427.5
              2.5                   151.5            466.25
              2.7                  -104.1            495.99
              2.95                 -111.1            469.09
              3.2                  -118.1            440.44
              3.45                 -125.1            410.04
              3.7                  -132.1            377.89
              3.95                 -139.1            343.99
              4.2                  -146.1            308.34
              4.45                 -153.1            270.94
              4.7                  -160.1            231.7901
              4.95                 -167.1            190.89
              5.2                  -174.1            148.2401
              5.45                 -181.1            103.84
              5.7                  -188.1             57.69
              5.95                 -195.1              9.790039

THE MAXIMUM SHEAR FORCE =   221.5   KN AT   0   M

THE MAXIMUM BENDING MOMENT =   495.99   KN-M AT   2.7   M
Ok
```

Sample Computer Problem 19 — BEAMS4.BAS

For any T-beam with one or both ends overhanging and any two concentrated loads, this program will:

(a)   Determine the centroid location of the cross section of the T-beam and the distances from the centroid to the extreme fibers
(b)   Determine the centroidal x-axis moment of inertia
(c)   Determine the support reactions
(d)   Determine the maximum bending moment and its location
(e)   Determine the maximum compressive and tensile stresses and their locations

(f) Determine the factors of safety in compression and in tension

(See also Sample Problem 3, Chap. 12.)

```
LIST
10 CLS: PRINT"BEAMS4:  DETERMINE STRESSES IN A T-BEAM"
20 PRINT"           SAMPLE PROBLEM 3, CHAPTER 12" : PRINT
30 PRINT"UNITS WILL BE 'FT' FOR SPAN & BEAM LENGTH, 'LB' FOR LOADS,"
40 PRINT"'IN' FOR SECTION LENGTHS, AND 'PSI' FOR STRESSES."
50 PRINT:PRINT"FOR THE T-SECTION, HOW MANY RECTANGLES WILL BE DESCRIBED?"
60 INPUT NR
70 PRINT:PRINT"GIVE DATA FOR EACH RECTANGLE ON ONE LINE, SEPARATED BY COMMAS."
80 PRINT"ENTER HORIZONTAL & VERTICAL LENGTHS & CENTROID DISTANCE FROM BASE
90 A = 0 : YS = 0 : IX = 0
100 FOR K = 1 TO NR
110 INPUT B(K), H(K), Y(K)
120 A(K) = B(K) * H(K)
130 A = A + A(K) : V(K) = A(K) * Y(K) : YS = YS + V(K)
140 IF K = NR THEN 160 ELSE 150
150 NEXT K
160 YO = YS/A
170 PRINT:PRINT"ENTER THE TOTAL VERTICAL DEPTH OF THE T-SECTION."
180 INPUT D
190 CC = D - YO : CT = YO
200 FOR K = 1 TO NR
210 A(K) = B(K) * H(K)
220 I(K) = (B(K) * (H(K)^3)/12) + (A(K) * (Y(K) - YO)^2)
230 IX = IX + I(K)
240 IF K = NR THEN 260 ELSE 250
250 NEXT K
260 PRINT:PRINT"T-SECTION CENTROID IS LOCATED ";YO;" IN. ABOVE THE BASE."
270 PRINT:PRINT"THE DISTANCES TO THE OUTER FIBERS ARE ";CC;" IN. (TOP) & ";CT;
280 PRINT" IN. (BOTTOM)"
290 PRINT:PRINT"T-SECTION CENTROIDAL X-AXIS MOMENT OF INERTIA = ";IX;" IN^4"
300 PRINT:PRINT"GIVE SPAN (FT), & LENGTHS (FT) OF LEFT OVERHANG & RIGHT OVERHANG
310 INPUT L, LO, RO
320 PRINT:PRINT"FOR EACH OF THE TWO LOADS, GIVE ITS VALUE (LB), & DISTANCE"
330 PRINT:PRINT"FROM THE LEFT END OF THE BEAM (FT)."
340 PRINT:PRINT"GIVE THE LOAD WHICH IS CLOSER TO THE LEFT END FIRST."
350 INPUT P1, X1
360 INPUT P2, X2
370 RL = (P1 * (L + LO - X1)/L) + (P2 * (L + LO - X2)/L)
380 RR = (P1 * (X1 - LO)/L) + (P2 * (X2 - LO)/L)
390 CLS:PRINT"REACTIONS ARE RL = ";RL;" LB AND RR = ";RR;" LB"
400 MMAX = 0
410 IF X1 < LO AND X2 < LO THEN 420 ELSE 450
420 MMAX = - P1 * (LO - X1) - P2 * (LO - X2)
430 XMAX = LO
440 GOTO 840
450 IF X1 < LO AND X2 >= LO THEN 460 ELSE 600
460 ML = - P1 * (LO - X1)
470 IF X2 <= (L + LO) THEN 480 ELSE 540
480 M2 = - P1 * (X2 - X1) + RL * (X2 - LO)
490 IF ABS(ML) > ABS(M2) THEN 500 ELSE 510
500 MMAX = ML : XMAX = LO
510 IF ABS(ML) <= ABS(M2) THEN 520 ELSE 530
520 MMAX = M2 : XMAX = X2
530 GOTO 840
540 MR = - PI * (L + LO - X1) + RL * L
550 IF ABS(ML) > ABS(MR) THEN 560 ELSE 570
560 MMAX = ML : XMAX = LO
570 IF ABS(ML) <= ABS(MR) THEN 580 ELSE 590
580 MMAX = MR : XMAX = L + LO
590 GOTO 840
600 IF X1 >= LO AND X1 <= (L + LO) THEN 610 ELSE 760
610 IF X2 >= LO AND X2 <= (L + LO) THEN 620 ELSE 690
620 M1 = RL * (X1 - LO)
630 M2 = RL * (X2 - LO) - P1 * (X2 - X1)
640 IF ABS(M1) > ABS(M2) THEN 650 ELSE 660
650 MMAX = M1 : XMAX = X1
660 IF ABS(M1) <= ABS(M2) THEN 670 ELSE 680
670 MMAX = M2 : XMAX = X2
680 GOTO 840
690 M1 = RL * (X1 - LO)
```

```
700 MR = RL * L - P1 * (L + LO - X1)
710 IF ABS(M1) > ABS(MR) THEN 720 ELSE 730
720 MMAX = M1 : XMAX = X1
730 IF ABS(M1) <= ABS(MR) THEN 740 ELSE 750
740 MMAX = MR : XMAX = L + LO
750 GOTO 840
760 IF X1 > (L + LO) THEN 770 ELSE 800
770 IF X2 > (L + LO) THEN 780 ELSE 800
780 MMAX = - RL * L : XMAX = L + LO
790 GOTO 840
800 PRINT:PRINT"IT SEEMS THAT YOUR FIRST LOAD IS FURTHER FROM THE LEFT END OF"
810 PRINT:PRINT"THE BEAM THAN YOUR SECOND LOAD.  START THE PROGRAM AGAIN AND"
820 PRINT:PRINT"BE SURE THAT THE FIRST LOAD ENTERED IS CLOSEST TO THE LEFT END."
830 END
840 PRINT:PRINT"MAXIMUM BENDING MOMENT = ";MMAX;" FT-LB AT ";XMAX;" FT"
850 PRINT:PRINT"      FROM THE LEFT END OF THE BEAM."
860 SC = MMAX * 12 * CC/IX : REM MAXIMUM COMPRESSION BENDING STRESS - TOP FIBER
870 ST = MMAX * 12 * CT/IX : REM MAXIMUM TENSION BENDING STRESS - BOTTOM FIBER
880 PRINT:PRINT"GIVE THE ULTIMATE TENSION & ULTIMATE COMPRESSION STRESSES (PSI)"
890 INPUT UT, UC
900 FT = UT/ST : FC = UC/SC
910 PRINT:PRINT"MAXIMUM COMPRESSIVE STRESS = ";SC;" PSI IN TOP FIBER AT"
920 PRINT:PRINT"      ";XMAX;" FT FROM THE LEFT END OF THE BEAM."
930 PRINT:PRINT"MAXIMUM TENSILE STRESS = ";ST;" PSI IN BOTTOM FIBER AT "
940 PRINT:PRINT"      ";XMAX;" FT FROM THE LEFT END OF THE BEAM."
950 PRINT:PRINT"THE FACTOR OF SAFETY IN COMPRESSION = ";FC
960 PRINT:PRINT"THE FACTOR OF SAFETY IN TENSION = ";FT
970 END
Ok
```

```
RUN
BEAMS4:  DETERMINE STRESSES IN A T-BEAM
         SAMPLE PROBLEM 3, CHAPTER 12

UNITS WILL BE 'FT' FOR SPAN & BEAM LENGTH, 'LB' FOR LOADS,
'IN' FOR SECTION LENGTHS, AND 'PSI' FOR STRESSES.

FOR THE T-SECTION, HOW MANY RECTANGLES WILL BE DESCRIBED?
? 2

GIVE DATA FOR EACH RECTANGLE ON ONE LINE, SEPARATED BY COMMAS.

ENTER HORIZONTAL & VERTICAL LENGTHS & CENTROID DISTANCE FROM BASE
? 6, 1, 3.5
? 1, 3, 1.5

ENTER THE TOTAL VERTICAL DEPTH OF THE T-SECTION.
? 4

T-SECTION CENTROID IS LOCATED  2.833333  IN. ABOVE THE BASE.

THE DISTANCES TO THE OUTER FIBERS ARE  1.166667  IN. (TOP) &  2.833333
 IN. (BOTTOM)

T-SECTION CENTROIDAL X-AXIS MOMENT OF INERTIA =  10.75   IN^4

GIVE SPAN (FT), & LENGTHS (FT) OF LEFT OVERHANG & RIGHT OVERHANG
? 2, 0, .4

FOR EACH OF THE TWO LOADS, GIVE ITS VALUE (LB), & DISTANCE

FROM THE LEFT END OF THE BEAM (FT).

GIVE THE LOAD WHICH IS CLOSER TO THE LEFT END FIRST.
? 9000, 1.5
? 4000, 2.4
REACTIONS ARE RL =  1450  LB AND RR =  11550  LB

MAXIMUM BENDING MOMENT =  2175  FT-LB AT  1.5  FT

     FROM THE LEFT END OF THE BEAM.

GIVE THE ULTIMATE TENSION & ULTIMATE COMPRESSION STRESSES (PSI)
? 65000, 65000
```

MAXIMUM COMPRESSIVE STRESS =   2832.558   PSI IN TOP FIBER AT

   1.5   FT FROM THE LEFT END OF THE BEAM.

MAXIMUM TENSILE STRESS =   6879.069   PSI IN BOTTOM FIBER AT

   1.5   FT FROM THE LEFT END OF THE BEAM.

THE FACTOR OF SAFETY IN COMPRESSION =   22.94746

THE FACTOR OF SAFETY IN TENSION =   9.448954
Ok

Sample Computer Program 20 — BEAMS5.BAS

For the design of a cantilever beam using the AISC code, this program will:

(a) Determine the maximum bending moment and its location

(b) Determine the minimum required section modulus

(c) Check a selected beam shape for deflection and web shear

See also Sample Problem 8, Chap. 12.

```
LIST
10 CLS: PRINT"BEAMS5:   CANTILEVER BEAM DESIGN -- SAMPLE PROBLEM 8, CHAPTER 12"
20 PRINT
30 PRINT" AISC CODE WILL APPLY.   PLACE FREE END OF BEAM AT THE LEFT."
40 PRINT:PRINT"GIVE LENGTH OF BEAM (FT) & VALUE OF FULL UNIFORM LOAD (LB/FT)."
50 INPUT L, W
60 PRINT:PRINT"GIVE VALUE OF CONCENTRATED LOAD & ITS DISTANCE (FT) FROM FREE";
70 PRINT" END."
80 INPUT P, XP
90 PRINT:PRINT"GIVE ALLOWABLE BENDING STRESS (PSI) & WEB SHEAR STRESS (PSI)"
100 PRINT:PRINT"AND THE MODULUS OF ELASTICITY (PSI).   SEPARATE WITH COMMAS."
110 INPUT BS, WS, E
120 VMAX = P + (W * L)
130 MMAX = - P * (L - XP) - (W * L) * (L/2)
140 S = ABS(MMAX) * 12/BS
150 CLS:PRINT"MAXIMUM BENDING MOMENT (AT WALL) = ";MMAX;" FT-LB"
160 PRINT:PRINT"MINIMUM REQUIRED SECTION MODULUS = ";S;" IN^3"
170 PRINT:PRINT"SELECT AN APPROPRIATE TRIAL BEAM BASED ON ECONOMY & SECTION ";
180 PRINT"MODULUS"
190 PRINT:PRINT"GIVE BEAM DESIGNATION & VALUES OF SECTION MODULUS & MOMENT OF ";
200 PRINT"INERTIA"
210 INPUT B$, SM, MI
220 YMAX = (L * 12)/360
230 J = (W * L) * ((L * 12)^3)/(8 * E)
240 K = P * ((2 * (L * 12)^3) - (3 *(XP*12)* (L * 12)^2) +(XP*12)^3)/(6 * E)
250 I = (J + K)/ YMAX
260 PRINT:PRINT"MINIMUM MOMENT OF INERTIA FOR PERMISSIBLE DEFLECTION = ";I;
270 PRINT" IN^4"
280 IF I > MI THEN 320 ELSE 290
290 PRINT:PRINT B$;" IS SATISFACTORY FOR DEFLECTION BECAUSE MINIMUM MOMENT OF "
300 PRINT:PRINT"INERTIA ";I;" IN^4 IS LESS THAN THE BEAM'S ";MI;" IN^4"
310 GOTO 450
320 PRINT:PRINT"THE MOMENT OF INERTIA OF THE ";B$;" (";MI;" IN^4) IS NOT "
330 PRINT:PRINT"SUFFICIENT FOR DEFLECTION BECAUSE IT IS LESS THAN ";I;" IN^4"
340 PRINT:PRINT"SELECT ANOTHER BEAM WITH AT LEAST I= ";I;" IN^4 & S= ";S;" IN^3
350 PRINT:PRINT"GIVE BEAM DESIGNATION & VALUES OF SECTION MODULUS & MOMENT ";
360 PRINT"OF INERTIA"
370 B$ = " " : SM = 0 : MI = 0 : D = 0 : T = 0
380 INPUT B$, SM, MI
390 IF I > MI THEN 430 ELSE 400
400 CLS:PRINT B$;" IS SATISFACTORY FOR DEFLECTION BECAUSE MINIMUM MOMENT OF"
410 PRINT:PRINT"INERTIA ";I;" IN^4 IS LESS THAN THE BEAM'S ";MI;" IN^4"
420 GOTO 450
430 PRINT:PRINT"SELECT ANOTHER BEAM WITH AT LEAST I= ";I;" IN^4 & S= ";S;" IN^3"
440 GOTO 350
450 PRINT:PRINT"GIVE THE BEAM DEPTH OF SECTION (IN) & WEB THICKNESS (IN)"
460 INPUT D, T
470 WB = VMAX/(D * T)
```

```
480 IF WB < WS THEN 490 ELSE 540
490 PRINT:PRINT"THE ";B$;" BEAM IS SATISFACTORY IN WEB SHEAR BECAUSE THE WEB"
500 PRINT:PRINT"STRESS ";WB;" PSI IS LESS THAN THE ALLOWABLE STRESS ";WS;" PSI"
510 PRINT:PRINT:PRINT"THE ";B$;" BEAM IS SAFE IN BENDING & WEB SHEAR AND"
520 PRINT:PRINT"      IS SATISFACTORY IN DEFLECTION."
530 END
540 PRINT:PRINT"THE ";B$;" BEAM IS NOT SAFE IN WEB SHEAR BECAUSE THE WEB STRESS"
550 PRINT:PRINT WB;" PSI IS GREATER THAN THE ALLOWABLE STRESS ";WS;" PSI"
560 PRINT:PRINT"SELECT ANOTHER BEAM WITH A THICKER WEB OR LARGER DEPTH OR BOTH,"
570 PRINT:PRINT"AND ALSO WITH LARGE ENOUGH I & S."
580 GOTO 350
Ok


RUN
BEAMS5:  CANTILEVER BEAM DESIGN -- SAMPLE PROBLEM 8, CHAPTER 12

 AISC CODE WILL APPLY.  PLACE FREE END OF BEAM AT THE LEFT.

GIVE LENGTH OF BEAM (FT) & VALUE OF FULL UNIFORM LOAD (LB/FT).
? 12, 670

GIVE VALUE OF CONCENTRATED LOAD & ITS DISTANCE (FT) FROM FREE END.
? 16500, 6

GIVE ALLOWABLE BENDING STRESS (PSI) & WEB SHEAR STRESS (PSI)

AND THE MODULUS OF ELASTICITY (PSI).  SEPARATE WITH COMMAS.
? 24000, 14500, 30000000
MAXIMUM BENDING MOMENT (AT WALL) = -147240  FT-LB

MINIMUM REQUIRED SECTION MODULUS =  73.62  IN^3

SELECT AN APPROPRIATE TRIAL BEAM BASED ON ECONOMY & SECTION MODULUS

GIVE BEAM DESIGNATION & VALUES OF SECTION MODULUS & MOMENT OF INERTIA
? W16X50, 80.8, 657

MINIMUM MOMENT OF INERTIA FOR PERMISSIBLE DEFLECTION =  677.7561  IN^4

THE MOMENT OF INERTIA OF THE W16X50 ( 657  IN^4) IS NOT

SUFFICIENT FOR DEFLECTION BECAUSE IT IS LESS THAN  677.7561  IN^4

SELECT ANOTHER BEAM WITH AT LEAST I=  677.7561  IN^4 & S=  73.62  IN^3

GIVE BEAM DESIGNATION & VALUES OF SECTION MODULUS & MOMENT OF INERTIA
? W18X50, 89.1, 802
W18X50 IS SATISFACTORY FOR DEFLECTION BECAUSE MINIMUM MOMENT OF

INERTIA  677.7561  IN^4 IS LESS THAN THE BEAM'S  802  IN^4

GIVE THE BEAM DEPTH OF SECTION (IN) & WEB THICKNESS (IN)
? 18, .358

THE W18X50 BEAM IS SATISFACTORY IN WEB SHEAR BECAUSE THE WEB

STRESS  3808.194  PSI IS LESS THAN THE ALLOWABLE STRESS  14500  PSI


THE W18X50 BEAM IS SAFE IN BENDING & WEB SHEAR AND

     IS SATISFACTORY IN DEFLECTION.
Ok
```

Sample Computer Program 21 — SHAFTS1.BAS

For any solid circular shaft as in Sample Problem 4, Chap. 13, this program will:

(a)   Determine the torque
(b)   Determine the angle of twist in 20 diameters
(c)   Determine the maximum safe power transmission

```
LIST
10 CLS:PRINT"SHAFTS1:  SHAFT POWER TRANSMISSION -- SAMPLE PROBLEM 4, CHAPTER 13"
20 PRINT
30 PRINT"UNITS WILL BE 'MM' FOR SHAFT DIAMETER, 'N/MM^2' FOR STRESS &"
40 PRINT:PRINT"SHEAR MODULUS OF ELASTICITY, & 'RPM' FOR SHAFT SPEED."
50 PRINT:PRINT"GIVE SHAFT DIAMETER, SHAFT RPM, & FACTOR OF SAFETY ON ULTIMATE";
60 PRINT" SHEAR STRESS."
70 INPUT D, N, FS
80 PRINT:PRINT"GIVE ULTIMATE SHEAR STRESS & SHEAR MODULUS OF ELASTICITY."
90 INPUT US, G
100 AS = US/FS : REM ALLOWABLE SHEAR STRESS
110 T = 3.141595 * D^3 * AS/(16 * 10^3) : REM TORQUE IN UNITS OF N-M
120 W = T * N/(9.5493 * 10^3) : REM MAXIMUM SAFE POWER TRANSMISSION IN KW
130 AT = AS * 20 * D/(G * D/2) : REM ANGLE OF TWIST (RADIANS) IN 20 DIAMETERS
140 AD = AT * 180/3.141593 : REM ANGLE OF TWIST (DEGREES)
150 CLS:PRINT:PRINT"SHAFT: ";D;" MM DIAM @ ";N;" RPM WITH ";
160 PRINT"FACTOR OF SAFETY (ULT) = ";FS
170 PRINT:PRINT TAB(10) "ALLOWABLE STRESS = ";AS;" N/MM^2, TORQUE = ";T;" N-M"
180 PRINT:PRINT TAB(10) "& ANGLE OF TWIST IN 20 DIAMETERS = ";AD;" DEGREES"
190 PRINT:PRINT TAB(10) "MAXIMUM SAFE POWER TRANSMISSION = ";W;" KW"
200 IF AD > 1 THEN 240 ELSE 210
210 PRINT:PRINT"ANGLE OF TWIST OF ";AD;" DEGREE FALLS WITHIN THE NORMALLY "
220 PRINT:PRINT"ACCEPTED LIMIT OF 1 DEGREE IN 20 DIAMETERS."
230 END
240 PRINT:PRINT"ANGLE OF TWIST OF ";AD;" DEGREES EXCEEDS THE USUAL LIMIT OF"
250 PRINT:PRINT"1 DEGREE IN 20 DIAMETERS.  IF THIS LIMIT WAS APPLIED TO THIS ";
260 PRINT"CASE,"
270 AT = 1 * 3.141593/180
280 AS = AT * (G * D/2)/(20 * D)
290 T = 3.141593 * D^3 * AS/(16 * 10^3)
300 W = T * N/(9.5493 * 10^3)
310 PRINT:PRINT"THE ALLOWABLE STRESS WOULD BE REDUCED TO ";AS;" N/MM^2"
320 PRINT:PRINT"AND THE SHAFT COULD SAFELY TRANSMIT ";W;" KW @ TORQUE = ";T;
330 PRINT" N-M"
340 END
Ok
```

```
RUN
SHAFTS1:  SHAFT POWER TRANSMISSION -- SAMPLE PROBLEM 4, CHAPTER 13

UNITS WILL BE 'MM' FOR SHAFT DIAMETER, 'N/MM^2' FOR STRESS &

SHEAR MODULUS OF ELASTICITY, & 'RPM' FOR SHAFT SPEED.

GIVE SHAFT DIAMETER, SHAFT RPM, & FACTOR OF SAFETY ON ULTIMATE SHEAR STRESS.
? 50, 1800, 10

GIVE ULTIMATE SHEAR STRESS & SHEAR MODULUS OF ELASTICITY.
? 690, 80000

SHAFT:  50  MM DIAM @  1800  RPM WITH FACTOR OF SAFETY (ULT) =  10

         ALLOWABLE STRESS =  69  N/MM^2, TORQUE =  1693.516  N-M

         & ANGLE OF TWIST IN 20 DIAMETERS =  1.976704  DEGREES

         MAXIMUM SAFE POWER TRANSMISSION =  319.2201  KW

ANGLE OF TWIST OF  1.976704  DEGREES EXCEEDS THE USUAL LIMIT OF

1 DEGREE IN 20 DIAMETERS.  IF THIS LIMIT WAS APPLIED TO THIS CASE,

THE ALLOWABLE STRESS WOULD BE REDUCED TO  34.90659  N/MM^2

AND THE SHAFT COULD SAFELY TRANSMIT  161.491  KW @ TORQUE =  856.7367  N-M
Ok
```

Sample Computer Program 22 — SHAFTS2.BAS

For a flange coupling connecting two solid circular shafts of any diameters as shown in Sample Problems 6 and 7, Chap. 13, this program will:

(a)   Determine the torque

(b)   Determine the shear force on each bolt

(c)   Determine the required bolt shear area and the required bearing area

(d)   Determine the shear force on the hub and the hub shear area

(e)   Determine the minimum safe diameter of the bolts

(f)   Check the hub for safety in shear

(g)   Determine the force on the key

(h)   Determine the minimum key shear area and the minimum key bearing area

(i)   Determine the maximum safe length of key

```
LIST
10 CLS : PRINT"SHAFTS2:  SHAFT FLANGE COUPLING POWER TRANSMISSION"
20 PRINT"              SAMPLE PROBLEMS 6 & 7, CHAPTER 13" : PRINT
30 PRINT"UNITS WILL BE 'MM' FOR DIAMETERS & THICKNESSES, 'RPM' FOR SHAFT SPEED,"
40 PRINT:PRINT"'KW' FOR POWER, & 'N/MM^2' FOR STRESSES"
50 PRINT:PRINT"ALL MATERIALS ARE AISI 1020 STEEL"
60 PRINT:PRINT"GIVE THE DIAMETER OF SHAFTS, HUB DIAMETER & FLANGE WEB THICKNESS
70 INPUT SD, HD, WT
80 PRINT:PRINT"GIVE THE BOLT CIRCLE DIAMETER (MM) & THE NUMBER OF BOLTS"
90 INPUT BC, NB
100 PRINT:PRINT"GIVE ULTIMATE SHEAR STRESS, ULTIMATE BEARING STRESS & FACTOR OF"
110 PRINT" SAFETY"
120 INPUT US, UC, FS
130 PRINT:PRINT"GIVE THE WIDTH & THICKNESS OF THE FLAT KEY (MM)"
140 INPUT B, KT
150 PRINT:PRINT"GIVE THE POWER TO BE TRANSMITTED (KW) & THE SHAFT SPEED (RPM)"
160 INPUT W, RPM
170 T = 9.5493 * W * (10^3)/RPM : REM TORQUE IN UNITS OF N-M
180 FB = T*(10^3)/(NB*(BC/2)) : REM SHEAR FORCE (N) ON EACH BOLT (SINGLE SHEAR)
190 SS = US/FS : SC = UC/FS : REM ALLOWABLE STRESSES (N/MM^2)
200 AS = FB/SS : AC = FB/SC : REM REQUIRED BOLT AREAS (MM^2)
210 DS = SQR(4 * AS/3.141593) : DC = AC/WT : REM MINIMUM DIAMETERS (MM)
220 IF DS >= DC THEN DM = DS ELSE DM = DC
230 FH = T * (10^3)/(HD/2) : REM SHEAR FORCE (N) ON HUB DIAMETER
240 AH = 3.141593 * HD * WT : REM HUB AREA RESISTING SHEAR
250 HS = FH/AH : REM SHEAR STRESS ON HUB
260 CLS:PRINT:PRINT"TORQUE = ";T;" N-M, SHEAR FORCE ON EACH BOLT = ";FB;" N"
270 PRINT"REQD BOLT SHEAR AREA = ";AS;" MM^2, REQD BEARING AREA = ";AC;" MM^2"
280 PRINT"SHEAR FORCE ON HUB = ";FH;" N, HUB AREA = ";AH;" MM^2"
290 PRINT:PRINT"MINIMUM DIAMETER FOR BOLTS IN SHEAR = ";DS;" MM"
300 PRINT"MINIMUM DIAMETER FOR BOLTS IN BEARING = ";DC;" MM"
310 PRINT:PRINT"MINIMUM SAFE DIAMETER OF BOLTS = ";DM;" MM"
320 IF HS > SS THEN 330 ELSE 340
330 PRINT:PRINT"THE HUB IS NOT SAFE IN SHEAR; STRESS ";HS;" > ALLOWABLE ";SS
340 PRINT:PRINT"THE HUB IS SAFE IN SHEAR; STRESS ";HS;" < ALLOWABLE ";SS
350 FK = T * (10^3)/(SD/2) : REM FORCE (N) ON KEY
360 KS = FK/SS : KC = FK/SC : REM REQUIRED SHEAR & BEARING AREAS OF KEY (MM^2)
370 LS = KS/B : LC = KC/(KT/2) :REM REQUIRED KEY LENGTHS FOR SHEAR & BEARING
380 IF LS >= LC THEN LM = LS ELSE LM = LC
390 PRINT:PRINT"FORCE ON KEY = ";FK;" N, MINIMUM KEY SHEAR AREA = ";KS;" MM^2"
400 PRINT:PRINT"MINIMUM KEY BEARING AREA = ";KC;" MM^2"
410 PRINT:PRINT"THE MINIMUM LENGTH OF KEY IN SHEAR = ";LS;" MM"
420 PRINT"THE MINIMUM LENGTH OF KEY IN BEARING = ";LC;" MM"
430 PRINT:PRINT"THE MINIMUM SAFE LENGTH OF KEY = ";LM;" MM"
440 END
Ok


RUN
SHAFTS2:  SHAFT FLANGE COUPLING POWER TRANSMISSION
           SAMPLE PROBLEMS 6 & 7, CHAPTER 13

UNITS WILL BE 'MM' FOR DIAMETERS & THICKNESSES, 'RPM' FOR SHAFT SPEED,

'KW' FOR POWER, & 'N/MM^2' FOR STRESSES

ALL MATERIALS ARE AISI 1020 STEEL

GIVE THE DIAMETER OF SHAFTS, HUB DIAMETER & FLANGE WEB THICKNESS
? 60, 110, 20
```

```
GIVE THE BOLT CIRCLE DIAMETER (MM) & THE NUMBER OF BOLTS
? 150, 4

GIVE ULTIMATE SHEAR STRESS, ULTIMATE BEARING STRESS & FACTOR OF
  SAFETY
? 345, 450, 4

GIVE THE WIDTH & THICKNESS OF THE FLAT KEY (MM)
? 16, 10

GIVE THE POWER TO BE TRANSMITTED (KW) & THE SHAFT SPEED (RPM)
? 19, 200

TORQUE =  907.1836  N-M, SHEAR FORCE ON EACH BOLT =  3023.945  N
REQD BOLT SHEAR AREA =  35.06024  MM^2, REQD BEARING AREA =  26.87951  MM^2
SHEAR FORCE ON HUB =  16494.25  N, HUB AREA =  6911.505  MM^2

MINIMUM DIAMETER FOR BOLTS IN SHEAR =  6.681323  MM
MINIMUM DIAMETER FOR BOLTS IN BEARING =  1.343976  MM
MINIMUM SAFE DIAMETER OF BOLTS =  6.681323  MM

THE HUB IS SAFE IN SHEAR; STRESS  2.386491  < ALLOWABLE  86.25

FORCE ON KEY =  30239.45  N, MINIMUM KEY SHEAR AREA =  350.6023  MM^2
MINIMUM KEY BEARING AREA =  268.7951  MM^2

THE MINIMUM LENGTH OF KEY IN SHEAR =  21.91264  MM
THE MINIMUM LENGTH OF KEY IN BEARING =  53.75902  MM

THE MINIMUM SAFE LENGTH OF KEY =  53.75902  MM
Ok
```

Sample Computer Program 23 — ECCENTR1.BAS

For an eccentrically loaded bolted joint as shown in Prob. 14-17, this program
will determine the minimum required bolt diameter.

```
LIST
10 CLS : PRINT"ECCENTR1:  ECCENTRIC LOAD ON BOLTED JOINT"
20 PRINT"                PROBLEM 14-17, CHAPTER 14" : PRINT
30 PRINT"UNITS WILL BE 'IN' FOR DIAMETER & LENGTH, 'LB' FOR LOAD, 'PSI' FOR";
40 PRINT" STRESS"
50 PRINT:PRINT"GIVE THE LOAD AND ITS ECCENTRIC DISTANCE FROM THE BOLT CENTROID"
60 INPUT P, E
70 PRINT:PRINT"THE 6 BOLTS ARE EQUALLY SPACED. GIVE THE DISTANCE BETWEEN BOLTS."
80 INPUT L
90 PRINT:PRINT"GIVE THE ALLOWABLE SHEAR STRESS ON THE BOLTS."
100 INPUT SS
110 FD = P/6 : REM VERTICAL FORCE ON EACH BOLT
120 REM HORIZ FORCE FROM MOMENTS ABOUT BOLT GROUP CENTROID & SIMILAR TRIANGLES
130 REM P*E =(2*(F3)*2.5*L) + (2*(1.5/2.5)*(F3)*1.5*L) + (2*(.5/2.5)*(F3)*.5*L)
140 F3 = P*E/((5 + 3*(1.5/2.5) + 1*(.5/2.5))*L) : REM HORIZ FORCE ON OUTER BOLT
150 R3 = SQR(F3^2 + FD^2) : REM RESULTANT FORCE ON EACH OUTER BOLT
160 AS = R3/SS : REM MINIMUM REQUIRED SHEAR AREA OF EACH OUTER BOLT
170 DB = SQR(4 * AS/3.141593) : REM MINIMUM REQUIRED BOLT DIAMETER
180 CLS:PRINT:PRINT"LOAD = ";P;" LB AT AN ECCENTRICITY OF ";E;" IN"
190 PRINT:PRINT"ALLOWABLE STRESS = ";SS;" PSI ON 6 BOLTS SPACED ";L;" IN APART"
200 PRINT:PRINT:PRINT"MINIMUM REQUIRED BOLT DIAMETER = ";DB;" IN"
210 PRINT:PRINT"SELECT THE STANDARD BOLT WHOSE DIAMETER IS AT LEAST ";DB;" IN."
220 END
Ok

RUN
ECCENTR1:  ECCENTRIC LOAD ON BOLTED JOINT
                PROBLEM 14-17, CHAPTER 14

UNITS WILL BE 'IN' FOR DIAMETER & LENGTH, 'LB' FOR LOAD, 'PSI' FOR STRESS

GIVE THE LOAD AND ITS ECCENTRIC DISTANCE FROM THE BOLT CENTROID
? 18000, 12

THE 6 BOLTS ARE EQUALLY SPACED. GIVE THE DISTANCE BETWEEN BOLTS.
? 4
```

```
GIVE THE ALLOWABLE SHEAR STRESS ON THE BOLTS.
? 10000

LOAD =  18000  LB AT AN ECCENTRICITY OF  12  IN

ALLOWABLE STRESS =  10000  PSI ON 6 BOLTS SPACED  4  IN APART

MINIMUM REQUIRED BOLT DIAMETER =  1.026583  IN

SELECT THE STANDARD BOLT WHOSE DIAMETER IS AT LEAST  1.026583  IN.
Ok
```

Sample Computer Program 24 — COMBINE1.BAS

For a solid circular shaft subjected to combined bending and torsion stresses as shown in Sample Problem 7, Chap. 14, this program will:

(a) Determine the minimum shaft diameter by using maximum shear stress for brittle materials

(b) Determine the minimum shaft diameter by using maximum principal (tensile) stress for ductile materials

```
LIST
10 CLS : PRINT"COMBINE1:  COMBINED BENDING & TORSION STRESSES"
20 PRINT"              SAMPLE PROBLEM 7, CHAPTER 14" : PRINT
30 PRINT"UNITS WILL BE 'KN-M' FOR TORQUE & MOMENT, 'N/MM^2' FOR STRESS,"
40 PRINT:PRINT"AND 'MM' FOR DIAMETER"
50 PRINT:PRINT"GIVE THE TRANSMITTED TORQUE AND THE MAXIMUM BENDING MOMENT."
60 INPUT T, M
70 PRINT:PRINT"GIVE THE ALLOWABLE SHEAR STRESS & THE ALLOWABLE TENSILE STRESS."
80 INPUT SS, ST
90 TS = (SQR(T^2 + M^2)) * (10^6)/SS : REM TORSION SECTION MODULUS
100 DS = (16 * TS/3.1415)^(1/3) : REM MIN SHAFT DIAMETER FOR MAX SHEAR STRESS
110 BS = (M/2 + (SQR(T^2 + M^2))/2) * (10^6)/ST : REM BENDING SECTION MODULUS
120 DT = (32 * BS/3.141593)^(1/3): REM MIN SHAFT DIAMETER FOR MAX TENSILE STRESS
130 CLS:PRINT:PRINT"FOR BRITTLE MATERIALS & MAXIMUM SHEAR STRESS,"
140 PRINT:PRINT"MINIMUM SHAFT DIAMETER = ";DS;" MM"
150 PRINT:PRINT:PRINT"FOR DUCTILE MATERIALS & MAXIMUM PRINCIPAL (TENSILE) STRESS
160 PRINT:PRINT"MINIMUM SHAFT DIAMETER = ";DT;" MM"
170 END
Ok

RUN
COMBINE1:  COMBINED BENDING & TORSION STRESSES
            SAMPLE PROBLEM 7, CHAPTER 14

UNITS WILL BE 'KN-M' FOR TORQUE & MOMENT, 'N/MM^2' FOR STRESS,

AND 'MM' FOR DIAMETER

GIVE THE TRANSMITTED TORQUE AND THE MAXIMUM BENDING MOMENT.
? 2.7, 2.0

GIVE THE ALLOWABLE SHEAR STRESS & THE ALLOWABLE TENSILE STRESS.
? 55, 75

FOR BRITTLE MATERIALS & MAXIMUM SHEAR STRESS,

MINIMUM SHAFT DIAMETER =  67.76245  MM


FOR DUCTILE MATERIALS & MAXIMUM PRINCIPAL (TENSILE) STRESS,

MINIMUM SHAFT DIAMETER =  71.39913  MM
Ok
```

Sample Computer Program 25 — COLUMN1.BAS

To design a circular linkage rod as a column as shown in Sample Problem 4, Chap. 15, this program will:

(a) Determine the critical slenderness ratio

(b) Determine the rod diameter from the intermediate machine design column formula

(c) Determine the corresponding radius of gyration

(d) Determine the slenderness ratio for the rod

(e) Compare this result with the acceptable range for intermediate columns

(f) Either accept the result or use the long slender column formula and repeat the design procedure

```
LIST
10 CLS : PRINT"COLUMN1:  DESIGN CIRCULAR LINKAGE ROD AS COLUMN"
20 PRINT"           SAMPLE PROBLEM 4, CHAPTER 15" : PRINT
30 PRINT"GIVE COMPRESSION YIELD STRESS & MODULUS OF ELASTICITY (PSI)"
40 INPUT SY, E
50 PRINT:PRINT"GIVE FACTOR OF SAFETY (YIELD) & UNSUPPORTED ROD LENGTH (IN)"
60 INPUT FS, L
70 PRINT:PRINT"GIVE AXIAL COMPRESSIVE LOAD & END CONDITION 'K' FACTOR"
80 INPUT P, K
90 SR = 3.141593 * SQR(2*E/(SY*(K^2))) : REM MIN SLENDER COL SLENDERNESS RATIO
100 REM SOLVE MACHINE DESIGN INTERMEDIATE COLUMN FORMULA FOR DIAMETER
110 DI = SQR(((P/.785398)*(FS/SY)) + (SY*((K*L/.25)^2)/((3.141593^2)*4*E)))
120 RI = .25 * DI : REM RADIUS OF GYRATION
130 CLS:PRINT:PRINT"CRITICAL SLENDERNESS RATIO = ";SR
140 PRINT:PRINT"DIAMETER FROM INTERMEDIATE COL FORMULA = ";DI;" IN"
150 PRINT:PRINT"CORRESPONDING RADIUS OF GYRATION = ";RI;" IN"
160 PRINT:PRINT"SLENDERNESS RATIO FROM LENGTH/RAD OF GYR = ";L/RI
170 IF L/RI < SR  AND  L/RI >= 40 THEN 180 ELSE 210
180 PRINT:PRINT:PRINT"USE OF INTERMEDIATE COLUMN FORMULA IS VALID."
190 PRINT:PRINT"MINIMUM ROD DIAMETER = ";DI;" IN WITH SLENDERNESS RATIO = ";L/RI
200 END
210 IF L/RI > SR THEN 230 ELSE 280
220 REM SOLVE LONG COLUMN MACHINE DESIGN FORMULA FOR DIAMETER
230 DL = (((P/.785398) * ((K * L/.25)^2) * FS)/(((3.141593)^2) * E))^(1/4)
240 RL = .25 * DL
250 PRINT:PRINT:PRINT"USE OF LONG SLENDER COLUMN FORMULA IS VALID."
260 PRINT:PRINT"MINIMUM ROD DIAMETER = ";DL;" IN WITH SLENDERNESS RATIO = ";L/RL
270 END
280 DS = SQR((P/.785398) * (FS/SY)):REM SOLVE SHORT MEMBER FORMULA FOR DIAMETER
290 RS = .25 * DS
300 PRINT:PRINT:PRINT"USE OF SHORT COMPRESSION MEMBER FORMULA IS VALID."
310 PRINT:PRINT"MINIMUM ROD DIAMETER = ";DS;" IN WITH SLENDERNESS RATIO = ";L/RS
320 END
Ok


RUN
COLUMN1:  DESIGN CIRCULAR LINKAGE ROD AS COLUMN
           SAMPLE PROBLEM 4, CHAPTER 15

GIVE COMPRESSION YIELD STRESS & MODULUS OF ELASTICITY (PSI)
? 60000, 30000000

GIVE FACTOR OF SAFETY (YIELD) & UNSUPPORTED ROD LENGTH (IN)
? 3, 18

GIVE AXIAL COMPRESSIVE LOAD & END CONDITION 'K' FACTOR
? 4500, 1

CRITICAL SLENDERNESS RATIO =  99.34589

DIAMETER FROM INTERMEDIATE COL FORMULA =  .7410151  IN

CORRESPONDING RADIUS OF GYRATION =  .1852538  IN

SLENDERNESS RATIO FROM LENGTH/RAD OF GYR =  97.16402
```

USE OF INTERMEDIATE COLUMN FORMULA IS VALID.

MINIMUM ROD DIAMETER = .7410151 IN WITH SLENDERNESS RATIO = 97.16402
Ok

RUN
COLUMN1: DESIGN CIRCULAR LINKAGE ROD AS COLUMN
              SAMPLE PROBLEM 4, CHAPTER 15

GIVE COMPRESSION YIELD STRESS & MODULUS OF ELASTICITY (PSI)
? 60000, 30000000

GIVE FACTOR OF SAFETY (YIELD) & UNSUPPORTED ROD LENGTH (IN)
? 3, 25

GIVE AXIAL COMPRESSIVE LOAD & END CONDITION 'K' FACTOR
? 4500, 1

CRITICAL SLENDERNESS RATIO = 99.34589

DIAMETER FROM INTERMEDIATE COL FORMULA = .8905531 IN

CORRESPONDING RADIUS OF GYRATION = .2226383 IN

SLENDERNESS RATIO FROM LENGTH/RAD OF GYR = 112.2898

USE OF LONG SLENDER COLUMN FORMULA IS VALID.

MINIMUM ROD DIAMETER = .8728824 IN WITH SLENDERNESS RATIO = 114.563
Ok

Sample Computer Program 26 — BEAMS6.BAS

For a beam with both ends fixed and a central concentrated load as shown in Sample Problem 3, Chap. 16, this program will:

(a)   Determine the maximum bending moment
(b)   Determine the minimum required x-axis centroidal moment of inertia
(c)   Determine the maximum deflection of the beam

```
LIST
10 CLS : PRINT"BEAMS6:  BEAM WITH BOTH ENDS FIXED AND CENTRAL LOAD"
20 PRINT"            SAMPLE PROBLEM 3, CHAPTER 16" : PRINT
30 PRINT"GIVE THE SPAN (FT) & THE CONCENTRATED LOAD (LB)"
40 INPUT L, P
50 PRINT:PRINT"GIVE THE ALLOWABLE BENDING STRESS & MODULUS OF ELASTICITY (PSI)"
60 INPUT SA, E
70 PRINT:PRINT"GIVE THE LIMITING DEPTH OF SYMMETRICAL CROSS-SECTION OF BEAM (IN)
80 INPUT D
90 MMAX = P * L/8 : REM MID-SPAN MOMENT (FT-LB) WITH CENTRAL LOAD
100 I = (MMAX * 12) * (D/2)/SA : REM MINIMUM REQUIRED MOMENT OF INERTIA (IN^4)
110 CLS:PRINT:PRINT"BEAM WITH FIXED ENDS, SPAN OF ";L;" FT, AND MID-SPAN"
120 PRINT:PRINT"CONCENTRATED LOAD OF ";P;" LB WITH LIMITED DEPTH OF ";D;" IN"
130 PRINT:PRINT"ALLOWABLE BENDING STRESS IS ";SA;" PSI & "
140 PRINT"MODULUS OF ELASTICITY = ";E;" PSI."
150 PRINT:PRINT:PRINT"MAXIMUM BENDING MOMENT (MID-SPAN & WALLS) = ";MMAX;" FT-LB
160 PRINT:PRINT"MINIMUM REQUIRED X-AXIS MOMENT OF INERTIA = ";I;"IN^4"
170 YMAX = P * ((L * 12)^3)/(192 * E * I) : REM MAXIMUM DEFLECTION
180 PRINT:PRINT:PRINT"MAXIMUM DEFLECTION OF BEAM (MID-SPAN) = ";YMAX;" IN."
190 END
Ok


RUN
BEAMS6:  BEAM WITH BOTH ENDS FIXED AND CENTRAL LOAD
            SAMPLE PROBLEM 3, CHAPTER 16

GIVE THE SPAN (FT) & THE CONCENTRATED LOAD (LB)
? 20, 50000
```

```
GIVE THE ALLOWABLE BENDING STRESS & MODULUS OF ELASTICITY (PSI)
? 9000, 30000000

GIVE THE LIMITING DEPTH OF SYMMETRICAL CROSS-SECTION OF BEAM (IN)
? 14

BEAM WITH FIXED ENDS, SPAN OF  20  FT, AND MID-SPAN

CONCENTRATED LOAD OF  50000  LB WITH LIMITED DEPTH OF  14   IN

ALLOWABLE BENDING STRESS IS  9000  PSI &

MODULUS OF ELASTICITY =  3E+07  PSI.


MAXIMUM BENDING MOMENT (MID-SPAN & WALLS) =  125000  FT-LB

MINIMUM REQUIRED X-AXIS MOMENT OF INERTIA =  1166.667 IN^4


MAXIMUM DEFLECTION OF BEAM (MID-SPAN) =  .1028572  IN.
Ok
```

A

Review of Statics

A-1 FUNDAMENTAL TERMS

The science of mechanics can be subdivided into *statics* and *dynamics*. Statics deals with bodies at rest; dynamics involves motion. This book is concerned with static systems.

Strength of materials is a study of the load-resisting ability of various machine and structural members.

The reader should have a clear understanding of the following technical terms: *length, area, volume, force, pressure, mass, weight, density, load, moment, torque, work,* and *power*. Typical units and definitions of these terms are given on pages 1 and 2.

A *scalar* quantity is specified by magnitude only, such as $40 or 46°F. A *vector* quantity is specified by both magnitude and direction. Force is a vector quantity.

A-2 FORCE SYSTEMS

Systems of forces acting on members can be classified as either *concurrent* or *nonconcurrent* and as either *coplanar* or *noncoplanar*. This gives four categories of force systems.

1. *Concurrent-coplanar* forces:
   (a) The lines of action of all forces pass through a common point.
   (b) All the forces lie in the same plane.
2. *Nonconcurrent-coplanar* forces:
   (a) The lines of action of all forces *do not* pass through a common point.
   (b) All the forces lie in the same plane.
3. *Concurrent-noncoplanar* forces:
   (a) The lines of action of all forces pass through a common point.
   (b) The forces *do not* all lie in the same plane.
4. *Nonconcurrent-noncoplanar* forces:
   (a) The lines of action of all forces *do not* pass through a common point.
   (b) The forces *do not* all lie in the same plane.

Two other types of force systems which frequently occur are special cases of those mentioned above. They are as follows.

1′.   *Collinear* forces:
    (a)   All forces act along the same line of action.
    (b)   This is a special case of *concurrent-coplanar* force systems.
2′.   *Parallel* forces:
    (a)   The lines of action of all forces are parallel.
    (b)   This is a special case of *nonconcurrent-coplanar* force systems.

A-3   CONCURRENT-COPLANAR FORCE SYSTEMS

A *resultant* is a single force which can replace two or more concurrent forces and produce the same effect as the original forces. The resultant of a collinear force system lies along the same line of action as the collinear forces. Since collinear forces lie along the same line, *their resultant is their algebraic sum.* However, *the resultant of concurrent forces which are not collinear is their vector sum.* Concurrent force vectors can be added by any of the following methods.

    1. Parallelogram method. This method is applied to *two* concurrent forces, as in Fig. A-1. Construct or sketch the parallelogram with the vectors as its sides, as shown in Fig. A-2. The *diagonal* of the parallelogram is the resultant of the two concurrent forces.

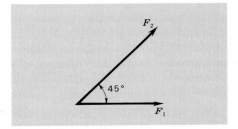

Fig. A-1 Force diagram.

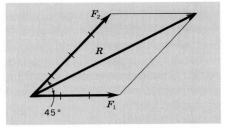

Fig. A-2 Parallelogram method for determining resultant.

    2. Triangle method. This method is applied to *two* concurrent forces, as in Fig. A-1. Draw or sketch one vector and connect the tail of the second vector to the head of the first, as shown in Fig. A-3. The line which forms the closing side of the triangle is the resultant of the two concurrent forces. Take note where the arrow on the resultant is placed.

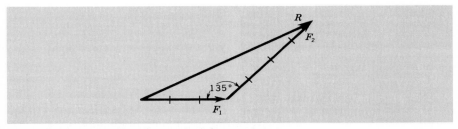

Fig. A-3 Triangle method for determining resultant.

3. Summation of rectangular components. This method is applied to *two or more concurrent forces*. All forces are converted to their vertical and horizontal (rectangular) components, as in Fig. A-4*a*. Both the vertical and horizontal components are algebraically added, resulting in one vertical vector and one horizontal vector. The resultant of these two vectors is the resultant of the original concurrent forces, as in Fig. A-4*b*.

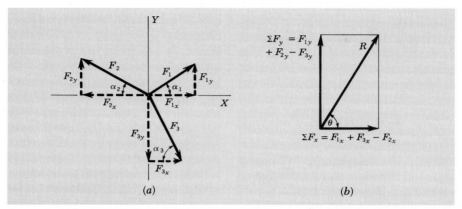

Fig. A-4 Component method for finding resultant: (*a*) vertical and horizontal components; (*b*) resultant of algebraic sum of vertical and horizontal components.

a. Determining vertical and horizontal components. Referring to Fig. A-4*a*, the components of force F_1 are F_{1y} and F_{1x}, which are determined from the equations

$$F_y = F \sin \alpha$$
$$F_x = F \cos \alpha$$

b. Determining the resultant. Referring to Fig. A-4*b*, the resultant is given by the equation

$$R = \sqrt{(\Sigma F_x)^2 + (\Sigma F_y)^2}$$

and the angle that the resultant makes with the *x* axis is given by

$$\tan \theta = \frac{\Sigma F_y}{\Sigma F_x}$$

4. Polygon method. This is applied to *more than two* concurrent forces, as in Fig. A-5*a*. This method is an extension of the triangle method in which all concurrent forces

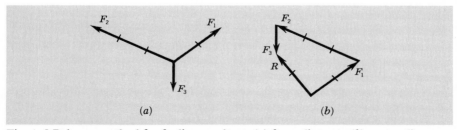

Fig. A-5 Polygon method for finding resultant: (*a*) force diagram; (*b*) vector diagram.

are successively joined head to tail. The line which closes the polygon is the resultant of the concurrent force system; see Fig. A-5b. Note that the arrowhead on the resultant meets the arrow of the last vector to be drawn.

Each of the four methods of determining the resultant of concurrent force systems can be used in either a graphical solution or a trigonometric solution. The graphical solution requires accurate layout of force vectors, with lengths proportional to force magnitudes and directions correctly aligned. The resultant is then measured from the drawing. The trigonometric solution for determining the resultant employs any of several mathematical relationships, the most common of which are given below.

1. Sine, cosine, tangent (for right triangles only):

$$\sin A = \frac{a}{c} \qquad \sin B = \frac{b}{c}$$

$$\cos A = \frac{b}{c} \qquad \cos B = \frac{a}{c}$$

$$\tan A = \frac{a}{b} \qquad \tan B = \frac{b}{a}$$

2. Pythagorean theorem (for right triangles only):

$$c^2 = a^2 + b^2$$

3. Law of sines (for any triangle):

$$\frac{a}{\sin A} = \frac{b}{\sin B} = \frac{c}{\sin C}$$

4. Law of cosines (for any triangle):

$$c^2 = a^2 + b^2 - 2ab \cos C$$

Note: If angle C is more than $90°$, substitute the expression $[-\cos(180° - C)]$ for cos C.

A-4 NONCONCURRENT-COPLANAR FORCE SYSTEMS

The resultant of a nonconcurrent-coplanar force system is a single force which usually does not pass through the axis of rotation of the body. The magnitude and direction of the resultant force can be determined by the rectangular component method using the equations

$$R = \sqrt{(\Sigma F_x)^2 + (\Sigma F_y)^2}$$

$$\tan \theta = \frac{\Sigma F_y}{\Sigma F_x}$$

and the perpendicular distance of the line of action of R from the axis of rotation of the body is given by

$$r = \frac{\Sigma M}{R}$$

where ΣM is the net unbalanced moment of the force system about the axis of rotation of the body.

The resultant of a nonconcurrent-coplanar force system is also equivalent to a single force passing through the axis of rotation and a couple.

For a complete discussion of nonconcurrent-coplanar force systems, refer to Sec. 4-2.

A-5 EQUILIBRANT

The *equilibrant* of a force system is a single force which would exactly balance the resultant of that system. The equilibrant must be equal in magnitude to the resultant, and it must act along the same line of action as the resultant (collinear with the resultant) but in the opposite direction.

A-6 STATIC EQUILIBRIUM

When a *concurrent-coplanar force system* is in equilibrium, the resultant of the system must be zero. In other words, the vector sum must be zero. In terms of the polygon method of adding force vectors, this means that concurrent force systems in equilibrium form closed polygons. That is, when all force vectors are connected head to tail graphically, the arrowhead of the last vector to be drawn will touch the tail (or starting point) of the first vector drawn. In terms of the component method of adding force vectors, since the force system is in equilibrium, there will be no unbalanced net component along any axis, including the x-axis and the y-axis. Thus, the algebraic sum of the rectangular components for each axis must equal zero, or in equation form,

$$\Sigma F_x = 0 \quad \text{and} \quad \Sigma F_y = 0$$

When a *nonconcurrent-coplanar force system* is in equilibrium, the resultant of the system must be zero. This means that there can be no unbalanced force and no unbalanced moment about any point or axis. These conditions for equilibrium can be expressed by the following equations:

$$\Sigma F_x = 0 \quad \Sigma F_y = 0 \quad \text{and} \quad \Sigma M = 0$$

B

Tables

TABLE 1A TYPICAL MECHANICAL PROPERTIES OF METALS*

| Metal | Ultimate Strength | | | | | | | | Tension Yield Stress | | Modulus of Elasticity | | Modulus of Rigidity | | Modulus of Rupture | |
| | Tension s_t | | Compression s_c | | Shear s_s | | | | s_y | | E | | G | | s_r | |
	10^3 psi	MPa†	10^3 psi	MPa	10^3 psi	MPa			10^3 psi	MPa	10^6 psi	10^3 MPa	10^6 psi	10^3 MPa	10^3 psi	MPa
Wrought iron	47	325	47	325	38	260			28	190	28	190	10.0	70		
AISI 1020 steel	65	450	65	450	50	345			45	310	30	200	11.5	80		
AISI 1045 steel	95	655	95	655	70	480			60	410	30	200	11.5	80		
AISI 1095 steel	142	980	142	980	105	725			83	570	30	200	11.5	80		
AISI 4130 steel	160	1105	160	1105	129	825			140	965	30	200	11.5	80		
302 Stainless steel	140	965	140	965	110	760			100	690	28	190	11.0	75		
Cast-iron Class 20	20	140	80	550	32	220					11	75	4.5	31	53	370
Cast-iron Class 40	40	275	125	860	55	380					16	110	5.5	38	78	540
Cast-iron Class 60	60	410	170	1170	65	450					19	130	8.0	55	98	680
Red brass	57	390			37	255			49	340	15	100	5.7	39		
Yellow brass	61	420			40	275			50	345	15	100	5.8	40		
6061-T6 Aluminum alloy	45	310			30	207			40	275	10.4	70	3.8	26		
Monel	80	550			56	385			40	275	26	180	9.5	65		
AZ 31B Magnesium alloy	40	275			19	130			30	200	6.5	45	2.4	16		

* This table is included to provide data for the problems in this book. These values will vary with heat treatment and degree of mechanical working. Allowable stresses for metals specified by the AISC Code for Connections and Welds are given in Chap. 9; the AISC allowable stresses for beams are given in Chapter 12.

† 1 MPa = 1 N/mm²

TABLE 1B   TYPICAL MECHANICAL PROPERTIES OF SELECTED PLASTICS

Polymer	Ultimate Strength Tension S_t 10³ psi	Tension MPa	Compression S_c 10³ psi	Compression MPa	Tension Yield Stress S_y 10³ psi	Tension Yield Stress MPa	Modulus of Elasticity E 10³ psi	Modulus of Elasticity 10³ MPa
Thermoset								
Phenolic	7	48	28	193			1200	8.3
Polyester (reinforced)	17	117	30	206			1700	11.7
Thermoplastic								
Nylon 6/6	12	83	15	103	8.5	58	420	2.9
Polyethylene (high-density)	5	34	3.2	22	3.4	23	100	0.7
Styrene-acrylonitrile (SAN)	11	76	14.5	100	11	76	520	3.6

TABLE 2   FACTORS OF SAFETY (N)

Loading Condition	Ductile Metals Steel, etc. N_u	Ductile Metals Steel, etc. N_y	Brittle Metals, Cast Iron, etc. N_u
Steady load	4	2	7
Varying load	6	3	10
Shock load	10	5	20

TABLE 3A   UNIFIED AND AMERICAN SCREW THREADS (ABSTRACTED FROM ANSI B1.1-1982)

Size	Basic Major Diameter, in	Coarse Threads, UNC		Fine Threads, UNF	
		Threads per in	Root (Minor) Diameter, in (External Thread)	Threads per in	Root (Minor) Diameter, in (External Thread)
0	0.0600	. . . .	. . . . . .	80	0.0465
1	0.0730	64	0.0561	72	0.0580
2	0.0860	56	0.0667	64	0.0691
3	0.0990	48	0.0764	56	0.0797
4	0.1120	40	0.0849	48	0.0894
5	0.1250	40	0.0979	44	0.1004
6	0.1380	32	0.1042	40	0.1109
8	0.1640	32	0.1302	36	0.1339
10	0.1900	24	0.1449	32	0.1561
12	0.2160	24	0.1709	28	0.1773
$\frac{1}{4}$	0.2500	20*	0.1959	28†	0.2113
$\frac{5}{16}$	0.3125	18	0.2524	24	0.2674
$\frac{3}{8}$	0.3750	16	0.3073	24	0.3299
$\frac{7}{16}$	0.4375	14	0.3602	20	0.3834
$\frac{1}{2}$	0.5000	13	0.4167	20	0.4459
$\frac{9}{16}$	0.5625	12	0.4723	18	0.5024
$\frac{5}{8}$	0.6250	11	0.5266	18	0.5649
$\frac{3}{4}$	0.7500	10	0.6417	16	0.6823
$\frac{7}{8}$	0.8750	9	0.7547	14	0.7977
1	1.0000	8	0.8647	12	0.9098
$1\frac{1}{8}$	1.1250	7	0.9704	12	1.0348
$1\frac{1}{4}$	1.2500	7	1.0954	12	1.1598
$1\frac{3}{8}$	1.3750	6	1.1946	12	1.2848
$1\frac{1}{2}$	1.5000	6	1.3196	12	1.4098
$1\frac{3}{4}$	1.7500	5	1.5335		
2	2.0000	$4\frac{1}{2}$	1.7594		
$2\frac{1}{4}$	2.2500	$4\frac{1}{2}$	2.0094		
$2\frac{1}{2}$	2.5000	4	2.2294		
$2\frac{3}{4}$	2.7500	4	2.4794		
3	3.0000	4	2.7294		
$3\frac{1}{4}$	3.2500	4	2.9794		
$3\frac{1}{2}$	3.5000	4	3.2294		
$3\frac{3}{4}$	3.7500	4	3.4794		
4	4.0000	4	3.7294		

* This thread is designated $\frac{1}{4}$-20 UNC.
† This thread is designated $\frac{1}{4}$-28 UNF.

TABLE 3B   METRIC SCREW THREADS (ABSTRACTED FROM ISO STANDARD 724-1978)

Nominal (Major) Diameter, mm	Coarse Threads		Fine Threads	
	Pitch, mm	Root (Minor) Diameter, mm	Pitch, mm	Root (Minor) Diameter, mm
1	0.25	0.729	0.2	0.783
1.2	0.25	0.929	0.2	0.983
1.6	0.35	1.221	0.2	1.383
2	0.4	1.567	0.25	1.729
2.5	0.45	2.013	0.35	2.121
3	0.5	2.459	0.35	2.621
4	0.7	3.242	0.5	3.459
5	0.8	4.134	0.5	4.459
6	1.	4.917	0.75	5.188
8	1.25	6.647	0.75	7.188
10	1.5	8.376	1.	8.917
12	1.75	10.106	1.	10.917
16	2.	13.835	1.	14.917
20	2.5*	17.294	1.5#	18.376
24	3.	20.752	1.5	22.376
30	3.5	26.211	1.5	28.376
36	4.	31.670	2.	33.835
42	4.5	37.129	2.	39.835
48	5.	42.587	2.	45.835
56	5.5	50.046	3.	52.752
64	6.	57.505	3.	60.752
72	—	—	4.	67.670
80	—	—	4.	75.670
90	—	—	4.	85.670
100	—	—	4.	95.670
110	—	—	4.	105.670

* This thread is designated M20. (Note that, for coarse threads, the pitch is not part of the designation.)
# This thread is designated M20 × 1.5.

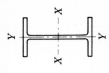

TABLE 4 W SHAPES: PROPERTIES FOR DESIGNING (SELECTED LISTINGS)

Designation	Area A, in²	Depth d, in	Flange Width b_f, in	Flange Thickness t_f, in	Web Thickness t_w, in	Elastic Properties Axis XX I, in⁴	S, in³	r, in	Axis YY I, in⁴	S, in³	r, in
W 36 × 300	88.3	36.74	16.655	1.680	0.945	20 300	1110	15.2	1 300	156	3.83
× 260	76.5	36.26	16.550	1.440	0.840	17 300	953	15.0	1 090	132	3.78
× 230	67.6	35.90	16.470	1.260	0.760	15 000	837	14.9	940	114	3.73
W 36 × 194	57.0	36.49	12.115	1.260	0.765	12 100	664	14.6	375	61.9	2.56
× 170	50.0	36.17	12.030	1.100	0.680	10 500	580	14.5	320	53.2	2.53
× 150	44.2	35.85	11.975	0.940	0.625	9 040	504	14.3	270	45.1	2.47
W 33 × 221	65.0	33.93	15.805	1.275	0.775	12 800	757	14.1	840	106	3.59
× 201	59.1	33.68	15.745	1.150	0.715	11 500	684	14.0	749	95.2	3.56
W 33 × 141	41.6	33.30	11.535	0.960	0.605	7 450	448	13.4	246	42.7	2.43
× 130	38.3	33.09	11.510	0.855	0.580	6 710	406	13.2	218	37.9	2.39
W 30 × 124	36.5	30.17	10.515	0.930	0.585	5 360	355	12.1	181	34.4	2.23
× 116	34.2	30.01	10.495	0.850	0.565	4 930	329	12.0	164	31.3	2.19
× 108	31.7	29.83	10.475	0.760	0.545	4 470	299	11.9	146	27.9	2.15
W 27 × 102	30.0	27.09	10.015	0.830	0.515	3 620	267	11.0	139	27.8	2.15
× 94	27.7	26.92	9.990	0.745	0.490	3 270	243	10.9	124	24.8	2.12
W 24 × 104	30.6	24.06	12.750	0.750	0.500	3 100	258	10.1	259	40.7	2.91

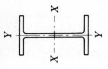

TABLE 4 (Continued)

Designation	Area A, in²	Depth d, in	Flange Width b_f, in	Flange Thickness t_f, in	Web Thickness t_w, in	Axis XX I, in⁴	Axis XX S, in³	Axis XX r, in	Axis YY I, in⁴	Axis YY S, in³	Axis YY r, in
W 24 × 94	27.7	24.31	9.065	0.875	0.515	2 700	222	9.87	109	24.0	1.98
× 84	24.7	24.10	9.020	0.770	0.470	2 370	196	9.79	94.4	20.9	1.95
× 76	22.4	23.92	8.990	0.680	0.440	2 100	176	9.69	82.5	18.4	1.92
W 21 × 73	21.5	21.24	8.295	0.740	0.455	1 600	151	8.64	70.6	17.0	1.81
× 68	20.0	21.13	8.270	0.685	0.430	1 480	140	8.60	64.7	15.7	1.80
× 62	18.3	20.99	8.240	0.615	0.400	1 330	127	8.54	57.5	13.9	1.77
W 18 × 60	17.6	18.24	7.555	0.695	0.415	984	108	7.47	50.1	13.3	1.69
× 55	16.2	18.11	7.530	0.630	0.390	890	98.3	7.41	44.9	11.9	1.67
× 50	14.7	17.99	7.495	0.570	0.355	800	88.9	7.38	40.1	10.7	1.65
W 16 × 57	16.8	16.43	7.120	0.715	0.430	758	92.2	6.72	43.1	12.1	1.60
× 50	14.7	16.26	7.070	0.630	0.380	659	81.0	6.68	37.2	10.5	1.59
× 45	13.3	16.13	7.035	0.565	0.345	586	72.7	6.65	32.8	9.34	1.57
× 40	11.8	16.01	6.995	0.505	0.305	518	64.7	6.63	28.9	8.25	1.57
× 36	10.6	15.86	6.985	0.430	0.295	448	56.5	6.51	24.5	7.00	1.52
W 14 × 426	125	18.67	16.695	3.035	1.875	6 600	707	7.26	2 360	283	4.34
× 342	101	17.54	16.360	2.470	1.540	4 900	559	6.98	1 810	221	4.24
× 257	75.6	16.38	15.995	1.890	1.175	3 400	415	6.71	1 290	161	4.13
W 14 × 233	68.5	16.04	15.890	1.720	1.070	3 010	375	6.63	1 150	145	4.10
× 176	51.8	15.22	15.650	1.310	0.830	2 140	281	6.43	838	107	4.02
× 145	42.7	14.78	15.500	1.090	0.680	1 710	232	6.33	677	87.3	3.98

Elastic Properties

448

W 14×132	38.8	14.66	14.725	1.030	0.645	1 530	209	6.28	548	74.5	3.76
×109	32.0	14.32	14.605	0.860	0.525	1 240	173	6.22	447	61.2	3.73
×90	26.5	14.02	14.520	0.710	0.440	999	143	6.14	362	49.9	3.70
W 14×74	21.8	14.17	10.070	0.785	0.450	796	112	6.04	134	26.6	2.48
W 14×48	14.1	13.79	8.030	0.595	0.340	485	70.3	5.85	51.4	12.8	1.91
W 14×34	10.0	13.98	6.745	0.455	0.285	340	48.6	5.83	23.3	6.91	1.53
×30	8.85	13.84	6.730	0.385	0.270	291	42.0	5.73	19.6	5.82	1.49
W 12×190	55.8	14.38	12.670	1.735	1.060	1 890	263	5.82	589	93.0	3.25
×136	39.9	13.41	12.400	1.250	0.790	1 240	186	5.58	398	64.2	3.16
×106	31.2	12.89	12.220	0.990	0.610	933	145	5.47	301	49.3	3.11
×96	28.2	12.71	12.160	0.900	0.550	833	131	5.44	270	44.4	3.09
×87	25.6	12.53	12.125	0.810	0.515	740	118	5.38	241	39.7	3.07
×79	23.2	12.38	12.080	0.735	0.470	662	107	5.34	216	35.8	3.05
×72	21.1	12.25	12.040	0.670	0.430	597	97.4	5.31	195	32.4	3.04
×65	19.1	12.12	12.000	0.605	0.390	533	87.9	5.28	174	29.1	3.02
W 12×26	7.65	12.22	6.490	0.380	0.230	204	33.4	5.17	17.3	5.34	1.51
W 10×112	32.9	11.36	10.415	1.250	0.755	716	126	4.66	236	45.3	2.68
×88	25.9	10.84	10.265	0.990	0.605	534	98.5	4.54	179	34.8	2.63
W 10×77	22.6	10.60	10.190	0.870	0.530	455	85.9	4.49	154	30.1	2.60
×68	20.0	10.40	10.130	0.770	0.470	394	75.7	4.44	134	26.4	2.59
×54	15.8	10.09	10.030	0.615	0.370	303	60.0	4.37	103	20.6	2.56
W 10×26	7.61	10.33	5.770	0.440	0.260	144	27.9	4.35	14.1	4.89	1.36
×22	6.49	10.17	5.750	0.360	0.240	118	23.2	4.27	11.4	3.97	1.33
W 8×67	19.7	9.00	8.280	0.935	0.570	272	60.4	3.72	88.6	21.4	2.12
×48	14.1	8.50	8.110	0.685	0.400	184	43.3	3.61	60.9	15.0	2.08
×35	10.3	8.12	8.020	0.495	0.310	127	31.2	3.51	42.6	10.6	2.03
×31	9.13	8.00	7.995	0.435	0.285	110	27.5	3.47	37.1	9.27	2.02

Data in Tables 4 through 9 from *Manual of Steel Construction* 8th ed., American Institute of Steel Construction, 1980.

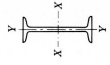

TABLE 5 S SHAPES: PROPERTIES FOR DESIGNING

Designation	Area A, in²	Depth d, in	Flange Width b_f, in	Flange Thickness t_f, in	Web Thickness t_w, in	Axis XX I, in⁴	Axis XX S, in³	Axis XX r, in	Axis YY I, in⁴	Axis YY S, in³	Axis YY r, in
S 24 × 121	35.6	24.50	8.050	1.090	0.800	3 160	258	9.43	83.3	20.7	1.53
× 106	31.2	24.50	7.870	1.090	0.620	2 940	240	9.71	77.1	19.6	1.57
S 24 × 100	29.3	24.00	7.245	0.870	0.745	2 390	199	9.02	47.7	13.2	1.27
× 90	26.5	24.00	7.125	0.870	0.625	2 250	187	9.21	44.9	12.6	1.30
× 80	23.5	24.00	7.000	0.870	0.500	2 100	175	9.47	42.2	12.1	1.34
S 20 × 96	28.2	20.30	7.200	0.920	0.800	1 670	165	7.71	50.2	13.9	1.33
× 86	25.3	20.30	7.060	0.920	0.660	1 580	155	7.89	46.8	13.3	1.36
S 20 × 75	22.0	20.00	6.385	0.795	0.635	1 280	128	7.62	29.8	9.32	1.16
× 66	19.4	20.00	6.255	0.795	0.505	1 190	119	7.83	27.7	8.85	1.19
S 18 × 70	20.6	18.00	6.251	0.691	0.711	926	103	6.71	24.1	7.72	1.08
× 54.7	16.1	18.00	6.001	0.691	0.461	804	89.4	7.07	20.8	6.94	1.14

S 15 × 50	14.7	15.00	5.640	0.622	0.550	486	64.8	5.75	15.7	5.57	1.03
× 42.9	12.6	15.00	5.501	0.622	0.411	447	59.6	5.95	14.4	5.23	1.07
S 12 × 50	14.7	12.00	5.477	0.659	0.687	305	50.8	4.55	15.7	5.74	1.03
× 40.8	12.0	12.00	5.252	0.659	0.462	272	45.4	4.77	13.6	5.16	1.06
S 12 × 35	10.3	12.00	5.078	0.544	0.428	229	38.2	4.72	9.87	3.89	0.980
× 31.8	9.35	12.00	5.000	0.544	0.350	218	36.4	4.83	9.36	3.74	1.00
S 10 × 35	10.3	10.00	4.944	0.491	0.594	147	29.4	3.78	8.36	3.38	0.901
× 25.4	7.46	10.00	4.661	0.491	0.311	124	24.7	4.07	6.79	2.91	0.954
S 8 × 23	6.77	8.00	4.171	0.426	0.441	64.9	16.2	3.10	4.31	2.07	0.798
× 18.4	5.41	8.00	4.001	0.426	0.271	57.6	14.4	3.26	3.73	1.86	0.831
S 7 × 20	5.88	7.00	3.860	0.392	0.450	42.4	12.1	2.69	3.17	1.64	0.734
× 15.3	4.50	7.00	3.662	0.392	0.252	36.7	10.5	2.86	2.64	1.44	0.766
S 6 × 17.25	5.07	6.00	3.565	0.359	0.465	26.3	8.77	2.28	2.31	1.30	0.675
× 12.5	3.67	6.00	3.332	0.359	0.232	22.1	7.37	2.45	1.82	1.09	0.705
S 5 × 14.75	4.34	5.00	3.284	0.326	0.494	15.2	6.09	1.87	1.67	1.01	0.620
× 10	2.94	5.00	3.004	0.326	0.214	12.3	4.92	2.05	1.22	0.809	0.643
S 4 × 9.5	2.79	4.00	2.796	0.293	0.326	6.79	3.39	1.56	0.903	0.646	0.569
× 7.7	2.26	4.00	2.663	0.293	0.193	6.08	3.04	1.64	0.764	0.574	0.581
S 3 × 7.5	2.21	3.00	2.509	0.260	0.349	2.93	1.95	1.15	0.586	0.468	0.516
× 5.7	1.67	3.00	2.330	0.260	0.170	2.52	1.68	1.23	0.455	0.390	0.522

TABLE 6   AMERICAN STANDARD CHANNELS: PROPERTIES FOR DESIGNING

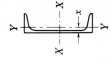

Designation	Area A, in²	Depth d, in	Flange Width b_f, in	Flange Thickness t_f, in	Web Thickness t_w, in	Axis XX I, in⁴	S, in³	r, in	Axis YY I, in⁴	S, in³	r, in	$\bar{x}$, in
C 15 × 50	14.7	15.00	3.716	0.650	0.716	404	53.8	5.24	11.0	3.78	0.867	0.798
× 40	11.8	15.00	3.520	0.650	0.520	349	46.5	5.44	9.23	3.37	0.886	0.777
× 33.9	9.96	15.00	3.400	0.650	0.400	315	42.0	5.62	8.13	3.11	0.904	0.787
C 12 × 30	8.82	12.00	3.170	0.501	0.510	162	27.0	4.29	5.14	2.06	0.763	0.674
× 25	7.35	12.00	3.047	0.501	0.387	144	24.1	4.43	4.47	1.88	0.780	0.674
× 20.7	6.09	12.00	2.942	0.501	0.282	129	21.5	4.61	3.88	1.73	0.799	0.698
C 10 × 30	8.82	10.00	3.033	0.436	0.673	103	20.7	3.42	3.94	1.65	0.669	0.649
× 25	7.35	10.00	2.886	0.436	0.526	91.2	18.2	3.52	3.36	1.48	0.676	0.617
× 20	5.88	10.00	2.739	0.436	0.379	78.9	15.8	3.66	2.81	1.32	0.692	0.606
× 15.3	4.49	10.00	2.600	0.436	0.240	67.4	13.5	3.87	2.28	1.16	0.713	0.634

452

C 9 × 20	5.88	9.00	2.648	0.413	0.448	60.9	13.5	3.22	2.42	1.17	0.642	0.583
× 15	4.41	9.00	2.485	0.413	0.285	51.0	11.3	3.40	1.93	1.01	0.661	0.586
× 13.4	3.94	9.00	2.433	0.413	0.233	47.9	10.6	3.48	1.76	0.962	0.668	0.601
C 8 × 18.75	5.51	8.00	2.527	0.390	0.487	44.0	11.0	2.82	1.98	1.01	0.599	0.565
× 13.75	4.04	8.00	2.343	0.390	0.303	36.1	9.03	2.99	1.53	0.854	0.615	0.553
× 11.5	3.38	8.00	2.260	0.390	0.220	32.6	8.14	3.11	1.32	0.781	0.625	0.571
C 7 × 14.75	4.33	7.00	2.299	0.366	0.419	27.2	7.78	2.51	1.38	0.779	0.564	0.532
× 12.25	3.60	7.00	2.194	0.366	0.314	24.2	6.93	2.60	1.17	0.703	0.571	0.525
× 9.8	2.87	7.00	2.090	0.366	0.210	21.3	6.08	2.72	0.968	0.625	0.581	0.540
C 6 × 13	3.83	6.00	2.157	0.343	0.437	17.4	5.80	2.13	1.05	0.642	0.525	0.514
× 10.5	3.09	6.00	2.034	0.343	0.314	15.2	5.06	2.22	0.866	0.564	0.529	0.499
× 8.2	2.40	6.00	1.920	0.343	0.200	13.1	4.38	2.34	0.693	0.492	0.537	0.511
C 5 × 9	2.64	5.00	1.885	0.320	0.325	8.90	3.56	1.83	0.632	0.450	0.489	0.478
× 6.7	1.97	5.00	1.750	0.320	0.190	7.49	3.00	1.95	0.479	0.378	0.493	0.484
C 4 × 7.25	2.13	4.00	1.721	0.296	0.321	4.59	2.29	1.47	0.433	0.343	0.450	0.459
× 5.4	1.59	4.00	1.584	0.296	0.184	3.85	1.93	1.56	0.319	0.283	0.449	0.457
C 3 × 6	1.76	3.00	1.596	0.273	0.356	2.07	1.38	1.08	0.305	0.268	0.416	0.455
× 5	1.47	3.00	1.498	0.273	0.258	1.85	1.24	1.12	0.247	0.233	0.410	0.438
× 4.1	1.21	3.00	1.410	0.273	0.170	1.66	1.10	1.17	0.197	0.202	0.404	0.436

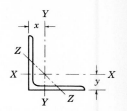

TABLE 7 ANGLES, EQUAL LEGS: PROPERTIES FOR DESIGNING (SELECTED LISTING)

Size and thickness, in	Weight per foot, lb	Area, in²	Axis XX and Axis YY				Axis ZZ
			I, in⁴	S, in³	r, in	x or y, in	r, in
L 8 × 8 × 1	51.0	15.0	89.0	15.8	2.44	2.37	1.56
$\frac{3}{4}$	38.9	11.4	69.7	12.2	2.47	2.28	1.58
$\frac{5}{8}$	32.7	9.61	59.4	10.3	2.49	2.23	1.58
$\frac{1}{2}$	26.4	7.75	48.6	8.36	2.50	2.19	1.59
L 6 × 6 × 1	37.4	11.0	35.5	8.57	1.80	1.86	1.17
$\frac{3}{4}$	28.7	8.44	28.2	6.66	1.83	1.78	1.17
$\frac{1}{2}$	19.6	5.75	19.9	4.61	1.86	1.68	1.18
$\frac{3}{8}$	14.9	4.36	15.4	3.53	1.88	1.64	1.19
L 5 × 5 × $\frac{3}{4}$	23.6	6.94	15.7	4.53	1.51	1.52	0.975
$\frac{1}{2}$	16.2	4.75	11.3	3.16	1.54	1.43	0.983
$\frac{3}{8}$	12.3	3.61	8.74	2.42	1.56	1.39	0.990
L 4 × 4 × $\frac{3}{4}$	18.5	5.44	7.67	2.81	1.19	1.27	0.778
$\frac{1}{2}$	12.8	3.75	5.56	1.97	1.22	1.18	0.782
$\frac{3}{8}$	9.8	2.86	4.36	1.52	1.23	1.14	0.788
$\frac{1}{4}$	6.6	1.94	3.04	1.05	1.25	1.09	0.795
L 3½ × 3½ × $\frac{3}{8}$	8.5	2.48	2.87	1.15	1.07	1.01	0.687
$\frac{1}{4}$	5.8	1.69	2.01	0.794	1.09	0.968	0.694
L 3 × 3 × $\frac{1}{2}$	9.4	2.75	2.22	1.07	0.898	0.932	0.584
$\frac{3}{8}$	7.2	2.11	1.76	0.833	0.913	0.888	0.587
$\frac{1}{4}$	4.9	1.44	1.24	0.577	0.930	0.842	0.592
L 2½ × 2½ × $\frac{3}{8}$	5.9	1.73	0.984	0.566	0.753	0.762	0.487
$\frac{1}{4}$	4.1	1.19	0.703	0.394	0.769	0.717	0.491
L 2 × 2 × $\frac{3}{8}$	4.7	1.36	0.479	0.351	0.594	0.636	0.389
$\frac{1}{4}$	3.19	0.938	0.348	0.247	0.609	0.592	0.391
$\frac{1}{8}$	1.65	0.484	0.190	0.131	0.626	0.546	0.398

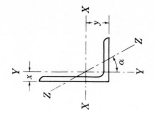

TABLE 8 ANGLES, UNEQUAL LEGS: PROPERTIES FOR DESIGNING (SELECTED LISTING)

Size and thickness, in	Weight per foot, lb	Area, in²	Axis XX				Axis YY				Axis ZZ	
			I, in⁴	S, in³	r, in	y, in	I, in⁴	S, in³	r, in	x, in	r, in	tan α
L 8 × 6 × 1	44.2	13.0	80.8	15.1	2.49	2.65	38.8	8.92	1.73	1.65	1.28	0.543
$\frac{3}{4}$	33.8	9.94	63.4	11.7	2.53	2.56	30.7	6.92	1.76	1.56	1.29	0.551
$\frac{1}{2}$	23.0	6.75	44.3	8.02	2.56	2.47	21.7	4.79	1.79	1.47	1.30	0.558
L 8 × 4 × 1	37.4	11.0	69.6	14.1	2.52	3.05	11.6	3.94	1.03	1.05	0.846	0.247
$\frac{3}{4}$	28.7	8.44	54.9	10.9	2.55	2.95	9.36	3.07	1.05	0.953	0.852	0.258
$\frac{1}{2}$	19.6	5.75	38.5	7.49	2.59	2.86	6.74	2.15	1.08	0.859	0.865	0.267
L 7 × 4 × $\frac{3}{4}$	26.2	7.69	37.8	8.42	2.22	2.51	9.05	3.03	1.09	1.01	0.860	0.324
$\frac{1}{2}$	17.9	5.25	26.7	5.81	2.25	2.42	6.53	2.12	1.11	0.917	0.872	0.335
$\frac{3}{8}$	13.6	3.98	20.6	4.44	2.27	2.37	5.10	1.63	1.13	0.870	0.880	0.340
L 6 × 4 × $\frac{3}{4}$	23.6	6.94	24.5	6.25	1.88	2.08	8.68	2.97	1.12	1.08	0.860	0.428
$\frac{1}{2}$	16.2	4.75	17.4	4.33	1.91	1.99	6.27	2.08	1.15	0.987	0.870	0.440
$\frac{3}{8}$	12.3	3.61	13.5	3.32	1.93	1.94	4.90	1.60	1.17	0.941	0.877	0.446
L 6 × 3½ × $\frac{3}{8}$	11.7	3.42	12.9	3.24	1.94	2.04	3.34	1.23	0.988	0.787	0.767	0.350
$\frac{5}{16}$	9.8	2.87	10.9	2.73	1.95	2.01	2.85	1.04	0.996	0.763	0.772	0.352

TABLE 8 (Continued)

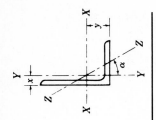

Size and thickness, in	Weight per foot, lb	Area, in²	Axis XX				Axis YY				Axis ZZ	
			I, in⁴	S, in³	r, in	y, in	I, in⁴	S, in³	r, in	x, in	r, in	tan α
L 5 × 3½ × ¾	19.8	5.81	13.9	4.28	1.55	1.75	5.55	2.22	0.977	0.996	0.748	0.464
½	13.6	4.00	9.99	2.99	1.58	1.66	4.05	1.56	1.01	0.906	0.755	0.479
⅜	10.4	3.05	7.78	2.29	1.60	1.61	3.18	1.21	1.02	0.861	0.762	0.486
L 5 × 3 × ½	12.8	3.75	9.45	2.91	1.59	1.75	2.58	1.15	0.829	0.750	0.648	0.357
¼	6.6	1.94	5.11	1.53	1.62	1.66	1.44	0.614	0.861	0.657	0.663	0.371
L 4 × 3½ × ½	11.9	3.50	5.32	1.94	1.23	1.25	3.79	1.52	1.04	1.00	0.722	0.750
¼	6.2	1.81	2.91	1.03	1.27	1.16	2.09	0.808	1.07	0.909	0.734	0.759
L 4 × 3 × ½	11.1	3.25	5.05	1.89	1.25	1.33	2.42	1.12	0.864	0.827	0.639	0.543
¼	5.8	1.69	2.77	1.00	1.28	1.24	1.36	0.599	0.896	0.736	0.651	0.558
L 3½ × 3 × ⅜	7.9	2.30	2.72	1.13	1.09	1.08	1.85	0.851	0.897	0.830	0.625	0.721
¼	5.4	1.56	1.91	0.776	1.11	1.04	1.30	0.589	0.914	0.785	0.631	0.727
L 3½ × 2½ × ⅜	7.2	2.11	2.56	1.09	1.10	1.16	1.09	0.592	0.719	0.660	0.537	0.496
¼	4.9	1.44	1.80	0.755	1.12	1.11	0.777	0.412	0.735	0.614	0.544	0.506
L 3 × 2½ × ⅜	6.6	1.92	1.66	0.810	0.928	0.956	1.04	0.581	0.736	0.706	0.522	0.676
¼	4.5	1.31	1.17	0.561	0.945	0.911	0.743	0.404	0.753	0.661	0.528	0.684
L 3 × 2 × ⅜	5.9	1.73	1.53	0.781	0.940	1.04	0.543	0.371	0.559	0.539	0.430	0.428
¼	4.1	1.19	1.09	0.542	0.957	0.993	0.392	0.260	0.574	0.493	0.435	0.440
L 2½ × 2 × ¼	3.62	1.06	0.654	0.381	0.784	0.787	0.372	0.254	0.592	0.537	0.424	0.626

TABLE 9 SECTION MODULUS TABLE FOR SHAPES USED AS BEAMS (SELECTED LISTING)

Section Modulus	Shape	Section Modulus	Shape	Section Modulus	Shape
1 110	W 36 × 300	143	W 14 × 90	48.6	W 14 × 34
				46.5	C 15 × 40
953	W 36 × 260	140	W 21 × 68	45.4	S 12 × 40.8
		131	W 12 × 96	43.3	W 8 × 48
837	W 36 × 230				
		128	S 20 × 75	42.0	W 14 × 30
757	W 33 × 221			42.0	C 15 × 33.9
684	W 33 × 201	127	W 21 × 62	38.2	S 12 × 35
664	W 36 × 194	126	W 10 × 112	36.4	S 12 × 31.8
580	W 36 × 170				
504	W 36 × 150	119	S 20 × 66	33.4	W 12 × 26
		118	W 12 × 87	31.2	W 8 × 35
448	W 33 × 141			29.4	S 10 × 35
406	W 33 × 130	112	W 14 × 74		
355	W 30 × 124			27.9	W 10 × 26
329	W 30 × 116	108	W 18 × 60	27.5	W 8 × 31
299	W 30 × 108	107	W 12 × 79	27.0	C 12 × 30
		103	S 18 × 70		
267	W 27 × 102			24.7	S 10 × 25.4
263	W 12 × 190	98.5	W 10 × 88	24.1	C 12 × 25
258	S 24 × 121				
		98.3	W 18 × 55	23.2	W 10 × 22
258	W 24 × 104	97.4	W 12 × 72		
		92.2	W 16 × 57	21.5	C 12 × 20.7
243	W 27 × 94			20.7	C 10 × 30
240	S 24 × 106	89.4	S 18 × 54.7	18.2	C 10 × 25
				16.2	S 8 × 23
222	W 24 × 94	88.9	W 18 × 50		
209	W 14 × 132	87.9	W 12 × 65	15.8	C 10 × 20
199	S 24 × 100	85.9	W 10 × 77		
				14.4	S 8 × 18.4
196	W 24 × 84	81.0	W 16 × 50		
187	S 24 × 90	75.7	W 10 × 68	13.5	C 10 × 15.3
186	W 12 × 136			13.5	C 9 × 20
		72.7	W 16 × 45	21.1	S 7 × 20
176	W 24 × 76	64.8	S 15 × 50		
175	S 24 × 80	64.7	W 16 × 40		
173	W 14 × 109	60.0	W 10 × 54	11.3	C 9 × 15
165	S 20 × 96			11.0	C 8 × 18.75
155	S 20 × 86	59.6	S 15 × 42.9		
				10.6	C 9 × 13.4
151	W 21 × 73	56.5	W 16 × 36	10.5	S 7 × 15.3
145	W 12 × 106	53.8	C 15 × 50	9.03	C 8 × 13.75
		50.8	S 12 × 50		

TABLE 10   TYPICAL PROPERTIES OF LUMBER*

| | Allowable Stress | | | | | | Modulus of Elasticity E | |
| | Compression Parallel to Grain | | Bending | | Horizontal Shear | | | |
Species	psi	MPa†	psi	MPa	psi	MPa	10^6 psi	10^3 MPa
Douglas fir	1 050	7.24	1 400	9.65	85	0.59	1.2	8.3
Eastern white pine	725	5.00	975	6.72	65	0.45	1.1	7.6
Mountain hemlock	925	6.38	1 250	8.62	90	0.62	1.1	7.6
Ponderosa pine	800	5.52	1 000	6.90	65	0.45	1.1	7.6
Sitka spruce	875	6.03	1 150	7.93	70	0.48	1.3	9.0

* This table is included to provide data for the problems in this book. These values will vary with moisture content, cross-sectional size, and grade. For further information see *National Design Specification for Stress-Grade Lumber,* National Forest Products Association, 1974.

† 1 MPa = 1 N/mm².

TABLE 11 LUMBER AND TIMBER, AMERICAN STANDARD SIZES*: PROPERTIES FOR DESIGNING, NATIONAL FOREST PRODUCTS ASSOCIATION

Nominal Size, in	American Standard Dressed Size, in	Area of Section, in²	Weight per Foot, lb	Moment of Inertia, in⁴	Section Modulus, in³
2 × 4	1½ × 3½	5.25	1.46	5.36	3.06
6	5½	8.25	2.29	20.8	7.56
8	7¼	10.9	3.03	47.6	13.1
10	9¼	13.9	3.86	98.9	21.4
12	11¼	16.9	4.69	178	31.6
14	13¼	19.9	5.53	291	43.9
16	15¼	22.9	6.36	443	58.1
3 × 4	2½ × 3½	8.75	2.43	8.93	5.10
6	5½	13.8	3.83	34.7	12.6
8	7¼	18.1	5.03	79.4	21.9
10	9¼	23.1	6.42	165	35.7
12	11¼	28.1	7.81	297	52.7
14	13¼	33.1	9.19	485	73.1
16	15¼	38.1	10.6	739	96.9
4 × 4	3½ × 3½	12.2	3.39	12.5	7.15
6	5½	19.2	5.33	48.5	17.6
8	7¼	25.4	7.06	111	30.7
10	9¼	32.4	9.00	231	49.9
12	11¼	39.4	10.9	415	73.8
14	13¼	46.4	12.9	678	102
16	15¼	53.4	14.8	1 034	136

Nominal Size, in	American Standard Dressed Size, in	Area of Section, in²	Weight per Foot, lb	Moment of Inertia, in⁴	Section Modulus, in³
10 × 10	9½ × 9½	90.3	25.0	679	143
12	11½	109	30.3	1 204	209
14	13½	128	35.6	1 948	289
16	15½	147	40.9	2 948	380
18	17½	166	46.1	4 243	485
20	19½	185	51.4	5 870	602
22	21½	204	56.7	7 868	732
24	23½	223	62.0	10 274	874
12 × 12	11½ × 11½	132	36.7	1 458	253
14	13½	155	43.1	2 358	349
16	15½	178	49.5	3 569	460
18	17½	201	55.9	5 136	587
20	19½	224	62.3	7 106	729
22	21½	247	68.7	9 524	886
24	23½	270	75.0	12 437	1 058
14 × 14	13½ × 13½	182	50.6	2 768	410
16	15½	209	58.1	4 189	541
18	17½	236	65.6	6 029	689
20	19½	263	73.1	8 342	856
22	21½	290	80.6	11 181	1 040
24	23½	317	88.1	14 600	1 243

TABLE 11 (Continued)

Nominal Size, in	American Standard Dressed Size, in	Area of Section, in²	Weight per Foot, lb	Moment of Inertia, in⁴	Section Modulus, in³
6 × 6	5½ × 5½	30.3	8.40	76.3	27.7
8	7½	41.3	11.4	193	51.6
10	9½	52.3	14.5	393	82.7
12	11½	63.3	17.5	697	121
14	13½	74.3	20.6	1 128	167
16	15½	85.3	23.6	1 707	220
18	17½	96.3	26.7	2 456	281
20	19½	107.3	29.8	3 398	349
8 × 8	7½ × 7½	56.3	15.6	264	70.3
10	9½	71.3	19.8	536	113
12	11½	86.3	23.9	951	165
14	13½	101.3	28.0	1 538	228
16	15½	116.3	32.0	2 327	300
18	17½	131.3	36.4	3 350	383
20	19½	146.3	40.6	4 634	475
22	21½	161.3	44.8	6 211	578
16 × 16	15½ × 15½	240	66.7	4 810	621
18	17½	271	75.3	6 923	791
20	19½	302	83.9	9 578	982
22	21½	333	92.5	12 837	1 194
24	23½	364	101	16 763	1 427
18 × 18	17½ × 17½	306	85.0	7 816	893
20	19½	341	94.8	10 813	1 109
22	21½	376	105	14 493	1 348
24	23½	411	114	18 926	1 611
26	25½	446	124	24 181	1 897
20 × 20	19½ × 19½	380	106	12 049	1 236
22	21½	419	116	16 150	1 502
24	23½	458	127	21 089	1 795
26	25½	497	138	26 945	2 113
28	27½	536	149	33 795	2 458
24 × 24	23½ × 23½	552	153	25 415	2 163
26	25½	599	166	32 472	2 547
28	27½	646	180	40 727	2 962
30	29½	693	193	50 275	3 408

* All properties and weights are given in dressed sizes only. The weights given are based on an assumed average weight of 40 lb/ft³.
Thickness up to 5 in is classified as lumber. Thickness of 5 in and above is classified as timber.
Lumber indicated is dry (19 percent moisture content or less). Timber indicated is green (more than 19 percent moisture content).

TABLE 12 RECTANGULAR AND POLAR MOMENT OF INERTIA, SECTION MODULUS, AND RADIUS OF GYRATION OF SIMPLE AREAS

	Rectangular	Polar
	$I_x = \dfrac{bh^3}{12}$ $S_x = \dfrac{bh^2}{6}$ $r_x = 0.289h$	$J = \dfrac{bh(b^2 + h^2)}{12}$
	$I_x = \dfrac{bh^3}{36}$ $S_x = \dfrac{bh^2}{24}$ $r_x = 0.236h$	
	$I_x = \dfrac{\pi d^4}{64}$ $S_x = \dfrac{\pi d^3}{32}$ $r_x = \dfrac{d}{4}$	$J = \dfrac{\pi d^4}{32}$ $S' = \dfrac{\pi d^3}{16}$
	$I_x = \dfrac{\pi(d_o^4 - d_i^4)}{64}$ $S_x = \dfrac{\pi(d_o^4 - d_i^4)}{32d_o}$ $r_x = \dfrac{\sqrt{d_o^2 + d_i^2}}{4}$	$J = \dfrac{\pi(d_o^4 - d_i^4)}{32}$ $S' = \dfrac{\pi(d_o^4 - d_i^4)}{16d_o}$
	$I_x = 0.11r^4$ $S_x = 0.191r^3$ $r_x = 0.264r$ $I_y = 0.393r^4$ $S_y = 0.393r^3$ $r_y = \dfrac{r}{2}$	$J = 0.503r^4$

TABLE 13   LIST OF SELECTED SI METRIC UNITS

Quantity	Unit Name	Unit Symbol
Length	meter	m
Mass	kilogram	kg
	metric ton	t (= 1000 kg)
Time	second	s
	hour	h
Temperature	kelvin*	K(≃ 273 + °C)
Plane angle	radian	rad (= 180/π ≃ 57.3°)
Force	newton	N (= 1 kg·m/s²)
Pressure, Stress	pascal, megapascal	Pa (= 1 N/m²), MPa (= 1 N/mm²)
Work, Energy	joule	J (= 1 N·m)
Power	watt	W (= 1 J/s)

* A difference of 1 K is equal to a difference of 1 °C.

TABLE 14   CONVERSION FACTORS: U.S. CUSTOMARY UNITS TO SI UNITS

Quantity	Multiply	By	To Get
Length	inch (in)	2.54×10^{-2}	meter (m)
	feet (ft)	3.048×10^{-1}	meter (m)
	mile (mi)	1.609×10^{3}	meter (m)
	yard (yd)	9.144×10^{-1}	meter (m)
Area	inch^2 (in^2)	6.452×10^{-4}	meter^2 (m^2)
	foot^2 (ft^2)	9.290×10^{-2}	meter^2 (m^2)
	yard^2 (yd^2)	8.361×10^{-1}	meter^2 (m^2)
Volume	inch^3 (in^3)	1.639×10^{-5}	meter^3 (m^3)
	foot^3 (ft^3)	2.832×10^{-2}	meter^3 (m^3)
	U.S. gallon (gal)	3.785×10^{-3}	meter^3 (m^3)
	U.S. gallon (gal)	3.785	liter (L or l)
Angle	degree (deg. or °)	1.745×10^{-2}	radian (rad)
Velocity	foot/minute (ft/min)	5.080×10^{-3}	meter/second (m/s)
	foot/second (ft/s)	3.048×10^{-1}	meter/second (m/s)
	mile/hour (mi/hr)	4.470×10^{-1}	meter/second (m/s)
	mile/hour (mi/hr)	1.609	kilometer/hour (km/h)
Acceleration	foot/second^2 (ft/s^2)	3.048×10^{-1}	meter/second^2 (m/s^2)
Density	pound-mass/foot^3 (lb/ft^3)	1.602×10^{1}	kilogram/meter^3 (kg/m^3)
Force	pound (lb)	4.448	newton (N)
Pressure, Stress	pound/inch^2 (lb/in^2)	6.895×10^{9}	megapascal (MPa)
	pound/inch^2 (lb/in^2)	6.895×10^{9}	newton/millimeter^2 (N/mm^2)
	pound/foot^2 (lb/ft^2)	4.788×10^{7}	megapascal (MPa)
	pound/foot^2 (lb/ft^2)	4.788×10^{7}	newton/millimeter^2 (N/mm^2)
Moment, Torque	foot-pound (ft·lb)	1.356	newton-meter (N·m)
	inch-pound (in·lb)	1.130×10^{-1}	newton-meter (N·m)
Energy, Work	foot-pound (ft·lb)	1.356	joule (J)
Power	foot-pound/minute (ft·lb/min)	2.260×10^{-2}	watt (W)
	horsepower (hp)	7.457×10^{2}	watt (W)
Mass	slug (lb·s^2/ft)	1.459×10^{1}	kilogram (kg)
	pound-mass (lbm)	4.536×10^{-1}	kilogram (kg)

TABLE 15   LIST OF FORMULAS

Pythagorean theorem $c^2 = a^2 + b^2$ $\qquad\qquad$ (2-1)

Law of sines $\dfrac{a}{\sin A} = \dfrac{b}{\sin B} = \dfrac{c}{\sin C}$ $\qquad\qquad$ (2-2)

Law of cosines $c^2 = a^2 + b^2 - 2ab \cos C$ $\qquad\qquad$ (2-3)

Components of a force $F_x = F \cos \alpha;\ F_y = F \sin \alpha$ $\qquad$ (2-4), (2-5)

Resultant of force components $R = \sqrt{(\Sigma F_x)^2 + (\Sigma F_y)^2}$ $\qquad\qquad$ (2-6)

Angle of resultant $\tan \theta = \dfrac{\Sigma F_y}{\Sigma F_x}$ $\qquad\qquad$ (2-7)

Equilibrium of concurrent forces $\Sigma F_x = 0;\ \Sigma F_y = 0$ $\qquad\qquad$ (2-8)

Equilibrium of a force system $\Sigma F = 0;\ \Sigma M = 0$ $\qquad$ (3-1), (3-2)

Resultant of noncoplanar force components
$$R = \sqrt{(\Sigma F_x)^2 + (\Sigma F_y)^2 + (\Sigma F_z)^2}$$ $\qquad\qquad$ (5-2)

Coefficient of friction $f = \dfrac{F_m}{N}$ $\qquad\qquad$ (6-1)

Simple stress $s = \dfrac{F}{A}$ $\qquad$ Strain $\epsilon = \dfrac{\delta}{l}$ $\qquad$ (7-1), (8-1)

Modulus of elasticity $E = \tan \theta = \dfrac{s}{\epsilon};\ \ E = \dfrac{Fl}{A\delta}$ $\qquad$ (8-2), (8-3)

Percent reduction in area $= \dfrac{\text{original area} - \text{final area}}{\text{original area}} (100)$ $\qquad\qquad$ (8-5)

Percent elongation $= \dfrac{\text{change in gage length}}{\text{original gage length}} (100)$ $\qquad\qquad$ (8-6)

Factor of safety $N_u = \dfrac{\text{ultimate stress}}{\text{allowable stress}};\ N_y = \dfrac{\text{yield stress}}{\text{allowable stress}}$ $\qquad$ (8-7), (8-8)

Modulus of rigidity $G = \dfrac{E}{2(1 + \mu)}$ $\qquad\qquad$ (8-9)

Thermal expansion $\delta = \alpha l\, \Delta t$ $\qquad\qquad$ (8-10)

Thermal stress $s = E\alpha\, \Delta t$ $\qquad\qquad$ (8-11)

Load on bolted and riveted joint $F = \left[n\!\left(\dfrac{\pi d^2}{4} \right) \right] s_s$ $\qquad\qquad$ (9-2a)

Load on bolted and riveted joint $F = [(b - nD)t] s_t$ $\qquad\qquad$ (9-3)

Load on bolted and riveted joint $F = [ntd] s_c$ $\qquad\qquad$ (9-4)

Efficiency of bolted joint

$$\eta = \dfrac{\text{strength of the joint}}{\text{tensile strength of the gross area}} \times 100$$ $\qquad\qquad$ (9-5)

Allowable fillet-weld force $F = F'(L)$ $\qquad\qquad$ (9-6)

Longitudinal section of pressure vessel $s_t = \dfrac{pD_c}{2t};\ F = \dfrac{pD_c l}{2}$ $\qquad$ (9-7), (9-8)

Transverse section of pressure vessel $s_t = \dfrac{pD_c}{4t};\ F = \dfrac{pD_c l}{4}$ $\qquad$ (9-9), (9-10)

TABLE 15 (Continued)

Centroid location $\quad \bar{x} = \dfrac{\Sigma Ax}{\Sigma A}; \; \bar{y} = \dfrac{\Sigma Ay}{\Sigma A}$ $\qquad\qquad$ (10-2), (10-1)

Formulas for centroid locations and moments of interia of simple
   areas $\qquad\qquad$ Tables 10-1, 10-2

Transfer formula $\quad I_{a-a} = I_x + Ad^2$ $\qquad\qquad$ (10-7)

Conditions for static equilibrium $\quad \Sigma F_x = 0; \; \Sigma F_y = 0; \; \Sigma M = 0$ $\qquad\qquad$ (11-1)

Stress due to bending $\quad s = \dfrac{Mc}{I}; \; s = \dfrac{M}{S}$ $\qquad\qquad$ (12-1), (12-2)

Modulus of rupture $\quad s_r = \dfrac{M_r c}{I}$ $\qquad\qquad$ (12-3)

Horizontal shear stress $\quad s_h = \dfrac{Va\bar{y}}{Ib}$ $\qquad\qquad$ (12-4)

Vertical web shear stress $\quad s_v = \dfrac{V}{td}$ $\qquad\qquad$ (12-7)

Formulas for beam deflections $\qquad\qquad$ Table 12-2

Allowable stress, laterally unsupported beam $\quad s = \dfrac{12\,000\,000 C_b}{ld/A_f}$ $\qquad\qquad$ (12-20)

Torsional shear stress $\quad s_s = \dfrac{Tc}{J}; \; s_s = \dfrac{T}{S'}$ $\qquad\qquad$ (13-1), (13-2)

Angle of twist $\quad \theta = \dfrac{s_s l}{Gc}; \; \theta = \dfrac{Tl}{JG}$ $\qquad\qquad$ (13-5), (13-6)

Horsepower $\quad \text{hp} = \dfrac{Tn}{63\,000}$ $\qquad\qquad$ (13-7a)

Metric Power $\quad \text{W} = \dfrac{Tn}{9.55}$ $\qquad\qquad$ (13-7b)

Combined axial and bending stress $\quad s = \dfrac{F}{A} \pm \dfrac{Mc}{I}$ $\qquad\qquad$ (14-1)

Combined stress, eccentrically loaded members $\quad s = \dfrac{F}{A} \pm \dfrac{Fec}{I}$ $\qquad\qquad$ (14-2)

Normal stress due to tension or compression

$$s_n = \dfrac{F}{2A} + \dfrac{F}{2A} \cos 2\alpha \qquad\qquad (14\text{-}3)$$

Shearing stress due to tension or compression

$$s_s = \dfrac{F}{2A} \sin 2\alpha \qquad\qquad (14\text{-}4)$$

Maximum resultant tensile stress $\quad (s_t)_{\max} = \dfrac{s_t}{2} + \sqrt{s_s^2 + \left(\dfrac{s_t}{2}\right)^2}$ $\qquad\qquad$ (14-5)

Maximum resultant shear stress $\quad (s_s)_{\max} = \sqrt{s_s^2 + \left(\dfrac{s_t}{2}\right)^2}$ $\qquad\qquad$ (14-7)

TABLE 15 (Continued)

Equivalent bending moment $\quad M_e = (s_t)_{max}S = \dfrac{M}{2} + \dfrac{1}{2}\sqrt{T^2 + M^2}$ $\qquad$ (14-8)

Equivalent torque $\quad T_e = (s_s)_{max}S' = \sqrt{T^2 + M^2}$ $\qquad$ (14-9)

Radius of gyration $\quad r = \sqrt{\dfrac{I}{A}}$ $\quad$ Euler's equation $\quad \dfrac{F}{A} = \dfrac{\pi^2 E}{(l/r)^2}$ $\qquad$ (15-2), (15-4)

Design equations for metal columns $\qquad\qquad\qquad\qquad\qquad\qquad\qquad\qquad$ Table 15-2

Design equation for wood columns (NFPA)

$\dfrac{F}{A} = \dfrac{0.30E}{\left(\dfrac{l}{b}\right)^2} \qquad \dfrac{F}{A} \leq s_{\text{allowable}} \qquad \dfrac{l}{b} \leq 50$ $\qquad$ (15-8)

Answers to Selected Problems

CHAPTER 2
2-1. $R = 23.3$ N; $\theta = 31°$ or 0.54 rad.　**2-2.** $R = 100$ lb; $\theta = 60°$ with either force.　**2-3.** $R = 1.73P$ lb; $\theta = 30°$ with either force.　**2-4.** $R = 87$ lb; $\theta = 16.7°$ or $16°42'$ with the 40-lb force, or $13.3°$ $(13°18')$ with the 50-lb force.　**2-5.** $F = 42.7$ lb; $\theta = 39°$ with the 60-lb force.　**2-7.** $m = 8.52$ kg.　**2-9.** $\mathbf{AB} = 60$ kN (tension); $\mathbf{BC} = 98$ kN (compression).

2-11.

$\theta°$	F_x (lb)	F_y (lb)
(a)　10	197	35
(b)　35	164	115
(c)　55	115	164
(d)　80	35	197
(e)　90	0	200

2-13. Any load greater than $W = 97$ lb.　**2-15.** $F = 1050$ lb.　**2-17.** Cable tensions $= 1.37$ tons at $30°$ and 1.065 tons at $40°$.　**2-19.** $\mathbf{BC} = 288$ lb (tension); $\mathbf{AB} = 577$ lb (compression).　**2-21.** $F = 200 + 0.0995W$ N.　**2-23.** Force in each arm $= 100$ lb (tension); $R_s = R_n = 173$ lb.　**2-25.** $F = 2.67$ kN.

CHAPTER 3
3-1. $R_l = 350$ lb; $R_r = 250$ lb.　**3-2.** $R = 775$ N; $F = 475$ N.　**3-3.** $R_1 = 8500$ lb; $R_2 = 13\,500$ lb.　**3-4.** $R_1 = 2280$ lb; $R_2 = 2160$ lb.　**3-5.** 58.3 in from rear axle.　**3-7.** $R_1 = 5520$ lb; $R_2 = 7080$ lb.　**3-9.** $R_1 = 400$ lb; $R_2 = 200$ lb; right

support 5 ft from right end.   **3-11.** $F = 2795$ lb.   **3-13.** $P = 4.69$ kN.   **3-15.**
$F = 44$ kN; $p = 318$ kPa.   **3-17.** $M = 180$ in·lb; $M = 180$ in·lb.

CHAPTER 4

4-1. $F = 17.33$ lb between upper and lower cylinders; $F = 8.66$ lb between lower cylinders and sides of box; $F = 45$ lb between lower cylinders and bottom of the box.
4-2. $R_a = 1667$ lb at 36.9° with horizontal; $R_b = 1333$ lb horizontal to left.   **4-3.**
BD $= 1.11$ kN (compression); $R_c = 1.93$ kN at 55° with horizontal.   **4-4.** $R_{ax} =$
190 lb; $R_{ay} = 289$ lb; $R_{bx} = 190$ lb; $R_{by} = 11$ lb; $R_{cx} = -190$ lb; $R_{cy} = 211$ lb.   **4-5.**
BC $= 6.5$ kN; $R_{ax} = 4.86$ kN; $R_{ay} = 0.18$ kN.   **4-7.** Upper hinge $R = 420$ lb horizontal to left; lower hinge $R = 638$ lb at 49° with horizontal.   **4-9.** $R_b = 2600$ lb;
$R_c = 2152$ lb; $R_d = 2810$ lb.   **4-11. CG** $= 1794$ lb; **DE** $= 1115$ lb; $R_{ax} = 1078$ lb;
$R_{ay} = 1889$ lb; $R_{bx} = 1794$ lb; $R_{by} = 1600$ lb.   **4-13. AB** $= 34$ k (compression);
AH $= 16$ k (tension); **BH** $= 10$ k (tension); **BC** $= 24$ k (compression); **HG** $= 16$ k (tension); **BG** $= 17$ k (tension); **CG** $= 10$ k (compression); $R_1 = R_2 = 30$ k.   **4-15.** $P =$
5550 lb (tension); $F_{bx} = 3330$ lb; $F_{by} = 8440$ lb; **EF** $= 4710$ lb (tension); **BF** $= 7770$ lb
(compression);   **BE** $= 950$ lb (compression);   **BC** $= 2660$ lb (compression);   **CE** $=$
1330 lb (tension);   **CD** $= 2660$ lb (compression);   **DE** $= 3760$ lb (tension).   **4-17.**
$R = 676$ lb acting at 102.8° from horizontal.   **4-19.** Upper horizontal members (left to right): 26 kip (compression); 46.22 kip (compression); 54.89 kip (compression); 46.22 kip (compression); 26 kip (compression). Lower horizontal members (left to right): 13 kip (tension); 36.11 kip (tension); 50.55 kip (tension); 50.55 kip (tension); 36.11 kip (tension); 13 kip (tension). Slanted members (left to right): 26 kip (compression); 26 kip (tension); 20.22 kip (compression); 20.22 kip (tension); 8.67 kip (compression); 8.67 kip (tension); 8.67 kip (tension); 8.67 kip (compression); 20.22 kip (tension); 20.22 kip (compression); 26 kip (tension); 26 kip (compression).

CHAPTER 5

5-1. AB $= 2.0$ tons (compression);   **BC** $=$ **BD** $= 1.14$ tons (compression).   **5-2.**
DE $= 50.6$ kN (tension);   **AD** $= 14.6$ kN (compression);   **CD** $= 19.9$ kN (compression).   **5-3. BD** $= 7500$ lb (tension).   **5-4. BD** $= 10\,600$ lb (tension).   **5-5.**
$P = 12.5$ tons (tension) in hoisting cable and in each pulley cable; $F = 48$ tons (compression) in each leg of crane; $F = 37.1$ tons (tension) in stay cable.   **5-7. CD** $=$
2415 lb.   **5-8. DF** $= 648$ lb (compression).   **5-9.** (*a*) **DE** $= 15.88$ tons (tension)
(*b*) **DF** $= 21.12$ tons (compression); (*c*) **AD** $= 11.68$ tons (tension); **BD** $= 15$ tons (tension).   **5-11. BC** $= 16\,450$ lb (tension).

CHAPTER 6

6-1. $P = 8.55$ lb.   **6-2.** Body will slide without additional force.   **6-7.** $f =$
0.118.   **6-9.** No. Total $F_m = 123.6$ lb; component of 60-lb body parallel to plane $=$
42.42 lb; tension in cord $= 20.82$ lb; $F = 166$ lb to start 200-lb body to the right.
6-11. $F = 1064$ lb; $P = 1684$ lb.   **6-13.** No. At base of ladder, $F_m = 35.4$ N, but horizontal force $= 66$ N.   **6-15.** Yes. Minimum F required $= 63.2$ lb.

CHAPTER 7

7-1. $s_t = 4000$ psi.   **7-2.** $s_c = 4.6$ MPa.   **7-3.** $F = 3890$ lb.   **7-4.** $s_c =$
4.17 MPa.   **7-5.** $d = 112$ mm.   **7-7.** $h = 0.358$ in (using root area in tension);
$h = 0.50$ in (using shank area in tension).   **7-9.** $F = 666$ kN.   **7-11.** Use 12

posts.   **7-13.** $s_t = 45\,400$ psi (root area).   **7-15.** (a) $d_o = 66$ mm, $d_i = 44$ mm; (b) $d_o = 106$ mm; $d_i = 94$ mm.

CHAPTER 8
8-1. $s_t = 15\,300$ psi; $\delta = 0.0734$ in.   **8-2.** $F = 238$ kN.   **8-3.** $\epsilon = 0.001088$ mm/mm; final length $= 993.08$ mm.   **8-4.** Original length $= 84$ in.   **8-5.** $s_t = 280$ MPa, $\epsilon = 1.40 \times 10^{-3}$ mm/mm, $s_t = 297$ MPa, $\epsilon = 1.48 \times 10^{-3}$ mm/mm, $s_t = 382$ MPa (above yield point), $\epsilon = 1.91 \times 10^{-3}$ mm/mm (not true strain).
8-9. $\delta = 0.00496$ in.   **8-11.** $E = 134$ GPa.   **8-13.** (a) $N_u = 3.15$; (b) $N_u = 6.30$.   **8-15.** (a) $d = 1.09$ in; (b) $d = 1.09$ in.   **8-17.** Use 12 in by 12 in.
8-19. 3.24 mm; 65.5 MPa (compression).   **8-21.** 0.0825 in, 330 psi.   **8-23.** (a) $\delta = 1.94$ mm; $s = 0$; (b) $\delta = 0$; $s = 108$ MPa (compression); (c) $F = 136$ kN.   **8-25.** $F = 158\,000$ lb.   **8-27.** (a) $s_c = 11\,800$ psi; (b) $s_c = 13\,300$ psi; (c) $s_c = 15\,200$ psi.

CHAPTER 9
9-1. $F = 26\,500$ lb (shear); ($F = 46\,200$ lb, tension; $F = 65\,300$ lb, bearing).   **9-2.** $F = 36\,800$ lb (shear); ($F = 49\,200$ lb, tension; $F = 94\,700$ lb, bearing).   **9-3.** Use 2 bolts.   **9-4.** $d = 0.56$ in; use two $\frac{9}{16}$ in bolts.   **9-5.** $F = 106\,000$ lb (shear); ($F = 179\,000$ lb tension).   **9-7.** Member A: The bolting is safe; ($s_s = 8300$ psi, $s_c = 16\,300$ psi, $s_t = 18\,400$ psi in angles at hole). Member B: $F = 67\,500$ lb (double shear); ($F = 198\,000$ lb, compression angles; $F = 299\,000$ lb, bearing).   **9-9.** Member A: Use 2 bolts; ($n = 1.4$, double shear; $n = 0.88$, bearing in gusset plate). Member B: Use 2 bolts; ($n = 1.09$, double shear; $n = 0.69$, bearing in gussett plate; tensile stress in angles at hole $= 9140$ psi).   **9-11.** Joint strength $F = 194\,000$ lb (row 1 tension); efficiency $= 89\%$; ($F = 251\,000$ lb, double shear; $F = 198\,000$ lb, row 2 tension; $F = 261\,000$ lb, bearing).   **9-13.** Joint strength $F = 167\,000$ lb (row 1 tension); efficiency $= 92\%$; ($F = 186\,000$ lb, double shear; $F = 177\,000$ lb row 2 tension; $F = 214\,000$ lb, row 3 tension; $F = 285\,000$ lb, bearing).   **9-15.** (a) Each side weld $= 1.43$ in $+ \frac{7}{8}$ in corner returns; (b) Yes, required $L = 2.86$ in $+ \frac{7}{8}$ in corner returns.
9-17. (a) $F = 24\,000$ lb (weld); ($F = 36\,300$ lb, plate); (b) $F = 36\,300$ lb (using plate allowable stress for butt weld).   **9-19.** (a) Overlap $= 4.3$ in; (b) $F = 48\,100$ lb (plate, AISC); ($F = 141\,000$ lb, weld).   **9-21.** (a) $s_t = 7840$ psi (tension, longitudinal section); (b) No, allowable $s_t = 5720$ psi for steady load.   **9-23.** $p = 1833$ psi (maximum).   **9-25.** $t = 0.273$ in (minimum).   **9-27.** Spacing $l = 5.3$ in (maximum).   **9-29.** $D = 44$ ft   **9-31.** $h = 34.5$ m (maximum).   **9-33.** (a) $t = \frac{3}{8}$ in (minimum); (single bead of weld); (b) $t = 0.155$ in (minimum).   **9-35.** (a) $h = 154$ ft; (b) $h = 387$ ft.

CHAPTER 10
10-1. $\bar{y} = 1.16$ in, $\bar{x} = 2.5$ in (by inspection) from point O.   **10-2.** $\bar{y} = 7.21$ in, $\bar{x} = 2.82$ in from lower right corner.   **10-3.** $\bar{y} = 131$ mm, $\bar{x} = 76.2$ mm from lower left corner.   **10-4.** $\bar{y} = 5.29$ in from base.   **10-5.** $\bar{y} = 7.12$ ft, $\bar{x} = 7.33$ ft from point D.   **10-7.** $\bar{y} = 0.58$ in, $\bar{x} = 4$ in (by inspection) from lower left corner.   **10-9.** $\bar{y} = 66$ mm, $\bar{x} = 207$ mm from lower left corner.   **10-11.** $\bar{y} = 10.58$ in from base on the vertical centerline.   **10-13.** Maximum $b = 3.72$ m, $(a = 3.04$ m).   **10-15.** Volume $= 589$ (10^3) mm^3.   **10-17.** Volume $= 6140$ π in^3 or 19 300 in^3.   **10-19.** (a) $I_x = 47.6$ in^4, $I_y = 24.3$ in^4; (b) $I_x = I_y = 7.36$ in^4; (c) $I_x = 1.03$ in^4, $I_y = 3.68$ in^4; (d) $I_x = I_y = 0.515$ in^4.   **10-21.** $I_x = 8.23$ in^4, $I_y = 7.92$ in^4.   **10-23.** $I_x = 1126$ in^4, $I_y = 1308$ in^4.   **10-25.** $I_x = 494$ in^4, $I_y = 186$ in^4.   **10-27.** $I_x =$

1864 in⁴. **10-29.** $I_x = 6010$ in⁴ (bolt holes deducted); $I_x = 6690$ in⁴, (bolt holes not deducted).

CHAPTER 11

11-1. $R_l = 1200$ lb, $R_r = 800$ lb; $V_l = 1200$ lb $= V_4$, $V_4' = -800$ lb $= V_r$. **11-2.** $R_1 = 12$ kN, $R_2 = 10$ kN; $V_l = 12$ kN $= V_1$, $V_1' = -1$ kN $= V_3$, $V_3' = -10$ kN $= V_r$. **11-3.** $R_l = 8.25$ kN, $R_r = 14.25$ kN; $V_l = 8.25$ kN, $V_{3.5} = 3$ kN, $V_{3.5}' = -12$ kN, $V_r = -14.25$ kN. **11-4.** $R_1 = 2040$ lb, $R_2 = 3760$ lb; $V_l = 2040$ lb, $V_{12} = 240$ lb $= V_{18}$; $V_{18}' = -3760$ lb $= V_r$. **11-5.** $R = 18$ kN; $V_l = -8$ kN $= V_2$, $V_2' = -18$ kN $= V_r$. **11-7.** $R_1 = 1735$ lb, $R_2 = 1640$ lb; $V_l = 1735$ lb, $V_5 = 610$ lb, $V_9 = 110$ lb, $V_{9.4} = 0$, $V_{14} = -1265$ lb, $V_r = -1640$ lb. **11-9.** $R_1 = 10.2$ kN, $R_2 = 16.8$ kN; $V_l = 0$, $V_1 = -1.4$ kN, $V_1' = 8.8$ kN, $V_2 = 7.4$ kN $= V_3$, $V_3' = -8.6$ kN $= V_{3.5}$, $V_{4.5} = -9.8$ kN $= V_5$, $V_5' = 7$ kN $= V_r$. **11-11.** $M_{max} = 4800$ ft·lb 4 ft from left support. **11-13.** $M_{max} = 19.7$ kN·m, 1.5 m from right support. **11-15.** $M_{max} = -43$ kN·m at wall; $(M_2 = -16$ kN·m). **11-17.** $M_{max} = 7320$ ft·lb 9.4 ft from left support; $(M_5 = 5870$ ft·lb, $M_9 = 7290$ ft·lb, $M_{14} = 4360$ ft·lb). **11-19.** $M_{max} = 14.8$ kN·m, 3 m from left end of beam; $(M_1 = -0.7$ kN·m, $M_2 = 7.4$ kN·m, $M_{3.5} = 10.5$ kN·m, $M_{4.5} = 1.3$ kN·m, $M_5 = -3.5$ kN·m). **11-21.** (a) $R_1 = (P/L)(L - a)$, $R_2 = Pa/L$; (b) $V_{max} = (P/L)(L - a)$ at all sections from R_1 to P, if $a < L/2$; (c) $M_{max} = Pa(L - a)/L$ located at P; (d) $M_x = Px(L - a)/L$. **11-23.** $R = 14\ 700$ lb, $M = -138\ 850$ ft·lb (at wall). **11-25.** $M_{max} = 735$ kN·m, occurs under rear axle (153 kN load) when it is 6.4 m from left support and front axle is 1.4 m from right support. $(M_{max} = 654$ kN·m for 81-kN load at 8.5 m from left support; $M_{max} = 451$ kN·m for 36-kN load at 10 m from left support.)

CHAPTER 12

12-1. $s_{0.9} = 16.2$ MPa; $s_{1.5} = 27$ MPa; $s_{2.4} = 10.8$ MPa. **12-2.** $W = 3910$ lb (maximum). **12-3.** $L = 10.8$ ft (maximum). **12-4.** $d_o = 140.8$ mm, $d_i = 70.4$ mm **12-5.** $N_u = 144$ (top fiber compression), $N_u = 75.4$ (bottom fiber tension); $(\bar{y} = 3.22$ in from base, $I_x = 155$ in⁴, $M_{max} = 3192$ ft·lb 3 ft from R_1). **12-7.** $L = 31.2$ ft (maximum, neglecting beam weight); $(L = 29.5$ ft, including beam weight). **12-9.** S8 × 18.4 (required $S = 14.4$ in³, neglecting beam weight); S8 × 23 (required $S = 14.8$ in³, including beam weight). **12-11.** $w = 268$ psf of floor area (maximum floor load in addition to beam weight); $(w = 272$ psf with beam weight neglected). **12-13.** $d = 3.57$ in (minimum; shaft weight included); $(d = 3.45$ in, neglecting shaft weight). **12-15.** $d = 2.47$ in (minimum, based on ultimate stress with $N_u = 6$); $(d = 2.22$ in, based on yield stress with $N_y = 3$). **12-17.** Yes, eastern white pine is satisfactory. $(s_b = 0.95$ MPa with beam weight). **12-19.** Use dressed 6 in × 14 in (required $S = 150$ in³ for bending, required $A = 68$ in² for shear; beam weight neglected). **12-21.** $F = 5760$ lb (maximum, for nominal size); $F = 4850$ lb (for dressed size). **12-23.** (a) $W = 16\ 800$ lb (maximum, for nominal size); $W = 15\ 300$ lb (for dressed size); (b) $s_b = 616$ psi (nominal size); $s_b = 637$ psi, (dressed size). **12-25.** Bolt spacing $= 11.3$ in center to center, two rows (maximum; beam weight neglected). **12-27.** $s_v = 3300$ psi (maximum web shear at supports). **12-29.** $s_h = 11\ 800$ psi at base of flange at supports for $W = 130\ 000$ lb; $(s_v = 12\ 300$ psi maximum web shear). **12-31.** $W = 8360$ lb (maximum, for dressed size; shear controls); $(W = 9600$ lb for nominal size). **12-33.** $s_b = 16\ 150$ psi at midspan. $s_v = 4130$ psi at supports, $y = 0.751$ in at midspan. **12-35.** (a) No, $y = 0.535$ in (beam weight neglected); $(y = 0.561$ in with beam weight); (b) $s_b = 14\ 500$ psi, $s_v = 1715$ psi (beam weight neglected); $(s_b = 15\ 100$ psi, $s_v = 1790$ psi with beam weight). **12-37.** (a) $y = 0.0923$ in (including shaft weight); (b) $s_h = 59.8$ psi at supports, $s_b = 4360$ psi at midspan. **12-39.** S15 $\times$ 42.9 (required $S = 50$ in³; beam weight neglected).

CHAPTER 13

13-1. $T = 7\ 950$ kN·mm (maximum). **13-2.** No; allowable $s_s = 18\ 350$ psi, actual $s_s = 29\ 400$ psi. **13-3.** (a) $d = 48.9$ mm (minimum); (b) $d_o = 50$ mm, $d_i = 25$ mm. **13-4.** $L = 411$ in or 34.2 ft (maximum). **13-5.** $\theta = 0.043$ rad. **13-7.** $d = 41.3$ mm (minimum, based on ultimate with $N_u = 10$). **13-9.** $n = 223$ rpm. **13-11.** $L = 53.3$ mm (bearing); ($L = 32$ mm, shear). **13-13.** $N_s = 1.79$ (shear), $N_c = 1.19$ (bearing). **13-15.** $s_s = 38$ MPa (shear), $s_c = 24$ MPa (bearing). **13-17.** $d = 38.4$ mm (based on angle of twist); $d = 23.1$ mm (based on shear stress).

CHAPTER 14

14-1. Bottom fiber maximum tension $= 21.9$ MPa, top fiber maximum compression $= 15.9$ MPa (both at midspan). **14-2.** $h = 10.24$ in (minimum depth). **14-3.** Right-side maximum tension $= 293$ MPa (at midspan). **14-4.** $F = 97\ 500$ lb. **14-5.** Right-side maximum compression $= 4550$ psi; left-side maximum tension $= 2050$ psi.

14-7.

BASE PRESSURES (psf)

	Footing weight	120-k load	85-k load	Total
Left-front corner	300	16 670	$-14\ 180$	2 790
Left-rear corner	300	16 670	4 720	21 690
Right-rear corner	300	$-3\ 330$	23 620	20 590
Right-front corner	300	$-3\ 330$	4 720	1 690

(Negative values indicate release of pressure)

14-9. $F = 1940$ lb (maximum, based on shock loading; tension controls).

14-11.

STRESSES (MPa)

	At A	At B
Due to 4.5-kN load	30	30
Due to 450-N load	36	-36
Due to 600-N load	21.8	-13.8
Total	87.8	-19.8

(Negative values are compression)

14-13. $N_u = 1.43$. **14-15.** $s_s = 13\ 580$ psi (no eccentricity). **14-17.** $d = 1.03$ in, use $1\frac{1}{8}$-in bolts. **14-19.** Maximum $s_s = 2940$ psi on lower right-hand bolt; (lower left $= 2760$ psi; center right $= 2250$ psi; center left $= 2015$ psi; upper right $= 1660$ psi;

upper left = 1330 psi).   **14-21.** (*a*) $F = 7350$ lb (bolt *A* controls); (*b*) $F = 8830$ lb (no eccentricity).   **14-23.** hp = 68.5   **14-25.** $d = 2.97$ in (minimum; from maximum principal stress); ($d = 2.96$ in, minimum; from maximum shear stress).   **14-27.** (*a*) $d = 75.8/\sqrt[3]{(s_s)_{max}}$; (*b*) $d = 90.6/\sqrt[3]{(s_t)_{max}}$.   **14-29.** $d_o = 90$ mm (minimum), $d_i = 45$ mm (from maximum principal stress).   **14-31.** $(s_t)_{max} = 16\,200$ psi, $(s_s)_{max} = 12\,100$ psi, $\theta = 35°$.

CHAPTER 15

15-1. $F = 103\,600$ lb.   **15-2.** $F = 213\,000$ lb.   **15-3.** $F = 69\,300$ lb.
15-4. Maximum safe load $F = 15.26$ kN (intermediate column; machine-design formula). The member is satisfactory.   **15-5.** $L = 13.95$ ft or 167.5 in.   **15-7.** $F = 690\,000$ lb ($I_y = 477.8$ in⁴).   **15-9.** $F = 51\,000$ lb.   **15-11.** Two $3\frac{1}{2}'' \times 3\frac{1}{2}'' \times \frac{1}{4}''$ angles (can carry 34 200 lb).   **15-13.** S 10 × 25.4 (can carry 49 000 lb).   **15-15.** 0.516×1.03 in (slender column; machine-design formula).   **15-17.** The column is safe. Maximum safe load $F = 24\,900$ lb (dressed size).   **15-19.** 11.2 in × 11.2 in. Use 12 × 12 (can carry 109,000 lb).   **15-21.** For left column, use 6 in × 6 in, (or 6 in diam.); for right column, use 6 in × 6 in, or 4 in × 8 in (or 4.7 in diam.).

CHAPTER 16

16-1. Force in rod $F = 4500$ lb; rod diameter $d = 0.594$ in for AISI 1020 with steady load (use $\frac{5}{8}$-in-diameter rod). Maximum bending stress $s_b = 12\,300$ psi at wall.
16-2. Support reaction $R_1 = 2170$ lb; maximum bending stress $s_b = 868$ psi at wall (for dressed size); ($R_1 = 2100$ lb; $s_b = 810$ psi for nominal size). If support does not settle, $R_1 = 2550$ lb; $s_b = 600$ psi at wall (for dressed size).   **16-3.** $s_b = 692$ psi at wall; $s_h = 96$ psi at wall. ($M_{max} = 8300$ ft·lb; $V_{max} = 4610$ lb).   **16-4.** $s_b = 1435$ psi at wall; $s_h = 117$ psi at wall. ($M_{max} = 17\,200$ ft·lb; $V_{max} = 5600$ lb).   **16-5.** $s_b = 720$ psi at section 4.58 ft from support; $s_h = 78.3$ psi at support. ($M_{max} = 8640$ ft·lb; $V_{max} = 3760$ lb).   **16-7.** $s_b = 9150$ psi (compression) at bottom fiber at wall; $y_{max} = 0.0307$ in at midspan.